# FIELD THEORY

A MODERN PRIMER

**FRONTIERS IN PHYSICS: A Lecture Note and Reprint Series**

David Pines, Editor

Volumes of the Series published from 1961 to 1973 are not officially numbered. The parenthetical numbers shown are designed to aid librarians and bibliographers to check the completeness of their holdings.

| | | |
|---|---|---|
| (1) | N. Bloembergen | Nuclear Magnetic Relaxation: A Reprint Volume, 1961 |
| (2) | G. F. Chew | *S*-Matrix Theory of Strong Interactions: A Lecture Note and Reprint Volume, 1961 |
| (3) | R. P. Feynman | Quantum Electrodynamics: A Lecture Note and Reprint Volume, 1961 (6th printing, 1980) |
| (4) | R. P. Feynman | The Theory of Fundamental Processes: A Lecture Note Volume, 1961 (6th printing, 1980) |
| (5) | L. Van Hove, N. M. Hugenholtz, and L. P. Howland | Problems in Quantum Theory of Many-Particle Systems: A Lecture Note and Reprint Volume, 1961 |
| (6) | D. Pines | The Many-Body Problem: A Lecture Note and Reprint Volume, 1961 (5th printing, 1979) |
| (7) | H. Frauenfelder | The Mössbauer Effect: A Review—with a Collection of Reprints, 1962 |
| (8) | L. P. Kadanoff and G. Baym | Quantum Statistical Mechanics: Green's Function Methods in Equilibrium and Nonequilibrium Problems, 1962 (3rd printing, 1976) |
| (9) | G. E. Pake | Paramagnetic Resonance: An Introductory Monograph, 1962 [cr. (42)—2nd edition] |
| (10) | P. W. Anderson | Concepts in Solids: Lectures on the Theory of Solids, 1963 (4th printing, 1978) |
| (11) | S. C. Frautschi | Regge Poles and *S*-Matrix Theory, 1963 |
| (12) | R. Hofstadter | Electron Scattering and Nuclear and Nucleon Structure: A Collection of Reprints with an Introduction, 1963 |
| (13) | A. M. Lane | Nuclear Theory: Pairing Force Correlations to Collective Motion, 1964 |
| (14) | R. Omnès and M. Froissart | Mandelstam Theory and Regge Poles: An Introduction for Experimentalists, 1963 |
| (15) | E. J. Squires | Complex Angular Momenta and Particle Physics: A Lecture Note and Reprint Volume, 1963 |
| (16) | H. L. Frisch and J. L. Lebowitz | The Equilibrium Theory of Classical Fluids: A Lecture Note and Reprint Volume, 1964 |
| (17) | M. Gell-Mann and Y. Ne'eman | The Eightfold Way: (A Review—with a Collection of Reprints), 1964 |
| (18) | M. Jacob and G. F. Chew | Strong-Interaction Physics: A Lecture Note Volume, 1964 |
| (19) | P. Nozières | Theory of Interacting Fermi Systems, 1964 |

**FRONTIERS IN PHYSICS: A Lecture Note and Reprint Series**

David Pines, Editor *(continued)*

| | | |
|---|---|---|
| (20) | J. R. Schrieffer | Theory of Superconductivity, 1964 (2nd printing, 1971) |
| (21) | N. Bloembergen | Nonlinear Optics: A Lecture Note and Reprint Volume, 1965 (3rd printing, with addenda and corrections, 1977) |
| (22) | R. Brout | Phase Transitions, 1965 |
| (23) | I. M. Khalatnikov | An Introduction to the Theory of Superfluidity, 1965 |
| (24) | P. G. de Gennes | Superconductivity of Metals and Alloys, 1966 |
| (25) | W. A. Harrison | Pseudopotentials in the Theory of Metals, 1966 (2nd printing, 1971) |
| (26) | V. Barger and D. Cline | Phenomenological Theories of High Energy Scattering: An Experimental Evaluation, 1967 |
| (27) | P. Choquard | The Anharmonic Crystal, 1967 |
| (28) | T. Loucks | Augmented Plane Wave Method: A Guide to Performing Electronic Structure Calculations—A Lecture Note and Reprint Volume, 1967 |
| (29) | T. Ne'eman | Algebraic Theory of Particle Physics: Hadron Dynamics in Terms of Unitary Spin Currents, 1967 |
| (30) | S. L. Adler and R. F. Dashen | Current Algebras and Applications to Particle Physics, 1968 |
| (31) | A. B. Migdal | Nuclear Theory: The Quasiparticle Method, 1968 |
| (32) | J. J. J. Kokkedee | The Quark Model, 1969 |
| (33) | A. B. Migdal and V. Krainov | Approximation Methods in Quantum Mechanics, 1969 |
| (34) | R. Z. Sagdeev and A. A. Galeev | Nonlinear Plasma Theory, 1969 |
| (35) | J. Schwinger | Quantum Kinematics and Dynamics, 1970 |
| (36) | R. P. Feynman | Statistical Mechanics: A Set of Lectures, 1972 (5th printing, 1979) |
| (37) | R. P. Feynman | Photo-Hadron Interactions, 1972 |
| (38) | E. R. Caianiello | Combinatorics and Renormalization in Quantum Field Theory, 1973 |
| (39) | G. B. Field, H. Arp, and J. N. Bahcall | The Redshift Controversy, 1973 (2nd printing, 1976) |
| (40) | D. Horn and F. Zachariasen | Hadron Physics at Very High Energies, 1973 |
| (41) | S. Ichimaru | Basic Principles of Plasma Physics: A Statistical Approach, 1973 (2nd printing, with revisions, 1980) |

**FRONTIERS IN PHYSICS: A Lecture Note and Reprint Series**

David Pines, Editor *(continued)*

Volumes published from 1974 onward are being numbered as an integral part of the bibliography:

| Number | | |
|---|---|---|
| 43 | R. C. Davidson | Theory of Nonneutral Plasmas, 1974 |
| 44 | S. Doniach and E. H. Sondheimer | Green's Functions for Solid State Physicists, 1974 |
| 45 | P. H. Frampton | Dual Resonance Models, 1974 |
| 46 | S. K. Ma | Modern Theory of Critical Phenomena, 1976 |
| 47 | D. Forster | Hydrodynamic Fluctuations, Broken Symmetry, and Correlation Functions, 1975 (2nd printing, 1980) |
| 48 | A. B. Migdal | Qualitative Methods in Quantum Theory, 1977 |
| 49 | S. W. Lovesey | Condensed Matter Physics: Dynamic Correlations, 1980 |
| 50 | L. D. Faddeev and A. A. Slavnov | Gauge Fields: Introduction to Quantum Theory, 1980 |
| 51 | P. Ramond | Field Theory: A Modern Primer, 1981 |
| 52 | R. A. Broglia and A. Winther | Heavy Ion Reactions: Lecture Notes<br>Vol. I: Elastic and Inelastic Reactions, 1981 |
| 53 | R. A. Broglia and A. Winther | Heavy Ion Reactions: Lecture Notes<br>Vol. II, *in preparation* |

*Other Numbers in preparation.*

# FIELD THEORY

## A MODERN PRIMER

**PIERRE RAMOND**

*Physics Department*
*University of Florida*
*Gainesville, Florida*

*Physics Department*
*California Institute of Technology*
*Pasadena, California*

1981
THE BENJAMIN/CUMMINGS PUBLISHING COMPANY, INC.
ADVANCED BOOK PROGRAM
Reading, Massachusetts

London • Amsterdam • Don Mills, Ontario • Sydney • Tokyo

CODEN: FRHPA

Pierre Ramond

Field Theory: A Modern Primer

**Library of Congress Cataloging in Publication Data**

Ramond, Pierre, 1943–
Field theory.

(Frontiers in physics; 51)
Bibliography: p.
Includes index.
1. Quantum field theory. 2. Perturbation (Quantum dynamics) 3. Gauge fields (Physics) 4. Integrals, Path. I. Title. II. Series.
QC174.45.R35 530.1'43 80-27067
ISBN 0-8053-7892-8
ISBN-0-8053-7893-6 (pbk.)

Reproduced by Benjamin/Cummings Publishing Company, Inc., Advanced Book Program, Reading, Massachusetts, from camera-ready copy prepared under the supervision of the author.



Manufactured in the United States of America

TO MY GIRLS

CONTENTS

## EDITOR'S FOREWORD

The problem of communicating in a coherent fashion recent developments in the most exciting and active fields of physics seems particularly pressing today. The enormous growth in the number of physicists has tended to make the familiar channels of communication considerably less effective. It has become increasingly difficult for experts in a given field to keep up with the current literature; the novice can only be confused. What is needed is both a consistent account of a field and the presentation of a definite "point of view" concerning it. Formal monographs cannot meet such a need in a rapidly developing field, and, perhaps more important, the review article seems to have fallen into disfavor. Indeed, it would seem that the people most actively engaged in developing a given field are the people least likely to write at length about it.

FRONTIERS IN PHYSICS has been conceived in an effort to improve the situation in several ways. Leading physicists today frequently give a series of lectures, a graduate seminar, or a graduate course in their special fields of interest. Such lectures serve to summarize the present status of a rapidly developing field and may well constitute the only coherent account available at the time. Often, notes on lectures exist (prepared by the lecturer himself, by graduate students, or by postdoctoral fellows) and are distributed in mimeographed form on a limited basis. One of the principal purposes of the FRONTIERS IN PHYSICS Series is to make such notes available to a wider audience of physicists.

It should be emphasized that lecture notes are necessarily rough and informal, both in style and content; and those in the series will prove no exception. This is as it should be. The point of the series is to offer new, rapid, more informal, and, it is hoped, more effective ways for physicists to teach one another. The point is lost if only elegant notes qualify.

The publication of collections of reprints of recent articles in very active fields of physics will improve communication. Such collections are themselves useful to people working in the field. The value of the reprints will, however, be enhanced if the collection is accompanied by an introduction of moderate length which will serve to tie the collection together and, necessarily, constitute a brief survey of the present status of the field. Again, it is appropriate that such an introduction be informal, in keeping with the active character of the field.

The informal monograph, representing an intermediate step between lecture notes and formal monographs, offers an author the opportunity to present his views of a field which has developed to the point where a summation might prove extraordinarily fruitful but a formal monograph might not be feasible or desirable.

Contemporary classics constitute a particularly valuable approach to the teaching and learning of physics today. Here one thinks of fields that lie at the heart of much of present-day research, but whose essentials are by now well understood, such as quantum electrodynamics or magnetic resonance. In such fields some of the best pedagogical material is not readily available, either because it consists of papers long out of print or lectures that have never been published.

---

The above words, written in August, 1961, seem equally applicable today. Pierre Ramond has made significant contributions to our understanding of quantum field theory and particle physics. As he has noted in his Preface to this volume, despite the dramatic progress which has been made during the past decade in developing both new methods and applications of quantum field theory, few textbooks have been written which cover this material.

The present volume should go a long way toward bridging this gap, since it emphasizes the structure and methods of perturbation field theories while discussing both renormalization theory and the evaluation of Feynman's diagram for gauge theories at a level suitable for the beginning graduate student, with no important steps left out of the key derivations and problems at the end of each chapter.

It is a pleasure to welcome him as a contributor to this Series.

David Pines

# PREFACE

Since 't Hooft proved the renormalizability of Yang-Mills theories, there has been a consequent dramatic increase in both the methods and the applications of Quantum Field Theory. Yet, few textbooks have been written in the intervening years, so that the student of Field Theory is left to depend on the original literature or on one of the many excellent reviews of the subject. Unfortunately these are often written for the specialist rather than for the neophyte. These notes, based on a one-year graduate course in Field Theory offered at Caltech between 1978 and 1980, aim to fill this gap by introducing in a straightforward, calculational manner some of the tools used by the modern Field Theorist.

It is no longer possible to present perturbative Field Theory pedagogically in one year — the days of the standard one-year course based on QED are gone forever. Thus these notes cover only a selected set of topics. A modern presentation must consist of at least three parts — a first course emphasizing the structure and methods of perturbative Field Theories, with the intent of acquainting the student with renormalization theory and the evaluation of Feynman diagrams for gauge theories, — a second course dealing with applications of gauge theories, centering around perturbative calculations in Quantum Chromodynamics (QCD), Flavor Dynamics (Glashow-Weinberg-Salam model), and possibly Grand Unified Theories, and finally a third course on non-perturbative techniques. These notes address themselves to the first part, concentrating at an elementary level first on Classical Field Theory including a detailed discussion of the Lorentz group, Dirac and Majorana masses and supersymmetry, followed by a presentation of regularization methods, renormalization theory and other formal aspects of the subject. The approach is calculational - no proof of renormalizability is given, only plausibility arguments. Renormalization is treated in great detail for $\lambda\phi^4$ theory, but only lightly for gauge theories. The passage from Classical to Quantum Field Theory is described in terms of the Feynman Path Integral, which is appropriate to both perturbative and non-perturbative treatments. Also the $\zeta$-function technique for evaluating functional determinants is presented for simple theories.

The material is discussed in sufficient detail to enable the reader to follow every step, but some crucial theoretical aspects are not covered such as the Infrared Structure of unbroken gauge theories, and the description of calculations in broken gauge theories. Still it is hoped that these notes will serve as an introduction to the perturbative evaluation of gauge theories. Problems are included at the end of each section, with asterisks to denote their degree of difficulty. A bibliography is included to provide a guide for further studies.

I would like to express my gratitude to Professor S. Frautschi and J. Harvey for their diligent reading of the manuscript and constructive criticisms. Also special thanks go to E. Corrigan and J. Harvey for teaching me $\zeta$-function techniques, as well as to the students of Phys. 230 for their patience and their numerous suggestions. Finally, these notes would not have seen the light of day had it not been for the heroic efforts of two modern day scribes, Roma Gaines and Helen Tuck, to whom I am deeply grateful.

P. Ramond
Pasadena, Summer 1980

# I - HOW TO BUILD AN ACTION FUNCTIONAL

## 1. The Action Functional: elementary considerations

It is a most beautiful and awe-inspiring fact that all the fundamental laws of Classical Physics can be understood in terms of one mathematical construct called the Action. It yields the classical equations of motion, and analysis of its invariances leads to quantities conserved in the course of the classical motion. In addition, as Dirac and Feynman have shown, the Action acquires its full importance in Quantum Physics. As such, it provides a clear and elegant language to effect the transition between Classical and Quantum Physics through the use of the Feynman Path Integral (FPI).

Thus our task is clear: we first study the art of building acceptable Action functionals (AF) and later derive the quantum properties of the system a given AF describes by evaluating the associated Feynman Path Integral (FPI). First, examine the AF for an elementary system: take a point particle, with position vector $x_i(t)$ $(i = 1,2,3)$, at time t, moving in a time independent potential $V(x_i)$. The corresponding AF is given by

$$S([x_i];t_1,t_2) \equiv \int_{t_1}^{t_2} dt\left(\frac{1}{2} m \frac{dx_i}{dt}\frac{dx_i}{dt} - V(x_i)\right) \quad . \tag{1.1}$$

It is a function of the initial and final times $t_1$ and $t_2$ and a functional of the path $x_i(t)$ for $t_1 < t < t_2$. Repeated latin indices are summed over. This means that to a given path $x_i(t)$, we associate a number called the functional (in this case S). Functional relationship will be indicated by square brackets [...]. For instance, the length of a path is a functional of the path.

Consider the response of S to a small deformation of the path

$$x_i(t) \to x_i(t) + \delta x_i(t) \quad . \tag{1.2}$$

Then

$$S[x_i + \delta x_i] = \int_{t_1}^{t_2} dt \left(\frac{1}{2} m \frac{d(x_i + \delta x_i)}{dt} \frac{d(x_i + \delta x_i)}{dt} - V(x_i + \delta x_i)\right) . \tag{1.3}$$

Neglect terms of $O(\delta x)^2$ and use the chain rule to obtain

$$\frac{d(x_i + \delta x_i)}{dt} \frac{d(x_i + \delta x_i)}{dt} \simeq \frac{dx_i}{dt} \frac{dx_i}{dt} - 2 \frac{d^2 x_i}{dt^2} \delta x_i + 2 \frac{d}{dt} \left(\delta x_i \frac{dx_i}{dt}\right) \tag{1.4}$$

$$V(x_i + \delta x_i) \simeq V(x_i) + \delta x_i \partial_i V \quad . \tag{1.5}$$

[Here $\partial_i \equiv \frac{\partial}{\partial x_i}$ ]. Thus

$$S[x_i + \delta x_i] \simeq S[x_i] + \int_{t_1}^{t_2} dt \ \delta x_i \left(-\partial_i V - m \frac{d^2 x_i}{dt^2}\right) + m \int_{t_1}^{t_2} dt \frac{d}{dt} \left(\delta x_i \frac{dx_i}{dt}\right) . \tag{1.6}$$

The last term is just a "surface" term. It can be eliminated by restricting the variations to paths which vanish at the end points: $\delta x_i(t_1) = \delta x_i(t_2) = 0$. With this proviso the requirement that S not change under an arbitrary $\delta x_i$ leads to the classical equations of motion for the system. We symbolically write it as the vanishing of the functional derivative introduced by

$$S[x_i + \delta x_i] = S[x_i] + \int dt \ \delta x_i \frac{\delta S}{\delta x_i} + \ldots \quad . \tag{1.7}$$

That is

$$\frac{\delta S}{\delta x_i} = - \left(m \frac{d^2 x_i}{dt^2} + \partial_i V\right) = 0 \quad . \tag{1.8}$$

Thus we have the identification between equations of motion and extremization of S. Note, however, that extremization of S only leads to a class of possible paths. Which of those is followed depends on the boundary conditions, given as initial values of $x_i$ and $\frac{dx_i}{dt}$ .

A further, and most important point to be made is the correspondence between the symmetries of S and the existence of quantities conserved in the course of the motion of the system. An example will suffice. Take $V(x_i)$ to be a function of the length of $x_i$, i.e., $r = (x_i x_i)^{1/2}$. Then S is manifestly invariant under a rotation of the three-vector $x_i$ since it depends only on its length. Under an infinitesimal arbitrary rotation

$$\delta x_i = \varepsilon_{ij} x_j \quad , \quad \varepsilon_{ij} = -\varepsilon_{ji}, \text{ with } \varepsilon_{ij} \text{ time independent.} \tag{1.9}$$

Now, since S is invariant, we know that $\delta S = 0$, but as we have seen above $\delta S$ consists of two parts: the functional derivative which vanishes for the classical path,and the surface term. For this particular variation, however, we cannot impose boundary conditions on $\delta x_i(t)$, so the invariance of S together with the equations of motion yield

$$0 = \delta S = \int_{t_1}^{t_2} dt \frac{d}{dt} \left(m \frac{dx_i}{dt} \delta x_i\right) = \varepsilon_{ij}\, m\, x_i \frac{dx_j}{dt} \Bigg|_{t_1}^{t_2} \quad . \tag{1.10}$$

As this is true for any $\varepsilon_{ij}$, it follows that the

$$\ell_{ij}(t) \equiv \frac{1}{2} m\left(x_i \frac{dx_j}{dt} - x_j \frac{dx_i}{dt}\right) \quad , \tag{1.11}$$

satisfy

$$\ell_{ij}(t_1) = \ell_{ij}(t_2) \quad , \tag{1.12}$$

and are therefore conserved during the motion. These are, as you know, the components of the angular momentum. An infinitesimal form of the conservation law can be obtained by letting $t_2$ approach $t_1$. We have just proved in a simple case the celebrated theorem of Emmy Noether, relating an invariance (in this case, rotational) to a conservation law (of angular momentum).

To summarize the lessons of this elementary example:

1) Classical equations of motion are obtained by extremizing S.

2) Boundary conditions have to be supplied externally.

3) The symmetries of S are in correspondence with conserved quantities and therefore reflect the basic symmetries of the physical system.

This example dealt with particle mechanics; it can be generalized to Classical Field Theory, as in Maxwell's Electrodynamics or Einstein's General Relativity.

The Action is just a mathematical construct, and therefore unlimited in its possibilities. Yet, it also affords a description of the physical world which we believe operates in a definite way. Hence there should be <u>one</u> very special AF out of many that describes correctly what is going on. The problem is to find ways to characterize this unique Action. Noether's theorem gives us a hint since it allows us to connect the symmetries of the system with those of the functional. Certain symmetries, such as those implied by the Special Theory of Relativity, are well documented. Thus, any candidate action must reflect this fact. Other symmetries, such as electric charge conservation, further restrict the form of <u>the</u> AF. It is believed that Nature is partial to certain types of actions which are loaded with all kinds of invariances that vary from point to point. These give rise to the gauge theories which will occupy us later in this course. For the time being, let us learn how to build AF's for systems that satisfy the laws of the Theory of Special Relativity. Technically, these systems can be characterized by their invariance under transformations generated by the inhomogeneous Lorentz group, a.k.a., the Poincaré group, which is what will concern us next.

Problems.

Notes: - problems are given in order of increasing complexity.

- in the following, use the Action Functional as the main tool, although you may be familiar with more elementary methods of solution.

A. i) Prove that linear momentum is conserved during the motion described by $S = \int dt \, \frac{1}{2} m \dot{x}^2$ , $\dot{x} \equiv \frac{dx}{dt}$ .

ii) Suppose $V(x_i) = v(1 - \cos \frac{r}{a})$; find the expression for the rate of change of the linear momentum.

B. For a point particle moving in an arbitrary potential, derive the expression for the rate of change in angular momentum.

*C. For a point particle moving in a potential $V = -\frac{a}{r}$ , find the invariances of the AF. Hint: recall that the Newtonian orbits do not precess, which leads to a non-trivial conserved quantity, the Runge-Lenz vector.

*D. Given an AF invariant under uniform time translations, derive the expression for the associated conserved quantity. Use as an example a point particle moving in a time-independent potential. What happens if the potential is time-dependent?

## 2. The Lorentz Group (a cursory look)

The postulates of Special Relativity tell us that the speed of light is the same in all inertial frames. This means that if $x_i$ is the position of a light signal at time t in one frame, and the same light ray is found at $x'_i$ at time t' in another frame, we must have

$$s^2 \equiv c^2t^2 - x_ix_i = c^2t'^2 - x'_ix'_i \qquad . \tag{2.1}$$

The set of linear transformations which relate $(x'_i, t')$ to $(x_i, t)$ while preserving the above expression form a group called the Lorentz group (see problem). Choose units such that $c = 1$, and introduce the notation

$$x^\mu \quad \mu = 0,1,2,3 \text{ with } x^0 = t,\ (x^1,x^2,x^3) = (x^i) = \vec{x},$$

i.e.,
$$x^\mu = (x^o, x^i) \qquad i = 1,2,3$$
$$= (t, \vec{x}) \qquad .$$

In this compact notation $s^2$ can be written as

$$s^2 = x^0x^0 - x^ix^i \equiv x^\mu x^\nu g_{\mu\nu} \qquad , \tag{2.2}$$

where the metric $g_{\mu\nu} = g_{\nu\mu}$ is zero except for $\mu = \nu$ when $g_{00} = -g_{11} = -g_{22} = -g_{33} = 1$. Repeated indices are summed except when otherwise indicated. Then eq. (2.1) becomes

$$g_{\mu\nu}x^\mu x^\nu = g_{\mu\nu}x'^\mu x'^\nu \qquad . \tag{2.3}$$

Now look for a set of linear transformations

$$x'^\mu = \Lambda^\mu{}_\nu x^\nu = \Lambda^\mu{}_0 x^0 + \Lambda^\mu{}_i x^i \qquad , \tag{2.4}$$

which preserves $s^2$. The $\Lambda^\mu{}_\nu$ must therefore satisfy

$$g_{\mu\nu}x'^\mu x'^\nu = g_{\mu\nu}\Lambda^\mu{}_\rho \Lambda^\nu{}_\sigma x^\rho x^\sigma = g_{\rho\sigma}x^\rho x^\sigma \qquad . \tag{2.5}$$

As (2.5) must hold for any $x^\mu$, we conclude that

$$g_{\rho\sigma} = g_{\mu\nu}\Lambda^\mu{}_\rho \Lambda^\nu{}_\sigma \qquad . \tag{2.6}$$

It is more convenient for certain purposes to use a matrix notation: regard $x^\mu$ as a column vector x and $g_{\mu\nu}$ as a matrix g. Then

$$s^2 = x^T g x \quad , \tag{2.7}$$

and

$$x' = Lx \quad , \tag{2.8}$$

where L is the matrix equivalent of the $\Lambda^\mu{}_\nu$ coefficients, and T means the transpose. The L's must obey

$$g = L^T g L \quad , \tag{2.9}$$

to be Lorentz transformations (LT). Examine the consequences of eq. (2.9). First take its determinant

$$\det g = \det L^T \det g \det L \quad , \tag{2.10}$$

from which we deduce that

$$\det L = \pm 1 \quad . \tag{2.11}$$

The case $\det L = 1(-1)$ corresponds to proper (improper) LT's. As an example, the LT given numerically by $L = g$ is an improper one; physically it corresponds to $x^0 \to x^0$, $x^i \to -x^i$, i.e., space inversion. Second, take the 00 entry of eq. (2.6)

$$1 = \Lambda^\rho{}_0 g_{\rho\sigma} \Lambda^\sigma{}_0 = (\Lambda^0{}_0)^2 - (\Lambda^i{}_0)^2 \quad , \tag{2.12}$$

which shows that

$$|\Lambda^0{}_0| \geq 1 \quad . \tag{2.13}$$

When $\Lambda^0{}_0 \geq 1$, the LT's are said to be orthochronous, while $\Lambda^0{}_0 \leq -1$ gives non-orthochronous LT's. It follows then that LT's can be put in four categories (see problem):

1) proper orthochronous, also called restricted ($L_+^\uparrow$) with $\det L = +1$, $\Lambda^0{}_0 \geq 1$

2) proper non-orthochronous ($L_+^\downarrow$) with $\det L = +1$, $\Lambda^0{}_0 \leq -1$

3) improper orthochronous ($L_-^\uparrow$) with $\det L = -1$, $\Lambda^0{}_0 \geq 1$

4) improper non-orthochronous ($L_-^\downarrow$) with $\det L = -1$, $\Lambda^0{}_0 \leq -1$.

Let us give a few examples:

a) Rotations: $x'^0 = x^0$, $x'^i = a^{ij}x^j$ with $a^{ij}$ an orthogonal matrix.

Then we can write L in the block form

$$L = \left(\begin{array}{c|c} 1 & 0 \\ \hline 0 & a \end{array}\right) \tag{2.14}$$

so that $\det L = \det a$. We can have $\det a = \pm 1$, corresponding to proper and improper rotations, with L belonging to $L_+^\uparrow$ and $L_-^\uparrow$ respectively.

b) Boosts: the transformations

$$\begin{aligned} x'^0 &= x^0 \cosh\eta - x^1 \sinh\eta \\ x'^1 &= -x^0 \sinh\eta + x^1 \cosh\eta \\ x'^{2,3} &= x^{2,3} \end{aligned} \tag{2.15}$$

describe a boost in the 1-direction. Then in 2 x 2 block form

$$L = \left(\begin{array}{cc|c} \cosh\eta & -\sinh\eta & \\ & & 0 \\ -\sinh\eta & \cosh\eta & \\ \hline & 0 & 1 \end{array}\right) , \tag{2.16}$$

$$\det L = \cosh^2\eta - \sinh^2\eta = 1 , \tag{2.17}$$

$$\Lambda^0{}_0 = \cosh\eta \geq 1 . \tag{2.18}$$

This transformation therefore belongs to $L_+^\uparrow$. Note that the identification

$$\cosh\eta = (1 - v^2)^{-1/2} , \quad \sinh\eta = v(1 - v^2)^{-1/2} , \tag{2.19}$$

where v is the velocity of the boosted frame leads to the more familiar form.

c) Time inversion: defined by $x'^0 = -x^0$, $x'^i = x^i$. It has $\det L = -1$, $\Lambda^0{}_0 = -1$, and therefore belongs to $L_-^{\downarrow}$.

d) Full inversion: defined by $x'^\mu = -x^\mu$. It has $\det L = +1$, $\Lambda^0{}_0 = -1$, and belongs to $L_+^{\downarrow}$. Full inversion can be obtained as the product of a space and a time inversion.

Any Lorentz transformation can be decomposed as the product of transformations of these four types (see problem). Thus it suffices to concentrate on rotations and boosts. Since there are three rotations and three boosts, one for each space direction, the Lorentz transformations are described in terms of six parameters. We now proceed to build the six corresponding generators. Consider an infinitesimal LT

$$\Lambda^\mu{}_\nu = \delta^\mu{}_\nu + \varepsilon^\mu{}_\nu \quad , \tag{2.20}$$

where $\delta^\mu{}_\nu$ is the Kronecker delta which vanishes when $\mu \neq \nu$ and equals +1 otherwise. Evaluation of eq. (2.6) yields to $O(\varepsilon)$

$$0 = g_{\nu\rho}\varepsilon^\rho{}_\mu + g_{\mu\rho}\varepsilon^\rho{}_\nu \quad . \tag{2.21}$$

We use the metric $g_{\mu\nu}$ to lower indices, for example

$$x_\mu \equiv g_{\mu\nu}x^\nu = (x^0, -\vec{x}) \quad . \tag{2.22}$$

Eq. (2.21) becomes

$$0 = \varepsilon_{\nu\mu} + \varepsilon_{\mu\nu} \quad , \tag{2.23}$$

that is, $\varepsilon_{\mu\nu}$ is an antisymmetric tensor with (as advertised) $\frac{4.3}{2} = 6$ independent entries. Introduce the hermitean generators

$$L_{\mu\nu} \equiv i(x_\mu\partial_\nu - x_\nu\partial_\mu) \quad , \tag{2.24}$$

where

$$\partial_\mu \equiv \frac{\partial}{\partial x^\mu} = \left(\frac{\partial}{\partial t}\ ,\ \vec{\nabla}\right) \quad . \tag{2.25}$$

In terms of these we can write

$$\delta x^\mu = i\,\frac{1}{2}\,\varepsilon^{\rho\sigma} L_{\rho\sigma} x^\mu = \varepsilon^{\mu\rho} x_\rho \quad . \tag{2.26}$$

It is easy to see that the $L_{\mu\nu}$'s satisfy a Lie algebra

$$[L_{\mu\nu}, L_{\rho\sigma}] = ig_{\nu\rho} L_{\mu\sigma} - ig_{\mu\rho} L_{\nu\sigma} - ig_{\nu\sigma} L_{\mu\rho} + ig_{\mu\sigma} L_{\nu\rho} \ , \tag{2.27}$$

to be identified with the Lie algebra of SO(3,1). The most general representation of the generators of SO(3,1) that obeys the commutation relations (2.27) is given by

$$M_{\mu\nu} \equiv i(x_\mu \partial_\nu - x_\nu \partial_\mu) + S_{\mu\nu} \quad , \tag{2.28}$$

where the hermitean $S_{\mu\nu}$ satisfy the same Lie algebra as the $L_{\mu\nu}$ and commute with them. The hermitean generators $M_{ij}$, for i,j = 1,2,3, form an algebra among themselves

$$[M_{ij}, M_{k\ell}] = -i\delta_{jk} M_{i\ell} + i\delta_{ik} M_{j\ell} + i\delta_{j\ell} M_{ik} - i\delta_{i\ell} M_{jk} \quad , \tag{2.29}$$

which is that of the rotation group SU(2). A more familiar expression can be obtained by introducing the new operators

$$J_i \equiv \frac{1}{2}\,\varepsilon_{ijk} M_{jk} \quad , \tag{2.30}$$

where $\varepsilon_{ijk}$ is the Levi-Civitá symbol, totally antisymmetric in all of its three indices, and with $\varepsilon_{123} = +1$. (Repeated Latin indices are summed over.) Then, we find

$$[J_i, J_j] = i\varepsilon_{ijk} J_k \quad . \tag{2.31}$$

Define the boost generators

$$K_i \equiv M_{0i} \quad . \tag{2.32}$$

It follows from the Lie algebra that

$$[K_i, K_j] = -i\varepsilon_{ijk} J_k \quad , \tag{2.33}$$

$$[J_i, K_j] = i\varepsilon_{ijk} K_k \quad . \tag{2.34}$$

Here, both $K_i$ and $J_i$ are hermitean generators ; the $K_i$ are non-compact generators. We can disentangle these commutation relations by introducing the new linear combinations

$$N_i \equiv \frac{1}{2}(J_i + iK_i) \quad . \tag{2.35}$$

Although not hermitean, $N_i \neq N_i^\dagger$, they have the virtue of yielding simple commutation relations

$$[N_i, N_j^\dagger] = 0 \quad , \tag{2.36}$$

$$[N_i, N_j] = i\varepsilon_{ijk} N_k \quad , \tag{2.37}$$

$$[N_i^\dagger, N_j^\dagger] = i\varepsilon_{ijk} N_k^\dagger \quad . \tag{2.38}$$

This means that the $N_i$ and the $N_i^\dagger$ obey the Lie algebra of SU(2). We can therefore appeal to its well-known representation theory. In particular, we have two Casimir operators (operators that commute with all the generators)

$$N_i N_i \text{ with eigenvalues } n(n+1)$$

and

$$N_i^\dagger N_i^\dagger \text{ with eigenvalues } m(m+1) \quad ,$$

where m,n = 0, 1/2, 1, 3/2,..., using well-known results from the representation theory of the SU(2) (spin) group. These representations are labelled by the pair (n,m) while the states within a representation are further distinguished by the eigenvalues of $N_3$ and $N_3^\dagger$. Observe that the two SU(2)'s are not independent as they can be interchanged by the operation of parity, P,

$$J_i \to J_i \quad , \quad K_i \to -K_i \quad ,$$

and the operation of hermitean conjugation which changes the sign of i and therefore switches $N_i$ to $N_i^\dagger$. In general representations of the Lorentz group are neither parity nor (hermitean) conjugation eigenstates. Since $J_i = N_i + N_i^\dagger$, we can identify the spin of the representation with $m+n$. As an example consider the following representations:

a) (0,0) with spin zero is the scalar representation, and it has a well-defined parity (can appear as scalar or pseudoscalar);

b) (1/2,0) has spin 1/2 and represents a left-handed spinor (the handedness is a convention);

c) (0,1/2) describes a right-handed spinor.

These spinors have two components ("spin up" and "spin down"); they are called Weyl spinors. When parity is relevant, one considers the linear combination $(0,1/2) \oplus (1/2,0)$, which yields a Dirac spinor.

The fun thing is that given these spinor representations, we can generate any other representation by multiplying them together. This procedure is equivalent to forming higher spin states by taking the (Kronecker) product of many spin 1/2 states in the rotation group. Let us give a few examples:

a) $(1/2,0) \otimes (0,1/2) = (1/2,1/2)$ gives a spin 1 representation with four components. In tensor notation it will be written as a 4-vector.

b) $(1/2,0) \otimes (1/2,0) = (0,0) \oplus (1,0)$. Here the scalar representation is given by the antisymmetric product. The new representation (1,0) is represented by an antisymmetric, self-dual second rank tensor, i.e., a tensor $B_{\mu\nu}$ that obeys

$$B_{\mu\nu} = -B_{\nu\mu} \tag{2.39}$$

$$B_{\mu\nu} = \frac{i}{2} \varepsilon_{\mu\nu}{}^{\rho\sigma} B_{\rho\sigma} \quad , \tag{2.40}$$

where $\varepsilon^{\mu\nu\rho\sigma}$ is the Levi-Civitá symbol in four dimensions with $\varepsilon^{0123} = +1$, and total antisymmetry in its indices. The (0,1) representation would correspond to a tensor that is antiself-dual

$$B_{\mu\nu} = -\frac{i}{2} \varepsilon_{\mu\nu}{}^{\rho\sigma} B_{\rho\sigma} \quad . \tag{2.41}$$

For example, Maxwell's field strength tensor $F_{\mu\nu}$ transforms as $(0,1) \oplus (1,0)$ under the Lorentz group.

Finally, let us emphasize an important point. Suppose that we had considered LT's in the so-called "Euclidean space," where t is replaced by $\sqrt{-1}$ t. Then the commutation relations would have gone through except that $g_{\mu\nu}$ would have been replaced by $-\delta_{\mu\nu}$, the Kronecker delta, giving the Lie algebra of SO(4), the rotation group in four dimensions. The split-up of two commuting SU(2) groups is now achieved with the hermitean combinations $J_i \pm K_i$. These two SU(2)'s are completely independent since they cannot be switched by conjugation. Parity can still relate the two, but it loses much of its interest in Euclidean space where all directions are equivalent.

Problems.

A. Show that the Lorentz transformations satisfy the group axioms, i.e., if $L_1$ and $L_2$ are two LT's so is $L_1L_2$; the identity transformation exists, and if L is an LT, so is its inverse $L^{-1}$.

B. Show that det L and the sign of $\Lambda^0{}_0$ are Lorentz-invariant, and can therefore be used to catalog the Lorentz transformations.

C. Show that if L is a restricted LT (det L = +1, $\Lambda^0{}_0 \geq 1$), all Lorentz transformations can be written in the forms

L x space inversion for $L_-^\uparrow$ ,

L x time inversion for $L_-^\downarrow$ ,

L x space inversion x time inversion for $L_+^\downarrow$ .

D. Show that a restricted Lorentz transformation can be uniquely written as the product of a boost and a rotation.

*E. Index shuffling problem: Show that the components of a self-dual antisymmetric second rank tensor transform among themselves, i.e., irreducibly under the Lorentz group.

## 3. The Poincaré Group

Another fundamental principle is the invariance of the behavior of an isolated physical system under uniform translation in space and time. (This principle has to be generalized to arbitrary translations to generate gravitational interactions.) Such a transformation is given by

$$x^\mu \to x'^\mu = x^\mu + a^\mu \quad , \tag{3.1}$$

where $a^\mu$ is an arbitrary constant four-vector. Hence the general invariance group is a ten-parameter group called the Poincaré Group, under which

$$x^\mu \to x'^\mu = \Lambda^\mu{}_\nu x^\nu + a^\mu \quad , \tag{3.2}$$

The translations (3.1) do not commute with the LT's. Indeed consider two successive PG transformations with parameters $(\Lambda_1, a_1)$ and $(\Lambda_2, a_2)$

$$x^\mu \to \Lambda^\mu_{1\nu} x^\nu + a^\mu_1 \to \Lambda^\mu_{2\rho} \Lambda^\rho_{1\nu} x^\nu + \Lambda^\mu_{2\rho} a^\rho_1 + a^\mu_2 \quad , \tag{3.3}$$

and we see that the translation parameters get rotated. Nothing surprising here since this is what four-vectors do for a living! Such a coupling of the translation and Lorentz groups is called a semi-direct product. Still, as indicated by their name, the PG transformations form a group (see problem). In order to obtain the algebra of the generators, observe that we can write the change in x under a small translation as

$$\delta x^\mu = i\varepsilon^\rho P_\rho x^\mu \quad , \tag{3.4}$$

$$= \varepsilon^\mu \quad , \tag{3.5}$$

where $\varepsilon^\mu$ are the parameters, and

$$P_\rho = -i\partial_\rho \quad , \tag{3.6}$$

are the hermitean generators of the transformation. They clearly commute with one another

$$[P_\mu, P_\nu] = 0 \quad , \tag{3.7}$$

but not with the Lorentz generators (how can they? they are four-vectors!)

$$[M_{\mu\nu}, P_\rho] = -i\, g_{\mu\rho} P_\nu + i g_{\nu\rho} P_\mu \quad . \tag{3.8}$$

The commutation relations (3.7), (3.8) and those among the $M_{\mu\nu}$ define the Lie algebra of the Poincaré group. The "length" $P_\mu P^\mu$ of the vector $P_\rho$ is obviously invariant under Lorentz transformations and in view of (3.7) is therefore a Casimir operator. The other Casimir operator is not so obvious to construct, but as we just remarked, the length of any four-vector which commutes with the $P_\mu$'s will do. The Pauli-Lubanskí four-vector, $W^\mu$, is such a thing; it is defined by

$$W^\mu = \frac{1}{2}\, \varepsilon^{\mu\nu\rho\sigma} P_\nu M_{\rho\sigma} \quad . \tag{3.9}$$

Use of (3.7), (3.8) and of the antisymmetry of the Levi-Civitá symbol gives

$$[W^\mu, P^\rho] = 0 \quad , \tag{3.10}$$

while $W_\rho$ transforms as a four-vector,

$$[M_{\mu\nu}, W_\rho] = -i\, g_{\mu\rho} W_\nu + i\, g_{\nu\rho} W_\mu \quad . \tag{3.11}$$

Its length $W^\mu W_\mu$ is therefore a Casimir invariant. The most general representation of the ten Poincaré group generators is

$$P_\rho = -i\partial_\rho$$

$$M_{\mu\nu} = i(x_\mu \partial_\nu - x_\nu \partial_\mu) + S_{\mu\nu} \quad ,$$

so that

$$W^\mu = -\frac{i}{2}\, \varepsilon^{\mu\nu\rho\sigma} S_{\rho\sigma} \partial_\nu \quad . \tag{3.12}$$

The representation theory of the P.G. was worked out long ago by E. Wigner. Its representations fall into three classes.

1) The eigenvalue of $P_\rho P^\rho \equiv m^2$ is a real positive number. Then the eigenvalue of $W_\rho W^\rho$ is $-m^2 s(s+1)$, where s is the spin, which assumes discrete values $s = 0, \frac{1}{2}, 1, \ldots$ . This representation is labeled by the mass m and the spin s. States within the representation are distinguished by the third component of the spin $s_3 = -s, -s+1, \ldots, s-1, s$, and the continuous eigenvalues of $P_i$. Physically a state corresponds to a particle of mass m, spin s, three-momentum $p_i$ and spin projection $s_3$. Massive particles of spin s have $2s+1$ degrees of freedom.

2) The eigenvalue of $P_\rho P^\rho$ is equal to zero, corresponding to a particle of zero rest mass. $W_\rho W^\rho$ is also zero and, since $P^\rho W_\rho = 0$, it follows that $W_\mu$ and $P_\mu$ are proportional. The constant of proportionality is called the helicity, and is equal to $\pm$ s, where $s = 0, \frac{1}{2}, 1, \frac{3}{2}, \ldots$ is the spin of the representation. Thus massless particles of spin $s \neq 0$ have two degrees of freedom. They are further distinguished by the three values of their momenta along the x, y and z directions. Examples of particles falling in this category are the photon, with spin 1 and two states with helicity $\pm$ 1, the neutrino with helicity $\pm$ 1/2, and the graviton with two states of polarization $\pm$ 2.

3) $P_\rho P^\rho = 0$, but the spin is continuous. The length of W is minus the square of a positive number. This type of representation describes a particle of zero rest mass with an infinite number of polarization states labeled by a continuous variable. These do not seem to be realized in nature.

For more details on these, see V. Bargmann and E. P. Wigner, Proceedings of the National Academy of Sciences, Vol. 34, No. 5, 211 (1946). There are also "tachyon" representations with $P_\rho P^\rho < 0$, which we do not consider.

There are other representations of the Poincaré group; however they are not unitary. Quantum Mechanics allows for the identification of only the unitary representations with particle states. The Wigner representations are infinite dimensional, corresponding to particles with unbounded momenta. The situation is to be compared with that of the Lorentz group where we discussed finite dimensional but non unitary representations. The introduction of fields will enable us to make use of there representations.

Problems.

A. Show that the transformations (3.2) form a group.

B. Show that when $P_\rho P^\rho = m^2 > 0$, the eigenvalue of $W_\rho W^\rho$ is indeed $-m^2 s(s+1)$.

*C. Find the representation of the Poincaré Group generators on the spacelike surface $x_0 = 0$ in the case $m^2 = 0$, $s = 0$.

Hint: by setting $x_0 = 0$, one has to re-express its conjugate variable $P_0$ in terms of the remaining variables. Use a Casimir operator to do this. Then re-express all of the P.G. generators in terms of $x_i, P_i$ and $m^2$. See P.A.M. Dirac, Rev. Mod. Phys. 21, 392 (1949).

*D. Repeat the previous problem on the spacelike surface $x^0 = x^3$.

**E. Repeat problem D when $m^2 > 0$ and $s \neq 0$.

4. Behavior of Local Fields under the Poincaré Group

Consider an arbitrary function of the spacetime point P. In a given inertial frame, where P is located at $x^\mu$, this function will be denoted by $f(x^\mu)$; in another where P is at $x'^\mu$ it will be written as $f'(x'^\mu)$ because the functional relationship will in general be frame-dependent. Write for an infinitesimal transformation the change in the function as

$$\begin{aligned}\delta f &\equiv f'(x') - f(x) \\ &= f'(x+\delta x) - f(x) \\ &= f'(x) - f(x) + \delta x^\mu \partial_\nu f' + O(\delta x^2) \ . \end{aligned} \tag{4.1}$$

To $O(\delta x^\mu)$, we can replace $\partial_\mu f'$ by $\partial_\mu f$. Then

$$\delta f = \delta_0 f + \delta x^\mu \partial_\mu f \ , \tag{4.2}$$

where we have introduced the functional change at the same x

$$\delta_0 f \equiv f'(x) - f(x) \ . \tag{4.3}$$

The second term in eq. (4.2) is called the transport term. We can formally write (4.2) as an operator equation

$$\delta = \delta_0 + \delta x^\mu \partial_\mu \ . \tag{4.4}$$

Under a translation in spacetime, there is no change in a local field, that is

$$\delta f = 0 = \delta_0 f + \varepsilon^\mu \partial_\mu f \ , \tag{4.5}$$

or

$$\delta_0 f = -\varepsilon^\mu \partial_\mu f = -i\, \varepsilon^\mu P_\mu f \ , \tag{4.6}$$

with $P_\mu$ defined by (3.6). However, under Lorentz transformations, the situation is more complicated and requires several examples for clarification.

a) <u>The scalar field</u>

We build (or imagine) a function of $x^\mu$, $\phi(x)$, which has the same value when measured in different inertial frames related by a Lorentz transformation

$$\phi'(x') = \phi(x) \quad . \tag{4.7}$$

This condition defines a scalar field (under LT's). Specializing to an infinitesimal transformation, we have, using (4.7) and (4.2)

$$0 = \delta\phi = \delta_0\phi + \delta x^\mu \partial_\mu \phi \quad , \tag{4.8}$$

with $\delta x^\mu$ given by (2.26). Setting

$$\delta_0\phi = -\frac{i}{2}\varepsilon^{\rho\sigma}M_{\rho\sigma}\phi \quad , \tag{4.9}$$

and comparing with (4.8) tells us that for a scalar field the representation of the Lorentz group generators $M_{\mu\nu}$ is just $i(x_\mu\partial_\nu - x_\nu\partial_\mu)$. That is, the operator $S_{\mu\nu}$ we had introduced earlier vanishes when acting on a scalar field. We can see how a non-trivial $S_{\mu\nu}$ can arise by considering the construct $\partial_\mu\phi(x)$. Note that it is a scalar under translations just as $\phi$ was, because the derivative operator is not affected by translations, (true for uniform translations only!). We have

$$\delta\partial_\mu\phi = [\delta,\partial_\mu]\phi + \partial_\mu\delta\,\phi \quad . \tag{4.10}$$

Now $\delta\,\phi$ vanishes since $\phi$ is a Lorentz-scalar. However, from (4.4) we see that

$$[\delta,\partial_\mu] = [\delta_0,\partial_\mu] + [\delta x^\nu\partial_\nu,\partial_\mu] \quad . \tag{4.11}$$

Since $\delta_0$ does not change $x^\mu$, it commutes with $\partial_\mu$, but $\delta x^\nu$ does not. Evaluation of the last commutator yields

$$[\delta,\partial_\mu] = \varepsilon_\mu{}^\nu\partial_\nu \quad . \tag{4.12}$$

Putting it all together we find

$$\delta_0 \partial_\mu \phi = -\frac{i}{2}\,\varepsilon^{\rho\sigma} L_{\rho\sigma} \partial_\mu \phi - \frac{i}{2}\,(\varepsilon^{\rho\sigma} S_{\rho\sigma})_\mu{}^\nu \partial_\nu \phi \quad , \tag{4.13}$$

where

$$(S_{\rho\sigma})_\mu{}^\nu = i(g_{\rho\mu} g_\sigma^\nu - g_{\sigma\mu} g_\rho^\nu) \quad . \tag{4.14}$$

One can check that they obey the same commutation relations as the $L_{\mu\nu}$'s. Comparison with the canonical form

$$\delta_0(\text{anything}) = -\frac{i}{2}\,\varepsilon^{\rho\sigma} M_{\rho\sigma}(\text{anything}) \quad , \tag{4.15}$$

yields the representation of the Lorentz generator on the field $\partial_\mu \phi$. A field transforming like $\partial_\mu \phi(x)$ is called a vector field. Note that the role of the "spin part" of $M_{\mu\nu}$ is to shuffle indices.

A tensor field with many Lorentz indices will transform like (4.13). The action of $S_{\rho\sigma}$ on it will be the sum of expressions like (4.14), one for each index. For instance, the action of $S_{\rho\sigma}$ on a second rank tensor $B_{\mu\nu}$ is given by

$$(S_{\rho\sigma} B)_{\mu\nu} = -i(g_{\sigma\mu} B_{\rho\nu} - g_{\rho\mu} B_{\sigma\nu} + g_{\sigma\nu} B_{\mu\rho} - g_{\rho\nu} B_{\mu\sigma}) \quad . \tag{4.16}$$

It is now easy to make Poincaré invariants out of scalar fields. Candidates are any scalar function of $\phi(x)$ such as $\phi^n$, $\cos\phi(x)$, etc...., $\partial_\mu \partial^\mu \phi(x)$, $(\partial_\mu \phi)(\partial^\mu \phi)$ (see problem) etc... . However, the expression $x^\mu \partial_\mu \phi$ is Lorentz invariant but not Poincaré invariant.

b) <u>The spinor fields</u>

The spinor representations of the Lorentz group $(\frac{1}{2},0)$ and $(0,\frac{1}{2})$ are realized by two-component complex spinors. Call these spinors $\psi_L(x)$ and $\psi_R(x)$ respectively. The two-valued spinor indices are not written explicitly. [In the literature L-like (R-like) spinor indices appear dotted (undotted).]

We write

$$\psi_L(x) \rightarrow \psi_L'(x') = \Lambda_L \psi_L(x) \qquad \text{for } (\tfrac{1}{2},0)$$

$$\psi_R(x) \rightarrow \psi_R'(x') = \Lambda_R \psi_R(x) \qquad \text{for } (0,\tfrac{1}{2}) \quad ,$$

where $\Lambda_{R,L}$ are 2 x 2 matrices with complex entries. When the transformation is a rotation we know the form of $\Lambda_{L,R}$ from the spinor representation of SU(2)

$$\Lambda_{L(R)} = e^{i\frac{\vec{\sigma}}{2}\cdot\vec{\omega}} \quad . \qquad \text{(rotation)} \tag{4.17}$$

The $\omega^i$ are the rotation parameters and the $\sigma^i$ are the hermitean 2 x 2 Pauli spin matrices given by

$$\sigma^1 = \begin{pmatrix} 0 & 1 \\ 1 & 0 \end{pmatrix} \quad , \sigma^2 = \begin{pmatrix} 0 & -i \\ i & 0 \end{pmatrix} \quad , \sigma^3 = \begin{pmatrix} 1 & 0 \\ 0 & -1 \end{pmatrix} \quad . \tag{4.18}$$

They obey

$$\sigma^i\sigma^j = \delta^{ij} + i\epsilon^{ijk}\sigma^k \quad . \tag{4.19}$$

After thus identifying the rotation generators $J_i$ with $\frac{1}{2}\sigma^i$, we have to write the boost generators in this 2 x 2 notation. We already know that the $K_i$ will not be represented unitarily because the split-up into two SU(2) groups was not unitary. The representation

$$\vec{K} = -\frac{i}{2}\vec{\sigma} \quad , \tag{4.20}$$

satisfies all the required commutation relations. So we write

$$\Lambda_L = e^{\frac{i}{2}\vec{\sigma}\cdot(\vec{\omega} - i\vec{\nu})} \quad , \tag{4.21}$$

where $\vec{\nu}$ are the boost parameters associated with $\vec{K}$. Since the $(\frac{1}{2},0)$ and $(0,\frac{1}{2})$ representations are related by parity, we construct $\Lambda_R$ from $\Lambda_L$ by changing the sign of the boost parameters:

$$\Lambda_R = e^{\frac{i}{2}\vec{\sigma}\cdot(\vec{\omega} + i\vec{\nu})} \quad . \tag{4.22}$$

These explicit forms for $\Lambda_L$ and $\Lambda_R$ enable us to derive important properties. First we see that $\Lambda_L$ and $\Lambda_R$ are related by

$$\Lambda_L^{-1} = \Lambda_R^{\dagger} \quad . \tag{4.23}$$

Secondly, the magic of the Pauli matrices

$$\sigma^2 \sigma^i \sigma^2 = - \sigma^{i*} \quad , \tag{4.24}$$

where the star denotes ordinary complex conjugation, enables us to write

$$\sigma^2 \Lambda_L \sigma^2 = e^{-\frac{i}{2} \vec{\sigma}^* \cdot (\vec{\omega} - i\vec{\nu})} = \Lambda_R^* \quad . \tag{4.25}$$

Thirdly, the hermitean conjugate equation of (4.24) with the hermiticity of the Pauli matrices yields

$$\Lambda_L^T = \sigma^2 \Lambda_L^{-1} \sigma^2 \quad , \tag{4.26}$$

whence

$$\sigma^2 \Lambda_L^T \sigma^2 \Lambda_L = 1 \quad \text{or} \quad \Lambda_L^T \sigma^2 \Lambda_L = \sigma^2 \quad . \tag{4.27}$$

The same equation holds for $\Lambda_R$. These relations will prove useful in the construction of Lorentz-invariant expressions involving spinor fields. As a first application, under a Lorentz transformation,

$$\begin{aligned} \sigma^2 \psi_L^* &\to \sigma^2 \Lambda_L^* \psi_L^* \\ &= \sigma^2 \Lambda_L^* \sigma^2 \sigma^2 \psi_L^* \\ &= \Lambda_R \sigma^2 \psi_L^* \quad , \end{aligned} \tag{4.28}$$

using the complex conjugate of (4.25). Eq. (4.28) indicates that given a spinor $\psi_L$ which transforms as $(\frac{1}{2},0)$, we can construct a related spinor $\sigma^2 \psi_L^*$ which transforms as $(0,\frac{1}{2})$. In a similar way we can see that $\sigma^2 \psi_R^*$ transforms as $(\frac{1}{2},0)$ if $\psi_R$ transforms as $(0,\frac{1}{2})$.

We noted earlier that by taking the antisymmetric product of two $(\frac{1}{2},0)$ representations we can construct the scalar representation. We can now show this explicitly. Let $\psi_L$ and $\chi_L$ be two spinors that transform as $(\frac{1}{2},0)$. As a consequence of (4.27), under a Lorentz transformation

$$\chi_L^T \sigma^2 \psi_L \to \chi_L^T \Lambda_L^T \sigma^2 \Lambda_L \psi_L = \chi_L^T \sigma^2 \psi_L \quad . \tag{4.29}$$

This is our scalar. The scalar representation appeared group-theoretically in the antisymmetric product, so by taking $\chi_L = \psi_L$ the scalar invariant should not exist. Explicitly we find

$$\psi_L^T \sigma^2 \psi_L = (\psi_{L_1} \; \psi_{L_2}) \begin{pmatrix} 0 & -i \\ i & 0 \end{pmatrix} \begin{pmatrix} \psi_{L_1} \\ \psi_{L_2} \end{pmatrix} = -i \; \psi_{L_1} \psi_{L_2} + i \, \psi_{L_2} \psi_{L_1} \quad , \tag{4.30}$$

which vanishes if $\psi_{L_1}$ and $\psi_{L_2}$ are regular numbers. However, if $\psi_{L_1}$ and $\psi_{L_2}$ are taken to be Grassmann numbers which anticommute among themselves this scalar invariant will take on a non-zero value. In fact, as we shall see later, spinor fields will be taken to be classical Grassmann (anticommuting) numbers.

We can also take $\chi_L = \sigma^2 \psi_R^*$. Then the invariant becomes

$$i(\sigma^2 \psi_R^*)^T \sigma^2 \psi_L = -i \psi_R^\dagger \psi_L \quad . \tag{4.31}$$

None of these invariants are real. By switching L to R their complex conjugates can be obtained.

We can build a four-vector representation out of two spinors. The simplest way is to start from a single $\psi_L \sim (\frac{1}{2},0)$ since out of it we can build a $(0,\frac{1}{2})$ spinor and then multiply them together. One knows that the quantity $\psi_L^\dagger \psi_L$ is invariant under rotations which are represented by unitary operators on the spinors. Not so however under the boosts for which

$$\psi_L^\dagger \psi_L \to \psi_L^\dagger e^{\vec{\sigma}\cdot\vec{\nu}} \psi_L \simeq \psi_L^\dagger \psi_L + \vec{\nu} \cdot \psi_L^\dagger \vec{\sigma} \psi_L + O(\nu^2) \quad . \tag{4.32}$$

The new quantity, fortunately, behaves as

$$\psi_L^\dagger \sigma^i \psi_L \to \psi_L^\dagger e^{\frac{\vec{\sigma}\cdot\vec{\nu}}{2}} \sigma^i e^{\frac{\vec{\sigma}\cdot\vec{\nu}}{2}} \psi_L \simeq \psi^\dagger \sigma^i \psi_L + \frac{1}{2} \nu^j \psi_L^\dagger \{\sigma^i, \sigma^j\} \psi_L + O(\nu^2)$$

$$\simeq \psi_L^\dagger \sigma^i \psi_L + \nu^i \psi_L^\dagger \psi_L + O(\nu^2) \quad , \tag{4.33}$$

where { , } denotes the anticommutator. Thus under a boost these quantities transform into one another

$$\delta\ \psi_L^\dagger \psi_L = \nu^i \psi_L^\dagger \sigma^i \psi_L$$

$$\delta\ \psi_L^\dagger \sigma^i \psi_L = \nu^i \psi_L^\dagger \psi_L \quad , \tag{4.34}$$

and $\psi_L^\dagger \sigma^i \psi_L$ behaves as a three-vector under rotations. Eq. (4.34) compares with the transformation law of a four-vector

$$\delta V^\mu = \varepsilon^\mu{}_\nu V^\nu \quad , \tag{4.35}$$

where $\varepsilon^{oi} = -\nu^i$ are the boost parameters. Thus the quantity

$$i\psi_L^\dagger \sigma^\mu \psi_L = i(\psi_L^\dagger \psi_L,\ \psi_L^\dagger \sigma^i \psi_L) \quad , \tag{4.36}$$

is a four-vector. We have identified $\sigma^o$ with the 2 x 2 unit matrix. Another four-vector can be obtained starting from $\psi_R$ and changing the sign of the space components

$$i\psi_R^\dagger \bar{\sigma}^\mu \psi_R \equiv i(\psi_R^\dagger \psi_R,\ -\psi_R^\dagger \vec{\sigma} \psi_R) \quad . \tag{4.37}$$

These two vectors are real since $\psi_L$ and $\psi_R$ are Grassmann variables: $(\psi_L^\dagger \psi_R)^* = \psi_L^T \psi_R^* = -\psi_R^\dagger \psi_L$, and their sum (difference) is even (odd) under parity.

Either of these combined with another four-vector can yield Lorentz invariants. The simplest four-vector, as we saw earlier, is the derivative operator $\partial_\mu$ which has the added virtue of preserving translational invariance. Since $\partial_\mu$ can act on any of the fields we have the following invariants that are bilinear in the spinor fields

$$\partial_\mu \psi_R^\dagger \bar{\sigma}^\mu \psi_R, \qquad \psi_R^\dagger \bar{\sigma}^\mu \partial_\mu \psi_R, \qquad \partial_\mu \psi_L^\dagger \sigma^\mu \psi_L, \qquad \psi_L^\dagger \sigma^\mu \partial_\mu \psi_L \quad . \tag{4.38}$$

The derivative operator is understood to act to the right and on its nearest neighbor only. These Lorentz invariants are no longer real; however, real linear combinations can be formed, such as

$$\frac{1}{2} \psi_L^\dagger \sigma^\mu \partial_\mu \psi_L - \frac{1}{2} \partial_\mu \psi_L^\dagger \sigma^\mu \psi_L \equiv \frac{1}{2} \psi_L^\dagger \sigma^\mu \overleftrightarrow{\partial}_\mu \psi_L \quad , \tag{4.39}$$

and a similar expression with L replaced by R and $\sigma^\mu$ by $\bar{\sigma}^\mu$.

If parity is a concern, one has to assemble $(\frac{1}{2},0)$ and $(0,\frac{1}{2})$ representations. Since we cannot equate $\psi_L$ with $\sigma_2 \psi_L^*$ without leading to a contradiction or to $\psi_L = 0$, we have to build a four-component spinor called a Dirac spinor

$$\Psi \equiv \begin{pmatrix} \psi_L \\ \psi_R \end{pmatrix} \quad , \tag{4.40}$$

on which the operation of parity is well defined.

$$P: \quad \Psi \to \Psi^P = \begin{pmatrix} \psi_R \\ \psi_L \end{pmatrix} = \begin{pmatrix} 0 & 1 \\ 1 & 0 \end{pmatrix} \Psi \equiv \gamma_0 \Psi \quad , \tag{4.41}$$

where we have defined the 4 x 4 matrix $\gamma_0$. We can project onto the left and right spinors by means of the projection operators

$$\frac{1}{2} (1 \pm \gamma_5) \quad , \tag{4.42}$$

where (in 2 x 2 block form)

$$\gamma_5 = \begin{pmatrix} 1 & 0 \\ 0 & -1 \end{pmatrix} \quad . \tag{4.43}$$

We can rewrite all the invariants we have built in terms of Dirac spinors.

For instance,

$$\psi_R^\dagger\psi_L + \psi_L^\dagger\psi_R = \Psi^\dagger\gamma^0\Psi \equiv \bar{\Psi}\Psi \quad , \tag{4.44}$$

where $\bar{\Psi} = \Psi^\dagger\gamma^0$ is the Pauli adjoint. Since (4.44) is Lorentz invariant it transforms contragrediently to $\Psi$. Similarly

$$\frac{1}{2}\,(\psi_L^\dagger\sigma^\mu\overleftrightarrow{\partial}_\mu\psi_L + \psi_R^\dagger\bar{\sigma}^\mu\overleftrightarrow{\partial}_\mu\psi_R) = \frac{1}{2}\,\bar{\Psi}\gamma^\mu\overleftrightarrow{\partial}_\mu\Psi \quad , \tag{4.45}$$

where we have introduced the 4 x 4 matrices

$$\gamma^i \equiv \begin{pmatrix} 0 & -\sigma^i \\ \sigma^i & 0 \end{pmatrix} \quad . \tag{4.46}$$

Since (4.45) is Lorentz invariant the $\mu$ index on the $\gamma$-matrices is a true four-vector index. They are, of course, the Dirac matrices in the Weyl representation. They obey

$$\{\gamma^\mu,\gamma^\nu\} = 2g^{\mu\nu} \quad . \tag{4.47}$$

The $\gamma_5$ matrix is related to the others by

$$\gamma_5 = i\gamma^0\gamma^1\gamma^2\gamma^3 \quad . \tag{4.48}$$

By means of the equivalence under Lorentz transformations of $\psi_L$ and $\sigma^2\psi_R{}^*$, we can build an associated Dirac spinor

$$\Psi^c \equiv \begin{pmatrix} \sigma^2\psi_R{}^* \\ -\sigma^2\psi_L{}^* \end{pmatrix} \quad . \tag{4.49}$$

Note that

$$(\Psi^c)^c = \Psi \quad . \tag{4.50}$$

$\Psi^c$ is called the (charge) conjugate spinor. Since $\sigma^2\psi_L{}^*$ transforms like $\psi_R$, we can construct a special type of four-component spinor called a Majorana spinor

$$\psi^M = \begin{pmatrix} \psi_L \\ -\sigma^2 \psi_L^* \end{pmatrix} \quad , \tag{4.51}$$

It is self-conjugate under (charge) conjugation. The Majorana spinor is a Weyl spinor in four-component form. Its physical interpretation will be discussed when we build Actions out of spinor fields. Suffice it to say that Majorana and/or Weyl spinors describe objects with half as many degrees of freedom as Dirac spinors.

As we remarked at the end of Section 2, one cannot relate in Euclidean space the two SU(2) groups that form the (Euclidean) Lorentz group. We can now see explicitly why. Since both are unitarily realized we have the new expressions

$$\Lambda_L \to \Lambda_L^E = e^{\frac{i}{2}\vec{\sigma}\cdot(\vec{\omega}+\vec{\nu})}$$

$$\Lambda_R \to \Lambda_R^E = e^{\frac{i}{2}\vec{\sigma}\cdot(\vec{\omega}-\vec{\nu})} \quad , \tag{4.52}$$

and there is no possible relation between $\Lambda_L^E$ and $\Lambda_R^E$. Thus Majorana spinors do not exist in Euclidean space because one cannot relate $\psi_L^E$ to $\psi_R^E$. However one is free to deal with $\psi_L^E$ and $\psi_R^E$ separately and even form Dirac spinors $\Psi^E$, with the understanding that the conjugation operation introduced earlier ceases to exist.

c) The vector field

The vector field transforms according to the $(\frac{1}{2}, \frac{1}{2})$ representation. We have already seen the effect of $S_{\rho\sigma}$ on an arbitrary vector field, $A_\mu(x)$. We might add that there is another representation of the vector field as a hermitean 2 x 2 matrix

$$A^\mu \to A = \begin{pmatrix} A^0 + A^3 & A^1 + iA^2 \\ A^1 - iA^2 & A^0 - A^3 \end{pmatrix} \quad . \tag{4.53}$$

Lorentz transformations are defined to be those that preserve the condition $A = A^\dagger$ and leave detA invariant. One can consider many invariants such as

$$A_\mu(x)A^\mu(x), \qquad \partial_\mu A_\nu(x)\partial^\nu A^\mu(x), \qquad \partial_\mu A_\nu(x)\partial^\mu A^\nu(x), \qquad \partial^\mu A_\mu(x), \text{ etc.} \ldots .$$

Since parity is defined for the vector representation we can define both vector and axial vector fields.

d) The spin-3/2 field

There are several ways to define a spin-3/2 field depending on the role we want parity to play. One procedure is to take the product of three $(\frac{1}{2},0)$ representations

$$(\tfrac{1}{2},0)\otimes(\tfrac{1}{2},0)\otimes(\tfrac{1}{2},0) = (\tfrac{3}{2},0)\oplus(\tfrac{1}{2},0)\oplus(\tfrac{1}{2},0) \quad . \tag{4.54}$$

The spin-3/2 corresponds to the completely symmetric product (the two $(\frac{1}{2},0)$ have mixed symmetry). Thus we can represent a spin-3/2 field by a field totally symmetric in the interchange of its three L-like spinor indices. Its transformation properties are obtained by a suitable generalization of the action of $S_{\rho\sigma}$ on an L-like index (see problem). The parity eigenstate is then a combination of $(\frac{3}{2},0)\oplus(0,\frac{3}{2})$. However this representation is rather awkward because of the many indices on the field symbol.

A more convenient representation of a spin-$\frac{3}{2}$ field is obtained through the product of a vector and a spinor

$$(\tfrac{1}{2},\tfrac{1}{2})\otimes(\tfrac{1}{2},0) = (1,\tfrac{1}{2})\oplus(0,\tfrac{1}{2}) \quad . \tag{4.55}$$

The corresponding field quantity has a four-vector and a spinor index. The parity eigenstate is the four-component "Rarita-Schwinger" field

$$\Psi_\mu = \begin{pmatrix} \psi_{\mu L} \\ \psi_{\mu R} \end{pmatrix} \quad , \tag{4.56}$$

where the spinor indices have been suppressed. As written $\Psi_\mu$ describes all the states in the product (4.55) together with their parity partners. Hence we must project out the extra $(\frac{1}{2},0)\oplus(0,\frac{1}{2})$ components, in a Lorentz invariant way. We impose on it the subsidiary conditions

$$\sigma^\mu \psi_{\mu L} = \bar{\sigma}^\mu \psi_{\mu R} = 0 \quad , \tag{4.57}$$

or in Dirac language

$$\gamma^\mu \Psi_\mu = 0 \quad . \tag{4.58}$$

The same kind of covariants and invariants can be assembled as in the spinor case except we have the extra vector index to play with. A sample of invariants is

$$\psi^T_{\mu L}\sigma_2 \psi^\mu_L, \quad \psi^T_{\mu R}\sigma_2 \psi^\mu_R, \quad \psi^\dagger_{\mu R}\psi^\mu_L \quad ,\ldots \tag{4.59}$$

We can use the set of vectors

$$\psi^\dagger_{\mu L}\sigma_\rho \psi_{\nu L}\varepsilon^{\mu\rho\nu\sigma}, \quad \psi^\dagger_{\mu R}\bar{\sigma}_\rho \psi_{\nu R}\varepsilon^{\mu\rho\nu\sigma} \quad , \tag{4.60}$$

in combination with $\partial_\rho$ to make invariants of the form

$$\partial_\mu \psi^\dagger_{\sigma L}\sigma_\rho \psi_{\nu L}\varepsilon^{\mu\rho\nu\sigma}, \text{ etc.} \quad . \tag{4.61}$$

The real scalar invariant is given by

$$\frac{1}{2}\left(\psi^\dagger_{\mu L}\sigma_\rho \overleftrightarrow{\partial}_\sigma \psi_{\nu L} - \psi^\dagger_{\mu R}\bar{\sigma}_\rho \overleftrightarrow{\partial}_\sigma \psi_{\nu R}\right)\varepsilon^{\mu\nu\rho\sigma} = \frac{1}{2}\bar{\Psi}_\mu \gamma_5 \gamma_\rho \overleftrightarrow{\partial}_\sigma \Psi_\nu \varepsilon^{\mu\nu\rho\sigma} \quad . \tag{4.62}$$

The presence of the relative minus sign i.e., of the $\gamma_5$, is dictated by the parity properties of the $\varepsilon$-operation. Finally, note that we can, as in the spin-$\frac{1}{2}$ case, impose a Majorana condition on the R-S fields.

e) <u>The spin-2 field</u>

Again there are many possible ways to describe a spin-2 field: (2,0), (0,2), (1,1). We choose the latter. It appears in the product

$$(\tfrac{1}{2},\tfrac{1}{2}) \otimes (\tfrac{1}{2},\tfrac{1}{2}) = [(0,0) \oplus (1,1)]_S \oplus [(0,1) \oplus (1,0)]_A \quad , \tag{4.63}$$

where S(A) denotes the symmetric (antisymmetric) part. Thus a spin-2 field can be described by a 2nd-rank symmetric tensor $h_{\mu\nu}$. The scalar component corresponds to its trace which can be subtracted by the tracelessness condition

$$g^{\mu\nu} h_{\mu\nu}(x) = 0 \quad . \tag{4.64}$$

Invariants are easily constructed by saturating the vector indices, and by the use of $\partial_\rho$. Examples are

$$h_{\mu\nu} h^{\mu\nu}, \quad \partial_\rho h_{\mu\nu} \partial^\rho h^{\mu\nu}, \quad \partial_\rho h_{\mu\nu} \partial^\mu h^{\rho\nu}, \text{ etc.,} \quad . \tag{4.65}$$

This tensor field appears in General Relativity where it is used to describe the graviton.

To conclude this section we note that many other fields with definite Lorentz transformation properties can be constructed. However we have chosen to discuss in some detail only those which prove useful in the description of physical phenomena. They are the ones to which we can associate fundamental particles. To Dirac spinors we associate charged fermions such as electron, muon, tau, quarks; to Weyl spinors the neutrinos $\nu_e, \nu_\mu, \nu_\tau$. To vector fields the photon, the gluons that mediate Strong Interactions, and the W-bosons that mediate Weak Interactions; and finally, to a tensor field the graviton that mediates gravitation.

Problems.

A. Build explicitly the action of $S_{\rho\sigma}$ on $\psi_L$ and $\psi_R$.

B. Write the action of $S_{\rho\sigma}$ on a Dirac spinor in terms of the Dirac matrices, i.e., in a representation independent way.

C. Build explicitly a field bilinear in the spinor $\chi_L$ and $\psi_L$ that transforms as the (1,0) representation. Can you build the same field out of one $\psi_L$ field?

D. Find the form Lorentz transformations take acting on the matrix (4.53) in terms of $\Lambda_L$ and $\Lambda_R$.

E. Given $\psi_L(x)$, $A_\mu(x)$, build at least two invariants where the two fields appear.

F. Find a representation for the Dirac matrices where the components of a Majorana spinor are real. Such a representation is called the Majorana representation.

## 5. General Properties of the Action

In the previous sections we have learned how to build Poincaré invariant expressions out of fields which have well-defined transformation properties under the Poincaré group. Now comes the time to assemble these invariants into Actions that describe reasonable physical theories. The requirement of Poincaré invariance insures that these theories will obey the axioms of Special Relativity. Yet as we become more and more adept at this game we will learn that there are too many candidate theories and that the single prescription of Poincaré invariance is not sufficient to pinpoint the true Action of the world. In an attempt to narrow our search we try to enumerate certain ad hoc prescriptions which have been found sufficient to yield acceptable theories.

First we deal with Action Functionals of the form

$$S \equiv \int_{\tau_1}^{\tau_2} d^4x \, \mathcal{L} \quad , \tag{5.1}$$

where $\tau_1$ and $\tau_2$ denote the limits of integrations and

$$d^4x = dt dx^1 dx^2 dx^3 \quad , \tag{5.2}$$

is the integration measure in four-dimensional Minkowski space. Sometimes we might alter for mathematical purposes, the number of space-time dimensions, or even consider the measure in Euclidian space with $d^4x$ replaced by the Euclidean measure

$$d^4\bar{x} = d\bar{x}^0 d\bar{x}^1 d\bar{x}^2 d\bar{x}^3 \quad , \tag{5.3}$$

where $\bar{x}^0 = ix^0$, $\bar{x}^i = x^i$. The integrand, $\mathcal{L}$, is called the Lagrange density, Lagrangian for short. It is a function of the fields and their derivatives only so as to ensure translation invariance. Also it depends on the fields taken at one space-time point $x^\mu$ only, leading to a local field theory. This is clearly the simplest choice to make: one can easily imagine non-local field theories but they are necessarily more complicated in nature. In fact our faith in local field theory is such that we believe it to be sufficient even in the description of non-local phenomena!

Secondly we demand that S be real. It is found (in retrospect) that this is a crucial requirement in obtaining satisfactory quantum field theories where total probability is conserved. In Classical Physics a complex potential leads to absorption, i.e., disappearance of matter into nothing; it is not a satisfactory situation.

Third we demand that S leads to classical equations of motions that involve no higher than second-order derivatives. Classical systems described by higher order differential equations will typically develop non-causal solutions. A well-known example is the Lorentz-Dirac equation of Electrodynamics. It is a third order differential equation that incorporates the effects of radiation reaction and shows non-causal effects such as preacceleration of particles yet to be hit by radiation. To satisfy this requirement we take $\mathcal{L}$ to contain at most two $\partial_\mu$ operations. As a consequence the classical equations (or their square for spinor fields) will display the operator $\partial_\mu \partial^\mu$ acting on a field. When the equations of motion turn into an eigenvalue condition on this operator, we will say we are dealing with a free field theory because we can identify $\partial_\mu \partial^\mu$ with a Casimir operator of the Poincaré group, with the equations of motion restricting us to a (free) particle representation.

Fourth we take S to be invariant under the Poincaré group, as we have already discussed.

Fifth there may be further invariance requirements on S. In fact the phenomenological success of gauge theories suggests that the relevant Action Functional is invariant under peculiar new transformations which involve new degrees of freedom such as electric charge, weak charge, color charge and other charges yet to be discovered. Gauge theories are described by actions which are invariant under local (i.e., x-dependent) transformations among these internal degrees of freedom. We will be much more specific on this subject later on.

In classical theory the Action has the definite units of angular momentum $ML^2T^{-1}$ or equivalently units of $\hbar$. In a natural unit system where $\hbar = 1$, S is taken to be "dimensionless." Then in four dimensions the Lagrange density has natural dimensions of $L^{-4}$.

Consider the action

$$S(\tau_1,\tau_2,[\Phi]) \equiv \int_{\tau_1}^{\tau_2} d^4x\, \mathscr{L}(\Phi,\partial_\mu\Phi) \quad , \tag{5.4}$$

where $\Phi(x)$ is any local field or any collection of local fields (it could be scalars, spinors,...; we suppress all indices); $\tau_1$ and $\tau_2$ are the boundaries of integrations. Under an arbitrary change in $\Phi, \delta\Phi$, the resulting change in S is

$$\delta S = \int_{\tau_1}^{\tau_2} d^4x\, \delta\mathscr{L} \tag{5.5}$$

$$= \int_{\tau_1}^{\tau_2} d^4x \left[\frac{\partial\mathscr{L}}{\partial\Phi}\delta\Phi + \frac{\partial\mathscr{L}}{\partial[\partial_\mu\Phi]}\delta(\partial_\mu\Phi)\right] \quad . \tag{5.6}$$

Since x does not change in this variation

$$\delta(\partial_\mu\Phi) = \partial_\mu\delta\Phi \quad . \tag{5.7}$$

Then use of the chain rule yields

$$\delta S = \int_{\tau_1}^{\tau_2} d^4x \left[\frac{\partial\mathscr{L}}{\partial\Phi} - \partial_\mu\frac{\partial\mathscr{L}}{\partial[\partial_\mu\Phi]}\right]\delta\Phi + \int_{\tau_1}^{\tau_2} d^4x\, \partial_\mu\left[\frac{\partial\mathscr{L}}{\partial[\partial_\mu\Phi]}\delta\Phi\right] \quad . \tag{5.8}$$

The last term is just a surface term which can be rewritten as a surface integral

$$\oint_\sigma d\sigma_\mu \frac{\partial \mathcal{L}}{\partial[\partial_\mu \Phi]} \delta\Phi \qquad , \qquad (5.9)$$

where $\sigma$ is the boundary surface and $d\sigma_\mu$ the surface element. Finally we demand that $\delta\Phi$ vanishes on $\sigma$. By requiring that S be stationary under an arbitrary change $\delta\Phi$ vanishing on the boundaries we obtain the Euler-Lagrange equations

$$\partial_\mu \frac{\partial \mathcal{L}}{\partial[\partial_\mu \Phi]} - \frac{\partial \mathcal{L}}{\partial \Phi} = 0 \qquad , \qquad (5.10)$$

which are the classical field equations for the system described by S. We can identify (5.10) with the functional derivative of S with respect to $\Phi$. Again note that it is well defined only for variations that vanish on the boundaries of integration. As an important consequence of dropping the surface term observe that the same equations of motion would have been obtained if we had started from the new Lagrangian density

$$\mathcal{L}' = \mathcal{L} + \partial_\mu \Lambda^\mu \qquad , \qquad (5.11)$$

with $\Lambda^\mu$ arbitrary. Such a change in $\mathcal{L}$ produces a change in S that entirely depends on the choice of boundary conditions on the fields that enter in $\mathcal{L}'$ [this freedom is no longer tolerated in the presence of gravity]. In classical mechanics, the transformation between $\mathcal{L}$ and $\mathcal{L}'$ is called a canonical transformation. In addition note that the addition of a constant to $\mathcal{L}$ does not change the nature of the classical system although it affects the coupling of the system to gravity as it generates an infinite energy.

Next we consider the response of the Action to yet unspecified (but not arbitrary) changes in the coordinates and in the fields, $\delta x^\mu$ and $\delta\Phi$ respectively.

To the coordinate change corresponds the change in the integration measure given by the Jacobi formula

$$\delta(d^4x) = d^4x\ \partial_\mu \delta x^\mu \quad . \tag{5.12}$$

Thus it follows that

$$\delta S = \int_{\tau_1}^{\tau_2} d^4x\ [\partial_\mu \delta x^\mu \mathcal{L} + \delta\mathcal{L}\ ] \quad . \tag{5.13}$$

Use of (4.4) yields

$$\delta\mathcal{L} = \delta x^\mu \partial_\mu \mathcal{L} + \delta_0 \mathcal{L} \tag{5.14}$$

$$= \delta x^\mu \partial_\mu \mathcal{L} + \frac{\partial\mathcal{L}}{\partial\Phi} \delta_0 \Phi + \frac{\partial\mathcal{L}}{\partial[\partial_\mu \Phi]} \delta_0 \partial_\mu \Phi \quad . \tag{5.15}$$

Now $\delta_0$ is just a functional change, therefore

$$\delta_0 \partial_\mu \Phi = [\delta_0, \partial_\mu]\Phi + \partial_\mu \delta_0 \Phi \quad , \tag{5.16}$$

$$= \partial_\mu \delta_0 \Phi \quad . \tag{5.17}$$

Use of the chain rule yields

$$\delta\mathcal{L} = \delta x^\rho \partial_\rho \mathcal{L} + \left[\frac{\partial\mathcal{L}}{\partial\Phi} - \partial_\mu \frac{\partial\mathcal{L}}{\partial[\partial_\mu \Phi]}\right] \delta_0 \Phi + \partial_\mu \left(\frac{\partial\mathcal{L}}{\partial[\partial_\mu \Phi]}\ \delta_0 \Phi\right) . \tag{5.18}$$

By invoking the classical equations of motion the change in the Action is

$$\delta S = \int_{\tau_1}^{\tau_2} d^4x \left[\mathcal{L}\ \partial_\mu \delta x^\mu + \delta x^\mu \partial_\mu \mathcal{L} + \partial_\mu \left(\frac{\partial\mathcal{L}}{\partial[\partial_\mu \Phi]} \delta_0 \Phi\right)\right] \quad , \tag{5.19}$$

$$= \int_{\tau_1}^{\tau_2} d^4x\ \ \partial_\mu \left[\mathcal{L}\delta x^\mu + \frac{\partial\mathcal{L}}{\partial[\partial_\mu \Phi]}\ \delta_0 \Phi\right] \quad . \tag{5.20}$$

Alternatively we can, by re-expressing $\delta_0$ in terms of $\delta$, obtain

$$\delta S = \int_{\tau_1}^{\tau_2} d^4x \, \partial_\mu \left[ \left( \mathcal{L} g^\mu_\rho - \frac{\partial \mathcal{L}}{\partial [\partial_\mu \Phi]} \partial_\rho \Phi \right) \delta x^\rho + \frac{\partial \mathcal{L}}{\partial [\partial_\mu \Phi]} \delta \Phi \right] \quad . \tag{5.21}$$

Next we write the variations in the coordinates and the fields in terms of the global (i.e., x-independent) parameters of the transformation

$$\delta x^\rho = \frac{\delta x^\rho}{\delta \omega^a} \delta \omega^a \quad , \tag{5.22}$$

$$\delta \Phi = \frac{\delta \Phi}{\delta \omega^a} \delta \omega^a \quad . \tag{5.23}$$

Here a is an index which enumerates the parameters of the transformation. Consequently

$$\delta S = \int_{\tau_1}^{\tau_2} d^4x \, \partial_\mu \left[ \left( \mathcal{L} g^\mu_\rho - \frac{\partial \mathcal{L}}{\partial [\partial_\mu \Phi]} \partial_\rho \Phi \right) \frac{\delta x^\rho}{\delta \omega^a} + \frac{\partial \mathcal{L}}{\partial [\partial_\mu \Phi]} \frac{\delta \Phi}{\delta \omega^a} \right] \delta \omega^a . \tag{5.24}$$

If the Action is invariant under the transformations (5.22) and (5.23), it follows that the current density

$$j^\mu_a \equiv - \left[ \mathcal{L} g^\mu_\rho - \frac{\partial \mathcal{L}}{\partial [\partial_\mu \Phi]} \partial_\rho \Phi \right] \frac{\delta x^\rho}{\delta \omega^a} - \frac{\partial \mathcal{L}}{\partial [\partial_\mu \Phi]} \frac{\delta \Phi}{\delta \omega^a} \quad , \tag{5.25}$$

is conserved, i.e.,

$$\partial_\mu j^\mu_a = 0 \quad . \tag{5.26}$$

This conservation equation is a consequence of the validity of (5.24) for all $\delta\omega^a$. We have just proved E. Noether's theorem for classical field theory which relates a conservation equation to an invariance of the Action.

On the other hand, if the Action is not conserved the conservation equation is no longer valid. For example it has a particularly simple form when $\delta x^\rho = 0$

$$\partial_\mu j^\mu_a = - \frac{\partial \mathcal{L}}{\delta \omega^a} \quad . \tag{5.27}$$

Now suppose we have found a set of transformations (5.22) and (5.23) which leave the Action invariant. Integrate (5.26) over an infinite range in the space directions and a finite interval over the time direction. We get

$$0 = \int_{T_1}^{T_2} dx^0 \int_{-\infty}^{+\infty} d^3x \, \partial_\mu j^\mu_a = \int_{T_1}^{T_2} dx^0 \frac{\partial}{\partial x^0} \int_{-\infty}^{+\infty} d^3x \, j^0_a + \int_{T_1}^{T_2} dx^0 \int d^3x \, \partial_i j^i_a \; . \tag{5.28}$$

The last term vanishes if the space boundaries are suitably chosen. We are left with

$$0 = \int_{-\infty}^{+\infty} d^3x \, j^0_a(T_1,\vec{x}) - \int_{-\infty}^{+\infty} d^3x \, j^0_a(T_2,\vec{x}) \qquad . \tag{5.29}$$

Therefore the charges defined by

$$Q_a(T) \equiv \int_{-\infty}^{+\infty} d^3x \, j^0_a(T,\vec{x}) \qquad , \tag{5.30}$$

are time independent since the above argument does not depend on the choice of the time integration limits:

$$\frac{dQ_a}{dt} = 0 \qquad . \tag{5.31}$$

So, from $\delta S = 0$, we have been able to deduce the existence of conserved charges.

When the parameters of the transformations are dimensionless as in the case of Lorentz transformations and internal transformations (but not translations) the resulting currents always have the dimensions of $L^{-D+1}$ in D dimensions, so that the charges are dimensionless.

Further we remark that a conserved current does not have a unique definition since we can always add to it the four-divergence of an antisymmetric tensor $\partial_\rho t^{\rho\mu}_a$. This is most clearly seen in the light of (5.11). Also since $j^\mu_a$ is conserved only after use of the equations of motion we have the freedom to add to it any quantity which vanishes by virtue of the equations of motion. This is particularly relevant when a is a Lorentz index, as in the case of a translation

$$\delta x^\mu = \varepsilon^\mu \quad : \frac{\delta x^\mu}{\delta\omega^a} \to g^\mu_\rho \quad , \ (a = \rho) \qquad , \tag{5.32}$$

or a Lorentz transformation

$$\delta x^\mu = \varepsilon^{\mu\nu} x_\nu \ : \frac{\delta x^\mu}{\delta\omega^a} \to \frac{1}{2}\,(g^\mu_\rho x_\nu - g^\mu_\nu x_\rho) \qquad . \tag{5.33}$$

In the latter case the parameter a is replaced by the antisymmetric pair $[\rho\nu]$.

Finally we remark that a transformation that leaves S invariant may change $\mathcal{L}$ by a total divergence, which means that the symmetry operation is accompanied by a canonical transformation. In quantum theory where one cannot rely on the equations of motion the statement of current conservation will lose its significance but will be replaced by relations between Green's functions known as Ward identities.

Problems

A. Consider the conformal transformations

$$\delta x^{\mu} = (2x^{\mu}x^{\rho} - g^{\mu\rho}x_{\tau}x^{\tau})\, c_{\rho} \quad ,$$

where $c_{\rho}$ is an infinitesimal four-vector. Show that these transformations, together with the dilatations

$$\delta x^{\mu} = \alpha x^{\mu} \qquad , \alpha \text{ infinitesimal},$$

and the Poincaré group transformations form a fifteen-parameter group, called the Conformal Group.

Find the generators of this group and their commutation relations.

B. The dilatations and the Poincaré group form a group called the Weyl group. Under dilatations a field $\Phi$ of dimension d transforms as

$$\delta\Phi = + d\Phi \ .$$

Assuming that S is invariant under the Weyl group and contains $\Phi$, find the conserved current corresponding to dilatations.

## 6. The Action for Scalar Fields

The most general form of the Lagrange density containing only one scalar field $\phi(x)$ is

$$\mathcal{L} = \frac{1}{2} \partial_\mu \phi(x) \partial^\mu \phi(x) - V(\phi(x)) \quad , \tag{6.1}$$

where the $\frac{1}{2}$ is purely conventional and V is a scalar function. The first term is called the kinetic term, the second the potential. The kinetic term has a larger invariance group than the potential: it is invariant under a shift of the field $\phi \to \phi + a$, where a is a global constant. In four dimensions, $\phi(x)$ therefore has natural dimensions of $L^{-1}$ (or of mass). The form of $V(\phi(x))$ is unrestricted in the classical theory. Some special examples are

$$\mathcal{L}_0 = \frac{1}{2} \partial_\mu \phi \partial^\mu \phi - \frac{1}{2} m^2 \phi^2 \quad , \tag{6.2}$$

where m has the dimension of mass. This Action describes a free particle of mass m (as we shall deduce later from its path integral treatment). Note that $\mathcal{L}_0$ is also invariant under the discrete symmetry

$$\phi(x) \to - \phi(x) \quad . \tag{6.3}$$

A more complicated example is given by

$$\mathcal{L} = \mathcal{L}_0 - \frac{\lambda}{4!} \phi^4 \quad , \tag{6.4}$$

which describes a self-interacting theory. Observe that (in four dimensions) $\lambda$ is a dimensionless parameter. The minus sign ensures the positivity of V. This Action leads to an acceptable quantum field theory. Another popular example is the Sine-Gordon Lagrangian

$$\mathcal{L} = \frac{1}{2} \partial_\mu \phi \partial^\mu \phi + \frac{m^4}{\lambda} \left[\cos \frac{\sqrt{\lambda}\phi}{m} - 1\right] \quad , \tag{6.5}$$

where $\lambda$ is dimensionless. For $\frac{\sqrt{\lambda}\phi}{m} \ll 1$, it reproduces the previous examples, except for the sign of the $\phi^4$ term. Alas it is not known to lead to an acceptable quantum field theory in four dimensions; however it yields a healthy quantum theory in two dimensions!

Whatever the form of V the equations of motion are easily obtained

$$\partial_\mu \partial^\mu \phi = - V'(\phi) \quad , \tag{6.6}$$

where the prime denotes the derivative with respect to $\phi$. We can construct the conserved quantity using the procedures of the previous section.

a) Under an infinitesimal translation characterized by $\delta x^\mu = \varepsilon^\mu$ and $\delta\phi = 0$, eqs. (5.25) and (5.26) become

$$j_{\mu\nu} = - g_{\mu\nu}\mathcal{L} + \partial_\mu \phi \partial_\nu \phi \quad , \tag{6.7}$$

$$\partial^\mu j_{\mu\nu} = 0 \quad . \tag{6.8}$$

Observe that in this case $j_{\mu\nu}$ is a symmetric tensor; it is called the energy momentum tensor. The corresponding conserved charge is

$$P_\mu = \int d^3x \; j_{\mu 0} = \int d^3x \; (-g_{\mu 0}\mathcal{L} + \partial_0 \phi \partial_\mu \phi) \; . \tag{6.9}$$

Then since $P_0$ is the energy of the system the energy density is

$$j_{00} = - \mathcal{L} + \partial_0 \phi \partial_0 \phi \quad , \tag{6.10}$$

$$= \frac{1}{2} \partial_0 \phi \partial_0 \phi + \frac{1}{2} \vec{\nabla}\phi \vec{\nabla}\phi + V(\phi) \quad , \tag{6.11}$$

and is seen to be positive definite when $V > 0$. The ground state field configuration is that which gives the lowest value for $j_{00}$. Since the derivative terms give a positive contribution it always occurs for a static field $\phi_0$ ($\partial_0 \phi_0 = \partial_i \phi_0 = 0$), in which case the energy density is the value of the potential $V(\phi_0)$ for this particular field.

b) Under a Lorentz transformation the conserved Noether current is a three-indexed quantity given by

$$j_{\mu\nu\rho} = (-g_{\mu\lambda}\mathcal{L} + \partial_\mu\phi\partial_\lambda\phi)(g_\nu^{\ \lambda}x_\rho - g_\rho^{\ \lambda}x_\nu) \quad , \tag{6.12}$$

$$= j_{\mu\nu}x_\rho - j_{\mu\rho}x_\nu \quad . \tag{6.13}$$

The corresponding conserved charges are the generators of Lorentz transformations

$$M_{\nu\rho} = \int d^3x\ j_{0\nu\rho} = \int d^3x\ (j_{0\nu}x_\rho - j_{0\rho}x_\nu) \quad . \tag{6.14}$$

The conservation of these charges is a consequence of the invariance of the Action under Poincaré transformations.

As an example of the application of Noether's theorem to transformations which are not necessarily invariances of S consider an infinitesimal dilatation

$$\delta x^\mu = \alpha x^\mu \qquad \delta\phi = -\alpha\phi \quad . \tag{6.15}$$

The Noether current is

$$j^\mu_D = (-g^\mu_\rho\mathcal{L} + \partial^\mu\phi\partial_\rho\phi)x^\rho + \phi\partial^\mu\phi \quad , \tag{6.16}$$

$$= j^{\mu\rho}x_\rho + \frac{1}{2}\partial^\mu\phi^2 \quad . \tag{6.17}$$

We see that using (6.8)

$$\partial_\mu j^\mu_D = j^\mu_\mu + \frac{1}{2}\partial^\mu\partial_\mu\phi^2 \quad . \tag{6.18}$$

When $\mathcal{L} = \frac{1}{2}\partial_\mu\phi\partial^\mu\phi - \frac{\lambda}{4!}\phi^4$, it is easy to show that $j^\mu_D$ is divergenceless (in four dimensions : see problem). However had we added to $\mathcal{L}$ the "mass term" $-\frac{1}{2}m^2\phi^2$, its contribution would have been

$$\partial_\mu j^\mu_D = m^2\phi^2 \neq 0 \quad . \tag{6.19}$$

The reason for the failure of conservation of $j_D^\mu$ in this case is that a dimensionful parameter appears in $\mathcal{L}$.

Recall that there are ambiguities in the form of $j_{\mu\nu}$. As an example consider the new definition

$$\Theta_{\mu\nu} \equiv j_{\mu\nu} + a\,(\partial_\mu\partial_\nu - g_{\mu\nu}\partial_\rho\partial^\rho)\phi^2 \qquad , \qquad (6.20)$$

where a is a dimensionless number. We still have

$$\partial^\mu\,\Theta_{\mu\nu} = 0 \qquad . \qquad (6.21)$$

We fix a by demanding that, for a dilatation invariant theory, $\Theta_{\mu\nu}$ be traceless. Taking as an example the Lagrangian given by (6.4) with $m^2 = 0$, we find

$$\Theta^\mu_\mu = (1+6a)\,[-\partial_\rho\phi\partial^\rho\phi - \phi\partial_\rho\partial^\rho\phi] \qquad , \qquad (6.22)$$

which sets $a = -\frac{1}{6}$. Furthermore the difference between $\Theta_{\mu\nu}$ and $j_{\mu\nu}$ is a surface term which does not alter the conserved charges. Now we can define a new dilation current as

$$j_D^{\mu\prime} \equiv x_\rho\Theta^{\mu\rho} \qquad . \qquad (6.23)$$

Then using (6.21)

$$\partial_\mu j'^\mu_D = \Theta^\mu{}_\mu \qquad , \qquad (6.24)$$

which shows that dilatation invariance is equivalent to tracelessness of $\Theta_{\mu\nu}$. This new dilatation current is related to the old one by

$$\begin{aligned} j'^\mu_D &= x_\rho j^{\mu\rho} - \frac{1}{6}\,x_\rho(\partial^\mu\partial^\rho - g^{\mu\rho}\partial_\tau\partial^\tau)\phi^2 \\ &= j_D^\mu - \frac{1}{2}\,\partial^\mu\phi^2 - \frac{1}{6}\,x_\rho(\partial^\mu\partial^\rho - g^{\mu\rho}\partial_\tau\partial^\tau)\phi^2 \\ &= j_D^\mu + \frac{1}{6}\,\partial_\rho[x^\mu\partial^\rho - x^\rho\partial^\mu]\phi^2 \qquad , \qquad (6.25) \end{aligned}$$

using the form (6.17). They are seen to differ from one another by a total divergence and thus the dilatation charge is not affected. The tensor $\Theta_{\mu\nu}$ is called the "new improved energy momentum tensor" [see F. Gürsey, Annals of Physics 24, 211 (1963) and S. Coleman and R. Jackiw, Annals of Physics 67, 552 (1971)]. The differences between $\Theta_{\mu\nu}$ and $j_{\mu\nu}$ and between $j_{\mu D}$ and $j'_{\mu D}$ are all surface terms.

These new forms for the energy momentum tensor and the dilatation current can be obtained canonically if we add to the scalar field Lagrangian a surface term of the form $\partial_\mu \Lambda^\mu$ where

$$\Lambda^\mu = \frac{1}{6}(x^\mu \partial^\rho - x^\rho \partial^\mu)\partial_\rho \phi^2 \quad , \tag{6.26}$$

and therefore corresponds to a canonical transformation.

In field theories of higher spin fields the dilatation invariance is always linked to a traceless energy momentum tensor. As we shall see later, dilatation invariance even when present in the original Lagrangian is broken by quantum effects.

The field theory of many scalar fields goes in much the same way except that interesting new symmetries arise. As an example consider N real scalar fields $\phi_a$ a = 1,...N and the Lagrangian

$$\mathcal{L} = \frac{1}{2}\sum_{a=1}^{N} \partial_\mu \phi_a \partial^\mu \phi_a \quad . \tag{6.27}$$

Besides the usual invariances, $\mathcal{L}$ is obviously invariant under a global (i.e., x-independent) rotation of the N real scalar fields into one another

$$\delta\phi_a = \varepsilon_{ab}\phi_b \quad , \quad \varepsilon_{ab} = -\varepsilon_{ba} \quad . \tag{6.28}$$

As a result there are $\frac{1}{2}N(N-1)$ conserved Noether currents

$$j^\mu_{ab} = \phi_a \partial^\mu \phi_b - \phi_b \partial^\mu \phi_a \quad . \tag{6.29}$$

This constitutes an example of an internal symmetry that arises because of the presence of many fields of the same type. If the theory is supplemented by a potential that depends only on the rotation invariant "length" $\phi_a\phi_a$, the internal rotation invariance is preserved.

Problems

A. In four-dimensions show that the canonical dilatation current is divergenceless when $\mathcal{L} = \frac{1}{2}\partial_\mu\phi\partial^\mu\phi - \frac{\lambda}{4!}\phi^4$.

B. In D dimensions, derive the expression for the divergence of the dilatation current when $\mathcal{L} = \frac{1}{2}\partial_\mu\phi\partial^\mu\phi - V(\phi)$.

*C. In general the canonical energy momentum tensor need not be symmetric. Show that one can always find a term $B^{\rho\mu\nu}$ antisymmetric under $\rho \to \mu$ or $\nu$ such that the Belinfante tensor

$$j_B^{\mu\nu} = j^{\mu\nu} + \partial_\rho B^{\rho\mu\nu} \quad ,$$

is symmetric and the conserved Noether current for L.T.'s is written in the form

$$j^{\mu\nu\rho} = (j_B^{\mu\nu}x^\rho - j_B^{\mu\rho}x^\nu) \quad .$$

Hint: $B^{\rho\mu\nu} = 0$ for scalar fields so it has to do with $S^{\mu\nu}$.

*D. Find $\delta\phi$ for a conformal transformation. Show that $S = \int d^4x \, \frac{1}{2}\partial_\mu\phi\partial^\mu\phi$ is invariant under a conformal transformation. Construct the conserved Noether current.

E. Derive the form of the conserved currents corresponding to the transformations (6.27) when $\mathcal{L}$ is given by (6.26).

## 7. The Action for spinor fields

In this section we concern ourselves primarily with the construction of Actions that involve the spinor Grassmann fields $\psi_L$ and $\psi_R$. Using the results of section 4 the simplest candidates for a spinor kinetic term are

$$\mathcal{L}_L = \frac{1}{2}\,\psi_L^\dagger \sigma^\mu \overleftrightarrow{\partial}_\mu \psi_L \quad , \qquad \mathcal{L}_L = \mathcal{L}_L^* \quad , \tag{7.1}$$

$$\mathcal{L}_R = \frac{1}{2}\,\psi_R^\dagger \sigma^\mu \overleftrightarrow{\partial}_\mu \psi_R \quad , \qquad \mathcal{L}_R = \mathcal{L}_R^* \quad , \tag{7.2}$$

or if parity is of interest

$$\mathcal{L}_{Dirac} = \frac{1}{2}\,\bar{\Psi}\gamma^\mu \overleftrightarrow{\partial}_\mu \Psi \quad , \tag{7.3}$$

$$= \mathcal{L}_L + \mathcal{L}_R \quad . \tag{7.4}$$

In the special case $\psi_R = -\sigma_2 \psi_L^*$ it is easy to see that $\mathcal{L}_R$ is equivalent to $\mathcal{L}_L$ up to a total divergence (see problem). Thus if $\Psi_M$ is a four-component Majorana spinor the Lagrangian is written as

$$\mathcal{L}_{Maj} = \frac{1}{4}\,\bar{\Psi}_M \gamma^\mu \overleftrightarrow{\partial}_\mu \Psi_M \quad , \tag{7.5}$$

and is equal to $\mathcal{L}_L$ as can be seen by using the Grassmann properties of $\psi_L$. In the literature, one often sees the Dirac kinetic term (7.3) written with the derivative operator acting only to the right and without the factor of $\frac{1}{2}$. Although superficially different from (7.3) this form differs from it by a total divergence. The distinction is not important as long as the system is not coupled to gravity.

From these expressions it is evident that in D dimensions the spinor field has (engineering) dimension $L^{-\frac{D+1}{2}}$ : spinor fields have dimension - 3/2 in four-dimensions.

These possible kinetic terms are invariant under conformal transformations (see problem), just as the scalar field kinetic term was, but in addition have phase invariances of their own. For instance consider $\mathcal{L}_L$ (the same can be said for $\mathcal{L}_R$). Since $\psi_L$ is a complex spinor, one can perform on it a phase transformation

$$\psi_L \rightarrow e^{i\alpha}\psi_L \quad , \tag{7.6}$$

which leaves $\mathcal{L}_L$ invariant as long as $\alpha$ does not depend on x. The Dirac Lagrangian (7.3) has two of these invariances. They can be reshuffled in four-component language as an overall phase transformation

$$\Psi \rightarrow e^{i\beta}\Psi \quad , \tag{7.7}$$

and a chiral transformation

$$\Psi \rightarrow e^{i\beta'\gamma_5}\Psi \quad . \tag{7.8}$$

Lastly, as in the scalar case, the Action with $\mathcal{L} = \mathcal{L}_L$ (or $\mathcal{L}_R$) is invariant under a constant shift in the fields since

$$\mathcal{L}_L(\psi_L+\alpha_L) = \mathcal{L}_L + \frac{1}{2}\partial_\mu(\alpha_L^\dagger\sigma^\mu\psi_L - \psi_L^\dagger\sigma^\mu\alpha_L) \quad . \tag{7.9}$$

By means of Noether's theorem we can build the conserved currents corresponding to the transformations (7.7) and (7.8). They are

$$j^\mu = i\bar{\Psi}\gamma^\mu\Psi = i\psi_L^\dagger\sigma^\mu\psi_L + i\psi_R^\dagger\bar{\sigma}^\mu\psi_R \tag{7.10}$$

and

$$j_5^\mu = i\bar{\Psi}\gamma^\mu\gamma^5\Psi = i\psi_L^\dagger\sigma^\mu\psi_L - i\psi_R^\dagger\bar{\sigma}^\mu\psi_R \quad , \tag{7.11}$$

respectively, while the conserved charges are

$$Q = i\int d^3x\bar{\Psi}\gamma^0\Psi = i\int d^3x(\psi_L^\dagger\psi_L + \psi_R^\dagger\psi_R) \tag{7.12}$$

and

$$Q_5 = i\int d^3x\bar{\Psi}\gamma^0\gamma^5\Psi = i\int d^3x(\psi_L^\dagger\psi_L - \psi_R^\dagger\psi_R) \quad . \tag{7.13}$$

For a Majorana field only the chiral transformation exists since $\psi_R$ is the conjugate of $\psi_L$; therefore $\psi_L$ and $\psi_R$ have opposite phase transformation.

Other non-kinetic quadratic invariants can be constructed out of spinor fields (see section 4). Using $\psi_L$ only we have

$$\mathcal{L}_L^m = \frac{im}{2}(\psi_L^T\sigma^2\psi_L+\psi_L^\dagger\sigma^2\psi_L^*) \tag{7.14}$$

$$\mathcal{L}_{L5}^m = \frac{m}{2}(\psi_L^T\sigma^2\psi_L-\psi_L^\dagger\sigma^2\psi_L^*) \quad , \tag{7.15}$$

where m is a parameter with dimensions of mass (in any number of dimensions). These are known as mass terms. Since $\psi_L$ can be used to describe a Majorana spinor $\Psi_M$, it follows that (7.14) can serve as the mass term for a Majorana spinor. In four-component notation (7.14) is known as the Majorana mass. Thus having only $\psi_L$ does not guarantee masslessness as is so often wrongly stated (e.g., the Weinberg-Salam model of weak and electromagnetic interactions where the neutrino is represented by a two-component left-handed spinor without a right-handed partner. There the masslessness of the neutrino results from the absence of certain Higgs bosons, in which case the fermion number conservation keeps the neutrino massless even after radiative corrections). This remark assumes special relevance in the conventional description of the neutrino in terms of a left-handed field. Note that $\mathcal{L}_L^m$ breaks the continuous phase symmetry (7.6) leaving only the discrete remnant $\psi_L \to -\psi_L$. In Majorana notation

$$\mathcal{L}_L^m = -\frac{im}{2}\bar{\Psi}_M\Psi_M \quad , \tag{7.16}$$

$$\mathcal{L}_{L5}^m = -\frac{m}{2}\bar{\Psi}_M\gamma_5\Psi_M \quad . \tag{7.17}$$

When both $\psi_L$ and $\psi_R$ are present two more quadratic invariants are available, leading to

$$\mathcal{L}_D^m = im\bar{\Psi}\Psi = im(\psi_L^\dagger\psi_R+\psi_R^\dagger\psi_L) \quad , \tag{7.18}$$

$$\mathcal{L}_{D5}^m = m\bar{\Psi}\gamma_5\Psi = m(\psi_L^\dagger\psi_R-\psi_R^\dagger\psi_L) \quad . \tag{7.19}$$

Both are left invariant by the overall phase transformation (7.7), but not by the chiral transformation (7.8) under which

$$\Psi \to e^{i\beta\gamma_5}\Psi, \qquad \bar{\Psi} = \Psi^\dagger\gamma^0 \to \bar{\Psi}e^{i\beta\gamma_5} \quad , \tag{7.20}$$

and thus

$$\mathcal{L}_D^m \to im\bar{\Psi}e^{2i\beta\gamma_5}\Psi \quad . \tag{7.21}$$

Application of (5.27) yields

$$\partial_\mu j_5^\mu = -2m\bar{\Psi}\gamma_5\Psi \quad , \tag{7.22}$$

while $j^\mu$ of (7.10) is still divergenceless. This is not to say that we cannot have terms quadratic in Dirac fields, free of derivatives that respect chiral invariance as the following example will show. Consider the term

$$\sigma(x)\bar{\Psi}(x)\Psi(x) + i\pi(x)\bar{\Psi}(x)\gamma_5\Psi(x) \quad , \tag{7.23}$$

which is the sum of $\mathcal{L}_D^m$ and $\mathcal{L}_{D5}^m$, with the coefficients depending this time on x. To preserve chiral invariance, $\sigma$ and $\pi$ must transform under chiral transformations as

$$[\sigma(x)+i\gamma_5\pi(x)] \to [\sigma'(x)+i\gamma_5\pi'(x)] = e^{-i\beta\gamma_5}[\sigma(x)+i\gamma_5\pi(x)]e^{-i\beta\gamma_5}; \tag{7.24}$$

for infinitesimal $\beta$, the $\sigma$ and $\pi$ fields are rotated into one another

$$\left.\begin{aligned} \delta\sigma &= +2\beta\pi \\ \delta\pi &= -2\beta\sigma \end{aligned}\right\} \quad . \tag{7.25}$$

This transformation leaves invariant $\sigma^2 + \pi^2$. Hence the Lagrangian

$$\mathcal{L}_f = \frac{1}{2}\,\bar{\Psi}\gamma^\mu \overleftrightarrow{\partial}_\mu \Psi + ih\bar{\Psi}[\sigma + i\gamma_5\pi]\Psi \tag{7.26}$$

is chirally invariant. If $\sigma$ and $\pi$ are canonical fields, h is a dimensionless constant (usually called the Yukawa coupling constant). One can give $\sigma$ and $\pi$ a life of their own by adding to $\mathcal{L}$ their kinetic terms as well as self interactions that preserve (7.25), leading to the Lagrangian

$$\mathcal{L} = \frac{1}{2}\,\bar{\Psi}\gamma^\mu \overleftrightarrow{\partial}_\mu \Psi + ih\bar{\Psi}\,[\sigma + i\gamma_5\pi]\Psi + \frac{1}{2}\,\partial_\mu\sigma\partial^\mu\sigma + \frac{1}{2}\,\partial_\mu\pi\partial^\mu\pi - V(\sigma^2+\pi^2) \quad . \tag{7.27}$$

This Lagrangian has the following symmetries (all global)

a) an overall Dirac phase symmetry $\Psi \to e^{i\alpha}\Psi$

$\sigma,\pi \to \sigma,\pi$

b) a chiral symmetry $\delta\Psi = i\beta\gamma_5\psi$

$\delta\sigma = -2\beta\pi$

$\delta\pi = 2\beta\sigma,$

which leaves $\sigma^2 + \pi^2$ invariant

c) a discrete parity transformation $\Psi \to \gamma_0\Psi$, $\sigma \to \sigma$, $\pi \to -\pi$; thus $\sigma(x)$ is a scalar field while $\pi(x)$ is a pseudoscalar field.

This type of Lagrangian was first constructed by Gell-Mann and Lévy. These theories are called $\sigma$-models and are relevant in describing the physical $\pi$-meson interactions. [In this example, the pion isospin was neglected.]

We see that the demand that the symmetry of the kinetic term be preserved in interaction leads to the introduction of extra fields. This is a general feature: enlargement of symmetries $\Rightarrow$ additional fields.

Note that in four dimensions invariant terms involving more than two spinor fields have dimensions of at least -6 so that dimensionful constants are needed to recover the dimension of $\mathcal{L}$. In two dimensions, however, terms like $(\bar{\Psi}\Psi)^2$ or $\bar{\Psi}\gamma^\mu\Psi\bar{\Psi}\gamma_\mu\Psi$ have the same dimensions as $\mathcal{L}$.

Since the 2-cpt spinor fields are always complex, the equations of motion are obtained by varying independently with respect to $\psi_L$ and $\psi_L^\dagger$. Extra care must be exercised because we treat $\psi_L$ and $\psi_L^\dagger$ as Grassmann fields and we cannot push a $\delta\psi$ past a $\psi$ without changing sign. For instance, we write

$$\delta\mathscr{L}_L = \frac{1}{2}\,\delta\psi_L^\dagger\sigma^\mu\partial_\mu\psi_L - \frac{1}{2}\,\partial_\mu\delta\psi_L^\dagger\sigma^\mu\psi_L + \frac{1}{2}\,\psi_L^\dagger\sigma^\mu\partial_\mu\delta\psi_L - \frac{1}{2}\,\partial_\mu\psi_L^\dagger\sigma^\mu\delta\psi_L \tag{7.28}$$

$$= \delta\psi_L^\dagger\sigma^\mu\partial_\mu\psi_L - (\partial_\mu\psi_L^\dagger\sigma^\mu)\delta\psi_L + \text{surface terms} \quad , \tag{7.29}$$

which leads to the conjugate equations

$$\sigma^\mu\partial_\mu\psi_L = 0 \text{ or } \partial_\mu\psi_L^\dagger\sigma^\mu = 0 \quad . \tag{7.30}$$

In the case of the Dirac spinor, independent variations for $\Psi$ and $\overline{\Psi}$ are performed to obtain the equations of motion.

Finally, let us note that one can build more complicated invariants involving spinor fields such as $\partial_\mu\overline{\Psi}\partial^\mu\Psi$. While there is nothing wrong with this type of term as far as invariance requirements, it does not lead to satisfactory theories in the sense that it violates the connection between spin and statistics. We will come back to this subject later, when we consider gauge theories.

Problems

A. Show that $\mathcal{L}_R$ with $\psi_R = \sigma^2\psi_L^*$ is equal to $\mathcal{L}_L$ plus a total divergence.

B. Find the Belinfante energy momentum tensor for $\mathcal{L}_{DIRAC}$.

C. Show that for $\mathcal{L} = \mathcal{L}_{DIRAC}$ the dilation current can be written as

$$j_D^\mu = x_\rho j_B^{\mu\rho} \quad ,$$

where $j_B^{\mu\rho}$ is the Belinfante form of the energy momentum tensor, thus showing that the Belinfante tensor coincides with the new improved energy momentum tensor for the Dirac field.

D. Given

$$\mathcal{L} = \frac{1}{2}\,\bar{\Psi}\gamma^\mu \overleftrightarrow{\partial}_\mu \Psi + im\bar{\Psi}\Psi + m'\bar{\Psi}\gamma_5\Psi \quad ,$$

use a chiral transformation to transform the pseudoscalar term away. What is the mass of the resultant Dirac field?

*E. Given a quadratic Lagrangian with both $\psi_L$ and $\psi_R$

$$\mathcal{L} = \mathcal{L}_L + \mathcal{L}_R + \mathcal{L}_L^m + \mathcal{L}_R^{m'} + iM(\psi_R^\dagger\psi_L + \psi_L^\dagger\psi_R) \quad ,$$

involving Dirac and Majorana masses. Rediagonalize the fields to obtain unmixed masses. What are the masses of the fields? What is the physical interpretation of the various degrees of freedom?

*F. For the $\sigma$-model Lagrangian, a) use Noether's theorem to derive the expression for the conserved chiral current; b) suppose we add to $\mathcal{L}$ a term linear in $\sigma$; find the divergence of the chiral current. This last equation embodies the PCAC (partially conserved axial current) hypothesis of pion physics.

*G. How does $\psi_L$ transform under a conformal transformation? Show that $\mathcal{L}_L$ is conformally invariant.

## 8. An Action with Scalar and Spinor Fields and Supersymmetry

There are several differences between the simplest kinetic term for spinor fields, $\mathcal{L}_L$ and its counterpart for a scalar field S. While $\mathcal{L}_L$ involves one derivative, the scalar kinetic term involves two; while $\psi_L$ is a Grassmann field, S is a normal field, and finally $\mathcal{L}_L$ has the phase invariance (7.6) while the kinetic term for one scalar field has none. Yet there are similarities since they both are conformally invariant. In this section we address ourselves to the possibility that there might exist a symmetry on the fields that relate the fermion and scalar kinetic terms. Such a symmetry is called a supersymmetry - it has the virtue of allowing non-trivial interactions between the scalar and spinor fields. To increase the odds we make the scalar field kinetic term resemble as much as possible $\mathcal{L}_L$. This is achieved by taking the kinetic term for two scalar fields, which we call S and P, and by comparing it with the kinetic term for a Majorana spinor field we call $\chi$. Then both kinetic terms have a phase invariance of their own. Indeed the Lagrangian,

$$\mathcal{L}_0^{WZ} = \frac{1}{2}\,\partial_\mu S \partial^\mu S + \frac{1}{2}\,\partial_\mu P \partial^\mu P + \frac{1}{4}\,\bar{\chi}\gamma^\mu \overleftrightarrow{\partial}_\nu \chi \qquad , \qquad (8.1)$$

besides being conformally invariant has two independent global phase invariances

$$\chi \to e^{i\alpha\gamma_5}\chi \quad , \qquad (S+iP) \to e^{i\beta}(S+iP) \qquad . \qquad (8.2)$$

Any further invariance will involve transformations that change the spinless fields S and P into the spinor field $\chi$. The general characteristics of this type of transformation are 1) the transformation parameter must be a Grassmann spinor field, call it $\alpha$, a global infinitesimal Majorana spinor parameter. 2) In its simplest form, the transformation of S and P must involve no derivative operator and that of $\chi$ must involve one since the fermion kinetic term has one less derivative than the scalar kinetic term. Thus we are led to

$$\delta(S \text{ or } P) = \bar{\alpha} M \chi \quad , \tag{8.3}$$

where M is some 4 x 4 matrix. Since no 4-vector indices are involved, it must only contain 1 or $\gamma_5$. Hence we fix it to be

$$\delta S = a\bar{\alpha}\chi \tag{8.4}$$

$$\delta P = ib\bar{\alpha}\gamma_5\chi \quad , \tag{8.5}$$

where a and b are unknown real coefficients. Here we have used the phase invariance (8.2) to define the variation of S to be along 1 and that of P along $i\gamma_5$. The right hand side of the variations is arranged to be real [In a Majorana representation for the Dirac matrices, all four components of the Majorana spinors are real and all the matrix elements of the $\gamma$-matrices are pure imaginary, so as to have real matrix elements for $i\gamma_5$.]. Then we have (assuming that $\partial_\mu$ does not change: see problem F)

$$\delta[\tfrac{1}{2}\partial_\mu S \partial^\mu S + \tfrac{1}{2}\partial_\mu P \partial^\mu P] = (a\partial^\mu S\bar{\alpha} + ib\partial^\mu P\bar{\alpha}\gamma_5)\chi \quad . \tag{8.6}$$

What can the variation of $\chi$ be? First note that

$$\delta[\tfrac{1}{4}\bar{\chi}\gamma^\mu\partial_\mu\chi] = \tfrac{1}{2}\delta\bar{\chi}\gamma^\mu\partial_\mu\chi - \tfrac{1}{2}\partial_\mu\bar{\chi}\gamma^\mu\delta\chi \tag{8.7}$$

up to a total divergence. Now we use the vector part of "Majorana-flip" properties: $\bar{\xi}\eta$, $\bar{\xi}\gamma_5\eta$, and $\bar{\xi}\gamma_\mu\gamma_5\eta$ are even as $\xi \rightarrow \eta$ while $\bar{\xi}\gamma_\mu\eta$ and $\bar{\xi}\sigma_{\mu\nu}\eta$ are odd. These hold for any two Majorana spinors $\xi$ and $\eta$ (see problem). Their application to (8.7) yields

$$\delta[\tfrac{1}{4}\bar{\chi}\gamma^\mu\partial_\mu\chi] = \delta\bar{\chi}\gamma^\mu\partial_\mu\chi \quad , \tag{8.8}$$

up to surface terms. Then putting it all together we see that

$$\delta\mathcal{L}_0^{WZ} = (\delta\bar{\chi}\gamma_\mu + a\partial_\mu S\bar{\alpha} + ib\partial_\mu P\bar{\alpha}\gamma_5)\partial^\mu\chi + \text{s.t.} \tag{8.9}$$

$$= -(\partial_\mu\delta\bar{\chi}\gamma^\mu + a\partial_\mu\partial^\mu S\bar{\alpha} + ib\partial_\mu\partial^\mu P\bar{\alpha}\gamma_5)\chi + \text{s.t}'. \quad , \tag{8.10}$$

where a partial integration has been performed to obtain (8.10) from (8.9). Thus $\mathcal{L}_0^{WZ}$ changes only by a total divergence if $\delta\bar{\chi}$ obeys the following equation ($\Box \equiv \partial_\mu\partial^\mu$)

$$\partial_\mu\delta\bar{\chi}\gamma^\mu + a\Box S\bar{\alpha} + ib\Box P\bar{\alpha}\gamma_5 = 0 \quad . \tag{8.11}$$

A solution is easily found to be

$$\delta\chi = a\gamma_\rho\alpha\partial^\rho S - ib\gamma_\rho\gamma_5\alpha\partial^\rho P \quad . \tag{8.12}$$

Here use of $\gamma_\rho\gamma_\sigma\partial^\rho\partial^\sigma = \partial_\rho\partial^\rho$ has been made. We have therefore achieved our goal: we have found a set of transformations between spinless and spin 1/2 fields which leaves the sum of their kinetic terms invariant (up to a canonical transformation). To further convince ourselves of the veracity of our find, we have to see if these transformations close among themselves and form a group.

As a starter examine the effect of two supersymmetry transformations on the fields. Explicitly

$$[\delta_1,\delta_2]S = a\bar{\alpha}_2\delta_1\chi - (1 \leftrightarrow 2)$$

$$= a\bar{\alpha}_2[a\gamma_\rho\alpha_1\partial^\rho S - ib\gamma_\rho\gamma_5\alpha\partial^\rho P] - (1 \leftrightarrow 2)$$

$$= 2a^2\bar{\alpha}_2\gamma_\rho\alpha_1\partial^\rho S \quad . \tag{8.13}$$

To get the last equation we have used the Majorana flip property of the axial vector part. Thus the effect of two supersymmetry transformations on S is to translate S by an amount $2ia^2\bar{\alpha}_2\gamma_\rho\alpha_1$. Let us see what happens to P:

$$[\delta_1,\delta_2]P = ib\bar{\alpha}_2\gamma_5\delta_1\chi - (1 \leftrightarrow 2)$$

$$= 2b^2\bar{\alpha}_2\gamma_\rho\alpha_1\partial^\rho P \quad , \tag{8.14}$$

where again the Majorana flip identity for axial vector has been used. Since transformations must be the same for S, P and X, we must have

$$b = \pm a \quad . \tag{8.15}$$

Let us finally verify that the action of two supersymmetry transformations on $\chi$ is itself a translation:

$$[\delta_1,\delta_2]\chi = a\gamma_\rho\alpha_2\partial^\rho\delta_1 S - ib\gamma_\rho\gamma_5\alpha_2\partial^\rho\delta_1 P - (1 \leftrightarrow 2)$$

$$= a^2\gamma_\rho\alpha_2\bar{\alpha}_1\partial^\rho\chi + b^2\gamma_\rho\gamma_5\alpha_2\bar{\alpha}_1\gamma_5\partial^\rho\chi - (1 \leftrightarrow 2). \tag{8.16}$$

We would like to rewrite the right hand side of this equation in a form similar to the others, that is involving $\bar{\alpha}_2\gamma_\rho\alpha_1$ and not the matrix $\alpha_2\bar{\alpha}_1$ that appears in (8.16). We do this using a trick due to Fierz: Take any two Dirac spinors (not necessarily Majorana), $\Psi$ and $\Lambda$. The 4 x 4 matrix $\Lambda\bar{\Psi}$ can be expanded as a linear combination of the 16 Dirac covariants 1, $\gamma_5$, $\gamma_5\gamma_\mu$, $\gamma_\mu$, $\sigma_{\mu\nu} = \frac{1}{4}[\gamma_\mu,\gamma_\nu]$. The coefficients are evaluated by taking the relevant traces. The result is

$$\Lambda\bar{\Psi} = -\frac{1}{4}\bar{\Psi}\Lambda - \frac{1}{4}\gamma_5\bar{\Psi}\gamma_5\Lambda + \frac{1}{4}\gamma_5\gamma_\rho\bar{\Psi}\gamma_5\gamma^\rho\Lambda - \frac{1}{4}\gamma_\rho\bar{\Psi}\gamma^\rho\Lambda + \frac{1}{2}\sigma_{\rho\sigma}\bar{\Psi}\sigma^{\rho\sigma}\Lambda \quad . \tag{8.17}$$

The numbers in front of the various terms constitute the first row of the celebrated Fierz matrix. They contain all the necessary information to generate the whole matrix. Application to our case yields

$$\alpha_2\bar{\alpha}_1 - \alpha_1\bar{\alpha}_2 = -\frac{1}{2}\bar{\alpha}_1\gamma^\rho\alpha_2\gamma_\rho + \bar{\alpha}_1\sigma^{\rho\sigma}\alpha_2\sigma_{\rho\sigma} \quad , \tag{8.18}$$

where we have used the Majorana flip properties. Use of (8.15), (8.18) leads to

$$[\delta_1,\delta_2]\chi = a^2\bar{\alpha}_2\gamma^\mu\alpha_1\gamma_\rho\gamma_\mu\partial^\rho\chi \quad . \tag{8.19}$$

By using the anticommutator of the $\gamma$-matrices, we rewrite it as

$$[\delta_1,\delta_2] = 2a^2\bar{\alpha}_2\gamma^\mu\alpha_1\partial_\mu\chi - a^2\bar{\alpha}_2\gamma^\mu\alpha_1\gamma_\mu\gamma^\rho\partial_\rho\chi \quad . \tag{8.20}$$

The first term on the right-hand side is the expected result, but unfortunately we have an extra term proportional to $\gamma^\rho\partial_\rho\chi$. This extra term vanishes only when the classical equations of motion are valid. In order to eliminate this term, we have to enlarge the definition of $\delta\chi$ and see where it leads us. Note that if we add to $\delta\chi$ of (8.12) an extra variation of the form

$$\delta_{extra}\chi = (F+i\gamma_5 G)\alpha \quad , \tag{8.21}$$

where F and G are functions of x, but not canonical fields since they have dimensions of $L^{-2}$, the relations (8.13) and (8.14) are not affected because of the Majorana flip conditions. For example

$$\begin{aligned}[\delta_1,\delta_2]_{extra}S &= a\bar{\alpha}_2\delta_{1\ extra}\chi - (1 \leftrightarrow 2)\\ &= a\bar{\alpha}_2(F+i\gamma_5 G)\alpha_1 - (1 \leftrightarrow 2)\\ &= 0 \quad .\end{aligned} \tag{8.22}$$

However, this extra variation gives a contribution on $\chi$, namely

$$[\delta_1,\delta_2]_{extra}\chi = (\delta_1 F+i\gamma_5\delta_1 G)\alpha_2 - (1 \leftrightarrow 2) \quad . \tag{8.23}$$

The extra term in (8.20) can be rewritten in a suggestive way by means of the Fierz rearrangement

$$\bar{\alpha}_2\gamma^\mu\alpha_1\gamma_\mu = \alpha_1\bar{\alpha}_2 - \gamma_5\alpha_1\bar{\alpha}_2\gamma_5 - (1 \leftrightarrow 2) \quad . \tag{8.24}$$

Comparison with (8.23) now shows that by choosing

$$\delta_1 F = a^2\bar{\alpha}_1\gamma^\rho\partial_\rho\chi \tag{8.25}$$

$$\delta_1 G = -ia^2\bar{\alpha}_1\gamma_5\gamma^\rho\partial_\rho\chi \quad , \tag{8.26}$$

we cancel the extra term and obtain the desired result. We leave it as an exercise (see problem) to show that the operator relation

$$[\delta_1,\delta_2] = 2a^2\bar{\alpha}_2\gamma^\mu\alpha_1\partial_\mu \tag{8.27}$$

is satisfied when acting on F and G.

Unfortunately the new $\delta\chi$ does not leave the original Action invariant because of $\delta_{extra}\chi$. But we observe that

$$\begin{aligned}\delta_{extra}\mathcal{L}_0^{WZ} &= \delta_{extra}\bar{\chi}\gamma^\mu\partial_\mu\chi \\ &= i\bar{\alpha}\gamma^\rho\partial_\rho\chi F - \bar{\alpha}\gamma_5\gamma^\rho\partial_\rho\chi G \\ &= -\frac{1}{2a^2}\,\delta(F^2+G^2) \quad .\end{aligned} \tag{8.28}$$

Therefore the Action

$$S_0^{WZ} = \int d^4x[\tfrac{1}{2}\,\partial_\mu S\partial^\mu S + \tfrac{1}{2}\,\partial_\mu P\partial^\mu P + \tfrac{1}{4}\,\bar{\chi}\gamma^\rho\partial_\rho\chi + \frac{1}{2a^2}\,(F^2+G^2)] \tag{8.29}$$

is invariant under the supersymmetry transformations

$$\begin{aligned}&\delta S = a\bar{\alpha}\chi\,;\ \delta P = ia\bar{\alpha}\gamma_5\chi\,;\ \delta F = a^2\bar{\alpha}\gamma^\rho\partial_\rho\chi\,; \\ &G = -ia^2\bar{\alpha}\gamma_5\gamma^\rho\partial_\rho\chi;\delta\chi \quad .\end{aligned} \tag{8.30}$$

These transformations now all satisfy the operator equation (8.27). This Action was first written down by Wess and Zumino, Nucl. Phys. B78 (1974) 1.

With the introduction of the auxiliary fields F and G, we now have the same number of spinless (S, P, F and G) and spinor (the 4 real components of $\chi$) fields irrespective of the equations of motion. The reader can convince himself that "on mass-shell" (i.e., on the classical path) where F and G are not necessary, the balance between spinless and spinor degrees of freedom is still true. This balance between the number of boson (even spin) and fermion (odd spin) degrees of freedom is a general feature of supersymmetric theories.

From (8.27) we see that the effect of two supersymmetry transformations is a translation. In addition since the supersymmetry parameters are spinors, it follows that the generators of the supersymmetry transform as spinors. Therefore we have an enlargement of the Poincaré group to include the supersymmetry generators (see problem). The F and G fields have no kinetic terms; they serve as auxiliary fields which are totally uncoupled for the free theory.

The beauty of the supersymmetry transformations (8.30) is their generalizability to interacting theories. For instance, one can introduce a supersymmetric Yukawa coupling term which leaves one global chiral invariance

$$\mathcal{L}_y^{WZ} = ih(\bar{\chi}\chi S - i\bar{\chi}\gamma_5\chi P + F(P^2 - S^2) - 2GSP) \tag{8.31}$$

or

$$\mathcal{L}_y^{WZ'} = ih'(\bar{\chi}\chi P + i\bar{\chi}\gamma_5\chi S + G(S^2 - P^2) - 2FSP) \quad . \tag{8.32}$$

Even mass terms can be written down

$$\mathcal{L}_m^{WZ} = -i\,\frac{m}{2}(\bar{\chi}\chi - \frac{1}{a}\,SF - \frac{1}{a}\,PG) \quad . \tag{8.33}$$

We can use this term to find an important (and fatal) property of supersymmetric theories. Consider the equations of motion for the Wess-Zumino Lagrangian with mass. They are

$$\not{\partial}\chi = i\,\frac{m}{2}\,\chi \tag{8.34}$$

$$\Box S = \frac{m}{2a}\,F \tag{8.35}$$

$$\Box P = \frac{m}{2a}\,G \tag{8.36}$$

$$0 = \frac{1}{a^2} F + \frac{m}{2a} S \tag{8.37}$$

$$0 = \frac{1}{a^2} G + \frac{m}{2a} P \quad . \tag{8.38}$$

The last two can be solved for F and G in terms of S and P without great difficulty and their result substituted in the equations for S and P. They become

$$\Box S = -\frac{m^2}{4} S \; ; \qquad \Box P = -\frac{m^2}{4} P \quad . \tag{8.39}$$

Hence the three fields $\chi$, S and P all have the same mass. This is a general feature of supersymmetry: all fields entering a supermultiplet have the same mass. This is because the mass operator $P_\mu P^\mu$ commutes with all supersymmetry generators. As an immediate consequence, we see that exact supersymmetry cannot exist in nature because particles of different spins show no mass degeneracy.

[Finally this little calculation hints at the role of the auxiliary fields when equations of motion can be solved. The following embryonic model of how auxiliary fields work will illustrate the point independently of the equations of motion: let $\phi(x)$ be a scalar field and A(x) be an auxiliary field. Take

$$\mathscr{L} = \frac{1}{2} \partial_\mu \phi \partial^\mu \phi + \frac{1}{2} A^2 + A\phi^2 \quad , \tag{8.40}$$

and complete the square to obtain

$$\mathscr{L} = \frac{1}{2} \partial_\mu \phi \partial^\mu \phi + \frac{1}{2} (A+\phi^2)^2 - \frac{1}{2} \phi^4 \quad . \tag{8.41}$$

Redefine the now uncoupled auxiliary field $A' = A + \phi^2$, and we are left with the interaction Lagrangian $\frac{1}{2} \partial_\mu \phi \partial^\mu \phi - \frac{1}{2} \phi^4$.]

This is the simplest example of a supersymmetric theory in 4 dimensions. Supersymmetry is at present a purely "theoretical symmetry" without any experimental support. However, we felt it was instructive to alert the reader to the existence of nontrivial symmetries among fields of different spins.

Problems

A. Prove the Majorana flip properties.

B. Verify by using $\gamma$-matrix identities the Fierz decomposition (8.17).

C. Identify the chiral invariance of $S_0^{WZ}$ and express its action on the fields.

D. Show that $\int (\frac{i}{2} \bar{\chi}\gamma_5\chi + \frac{1}{a} SG - \frac{1}{a} PF) d^4x$ is a supersymmetric invariant.

*E. Introduce the Majorana spinor generators of supersymmetry Q by writing a finite supersymmetry transformation as $e^{i\bar{\alpha}Q}$. Derive the expression for the anticommutator of two Q's and the commutator of Q with the Poincaré generators. The ensuing algebra involving both commutators and anticommutators forms a graded Lie algebra (superalgebra). As a consequence show that Q commutes with the mass.

*F. Find the change of the coordinate $x^\mu$ under a supersymmetry, and verify that $\partial_\mu$ is invariant.

**G. Use Noether's theorem to derive the expression for the conserved supersymmetric current. Use caution because $\mathcal{L}_0^{WZ}$ picks up a total divergence under a supersymmetric change.

## II - THE ACTION FUNCTIONAL IN QUANTUM MECHANICS: THE FEYNMAN PATH INTEGRAL

In the previous chapter we were concerned with the building of Action Functionals that yield (classical) theories conforming with the postulates of Special Relativity. This chapter deals with the use of the AF in Quantum Theory. For simplicity and clarity we first investigate the role of the Action in Quantum Mechanics, and then graduate to Quantum Field Theory in the next chapter.

Dirac and Feynman were the first to understand the role of the Action in Quantum Mechanics. Dirac's motivation stemmed from the desire to obtain a formulation of Quantum Mechanics where time and space variables were treated in an analogous fashion. Let me remind you that in the usual formulation of Quantum Mechanics, a quantum system is specified at an initial time to be in a certain state chosen among the eigenstates of a complete set of operators commuting with the Hamiltonian and among themselves. The Hamiltonian is then used to find in which state the system is at a later time t. One goes on to compute the transition amplitude from the state $S_0$ at $t_0$ to the state S at t, etc. As you can see, time plays a central role in this description, but for a relativistic system, one is uneasy because the manifest Lorentz invariance of the theory is lost even though the final answer turns out to be relativistically invariant. So Dirac was motivated to look for a formulation which did not take time as its centerpiece. To do this he went back to Classical Mechanics where there are two (analogous) descriptions: Hamilton's which singles out time *ab initio* and Lagrange's which does not. Specifically he looked for the meaning of the AF in classical mechanics with the intent of generalizing it to Quantum Mechanics. The answer was, of course, known, the Action being the generator of a canonical transformation which takes the system from one time to another. Hence it will be good to refresh your memory about canonical transformations:

## 1. Canonical Transformations in Classical and Quantum Mechanics

Consider a particle moving in one dimension. The state of motion of this particle at a time t is given by its coordinate q and momentum p, which are independent functions of t. Their time variation is given by a set of two first order differential equations (Hamilton's equations)

$$\frac{dq}{dt} = \frac{\partial H}{\partial p} \quad , \quad \frac{dp}{dt} = - \frac{\partial H}{\partial q} \quad , \tag{1.1}$$

where H, the Hamiltonian, is the energy of the system and depends on q, p and t. These equations can be neatly expressed in terms of the Poisson brackets

$$\{A,B\}_{q,p} \equiv \frac{\partial A}{\partial q}\frac{\partial B}{\partial p} - \frac{\partial A}{\partial p}\frac{\partial B}{\partial q} \quad , \tag{1.2}$$

defined here for two arbitrary functions A and B of q, p and t. Hamilton's equations are now

$$\frac{dq}{dt} = \{q,H\} \quad , \quad \frac{dp}{dt} = \{p,H\} \quad . \tag{1.3}$$

It follows that if F is any function of q, p and t, its time derivative is

$$\frac{dF}{dt} = \{F,H\} + \frac{\partial F}{\partial t} \quad . \tag{1.4}$$

The last term takes care of any explicit time dependence F may have. The Hamilton equation of motion can be derived from a variational principle

$$\delta \int_{t_1}^{t_2} dt\left(p\frac{dq}{dt} - H(p,q)\right) = 0 \quad , \tag{1.5}$$

where the independent variations $\delta p$ and $\delta q$ are taken to vanish at the end points.

We define a canonical transformation

$$p \to P \quad , \quad q \to Q \tag{1.6}$$

to be a transformation that leaves Hamilton's equations form invariant, i.e., in the new system $(Q,P)$ there exists a new Hamiltonian $\mathcal{H}(Q,P)$ such that

$$\frac{dQ}{dt} = \frac{\partial \mathcal{H}}{\partial P} \quad , \quad \frac{dP}{dt} = - \frac{\partial \mathcal{H}}{\partial Q} \quad . \tag{1.7}$$

It follows that these are also derivable from a variational principle

$$\delta \int_{t_1}^{t_2} dt(P \frac{dQ}{dt} - \mathcal{H}(Q,P)) = 0 \quad . \tag{1.8}$$

It implies that the integrands of (1.5) and (1.8) can differ at most by a total time derivative

$$p \frac{dq}{dt} - H(p,q) = P \frac{dQ}{dt} - \mathcal{H}(P,Q) + \frac{dG}{dt} \quad . \tag{1.9}$$

The function G is called the generating function of the canonical transformation. It can depend on t and on any "astride" pair of variables $(q,Q)$, $(q,P)$, $(p,Q)$ or $(p,P)$. Take G to depend on the independent variables $(q,Q)$. Then

$$\frac{dG}{dt} = \frac{\partial G}{\partial t} + \frac{\partial G}{\partial q}\frac{dq}{dt} + \frac{\partial G}{\partial Q}\frac{dQ}{dt} \quad . \tag{1.10}$$

Now consider (1.9) with q and Q as independent variables. It reads

$$(p - \frac{\partial G}{\partial q})\dot{q} - (P + \frac{\partial G}{\partial Q})\dot{Q} = H - \mathcal{H} + \frac{\partial G}{\partial t} \quad , \tag{1.11}$$

so that the remaining variables $(p,P)$ are now expressed by

$$p = \frac{\partial G}{\partial q} \quad , \quad P = - \frac{\partial G}{\partial Q} \tag{1.12}$$

and the new Hamiltonian is given by

$$\mathcal{H} = H + \frac{\partial G}{\partial t} \quad . \tag{1.13}$$

One could have equally well started by taking G to depend on the pair (q,P). A similar reasoning would have led to the equations

$$p = \frac{\partial G(q,P)}{\partial q} \quad , \quad Q = \frac{\partial G(q,P)}{\partial P} \quad . \tag{1.14}$$

A particular choice of G gives the identity transformation,

$$G = qP \tag{1.15}$$

as can be verified by means of eq. (1.14). An infinitesimal canonical transformation with parameter $\varepsilon << 1$

$$Q = q + O(\varepsilon) \quad , \quad P = p + O(\varepsilon) \tag{1.16}$$

will therefore be generated by a generating function $\varepsilon$ away from the form (1.15):

$$G(q,P) = qP + \varepsilon F(q,P) + O(\varepsilon^2) \quad , \tag{1.17}$$

$$= qP + \varepsilon F(q,p) + O(\varepsilon^2) \quad , \tag{1.18}$$

since p differs from P by $O(\varepsilon)$. The function F(q,p) which now depends entirely on the original system is called the <u>generator</u> of the canonical transformation. Substitution of (1.18) into (1.14) yields

$$p = P + \varepsilon \frac{\partial F}{\partial q} \tag{1.19}$$

$$Q = q + \varepsilon \frac{\partial F}{\partial p} \quad , \tag{1.20}$$

or

$$\delta q \equiv Q - q = \varepsilon \frac{\partial F}{\partial p} = \varepsilon\{q,F\} \tag{1.21}$$

$$\delta p \equiv P - p = - \varepsilon \frac{\partial F}{\partial q} = \varepsilon\{p,F\} \quad . \tag{1.22}$$

It follows that the infinitesimal change of any function f of q and p caused by an infinitesimal canonical transformation with generator $F(q,p)$ and parameter $\varepsilon$ is given by

$$\delta f(p,q) = \{f(p,q), \varepsilon F\} \quad . \tag{1.23}$$

In particular we see that the Hamiltonian is the generator of infinitesimal time translations, by comparing with (1.4).

Imagine now a very special type of canonical transformation which maps the variables q,p to another set Q,P which are time independent. In this case the knowledge of the transformation equations

$$q = q(Q,P,t)$$

$$p = p(Q,P,t)$$

is equivalent to the solution of the dynamical problem since Q and P are constants which can be identified in terms of initial conditions. It would be nice to know what generates such a transformation. Because we demand that

$$\frac{dQ}{dt} = \frac{dP}{dt} = 0 \quad , \tag{1.24}$$

the new Hamiltonian does not depend on Q and P. It can only be a constant with or without time dependence. For simplicity we take $\mathcal{H}$ to be zero. Then with q and Q as independent variables, (1.13) becomes

$$H\left(q,p = \frac{\partial S}{\partial q}, t\right) = \frac{\partial S}{\partial t} \tag{1.25}$$

where S(q,Q=constant, t) is the generating function. This is the Hamilton-Jacobi equation with S as a solution (S is called Hamilton's principal function). We know from (1.12) that

$$P = \text{constant} = -\frac{\partial S}{\partial Q}(q,Q,t)\Big|_{Q\,=\,\text{cst}} \quad , \tag{1.26}$$

which when inverted gives q as a function of Q, P and t, and therefore solves the dynamical problem.

Then, the time derivative of S

$$\frac{dS}{dt} = \frac{\partial S}{\partial t} + \frac{\partial S}{\partial q}\frac{dq}{dt} \tag{1.27}$$

$$= -H(q,p,t) + p\,\frac{dq}{dt} \tag{1.28}$$

is nothing but the Lagrangian. Integration yields

$$S = \int_{t_0}^{t} dt'L \quad , \tag{1.29}$$

which reveals S as the action regarded as a <u>function</u> of $q(t)$ and $q(t_0) = Q$ in the event that the solution of the problem has already been inserted in L and the time integration performed. Thus we have the central result that the action is the generating function of a canonical transformation which transforms the system variables from one time to another. Let us now see how to interpret this result in Quantum Mechanics.

Consider the Quantum description of one degree of freedom in one dimension. In terms of the operators $\hat{q}$ and $\hat{p}$ (^ will crown operators!) which obey the fundamental commutation relations

$$[\hat{q},\hat{q}] = [\hat{p},\hat{p}] = 0 \qquad ; \qquad [\hat{q},\hat{p}] = i\hbar \quad , \tag{1.30}$$

$\hbar$ being Planck's constant, the states of the system at a given time can be taken to be the position states $|q\rangle$ which satisfy

$$\hat{q}|q\rangle = q|q\rangle \quad , \tag{1.31}$$

$$\langle q|q'\rangle = \delta(q-q') \quad , \tag{1.32}$$

$$\int dq|q\rangle\langle q| = 1 \quad . \tag{1.33}$$

[Here q is a regular number or function; not an operator!] In Quantum Mechanics a canonical transformation between the operators $(\hat{q},\hat{p})$ and $(\hat{Q},\hat{P})$ is defined as a transformation that does not change the form of the fundamental commutation relations (1.30). Then the system will be described in terms of $|Q\rangle$ states with the same properties as the $|q\rangle$ states (with Q replaced by q, of course).

Following Dirac we focus on the "mixed" matrix element $\langle q|Q\rangle$. It is easy to see, using (1.31) that

$$\langle q|\hat{q}|Q\rangle = q\langle q|Q\rangle \tag{1.34}$$

and equivalently

$$\langle q|\hat{Q}|Q\rangle = Q\langle q|Q\rangle \quad . \tag{1.35}$$

Also since

$$\hat{p}|q\rangle = -i\hbar \frac{\partial}{\partial q} |q\rangle \quad , \tag{1.36}$$

it follows that

$$\langle q|\hat{p}|Q\rangle = i\hbar \frac{\partial}{\partial q} \langle q|Q\rangle \tag{1.37}$$

$$\langle q|\hat{P}|Q\rangle = -i\hbar \frac{\partial}{\partial Q} \langle q|Q\rangle \quad . \tag{1.38}$$

However, the operators $\hat{Q}$ and $\hat{q}$ need not commute so that the value of an arbitrary operator $F(\hat{q},\hat{Q})$ in the mixed representation may not be well defined until more demands are made on the form of F. For instance, it is clear from (1.34) and (1.35) that

$$\langle q|f_1(\hat{q})f_2(\hat{Q})|Q\rangle = f_1(q)f_2(Q)\langle q|Q\rangle \quad . \tag{1.39}$$

Hence we shall consider only "well-ordered" functions for which

$$\langle q|F(\hat{q},\hat{Q})|Q\rangle = F(q,Q)\langle q|Q\rangle \quad . \tag{1.40}$$

[Well ordered means they are separable as a function of $\hat{q}$ times a function of $\hat{Q}$.] It follows that if we set (with Dirac's forethought!)

$$\langle q|Q\rangle = e^{-i/\hbar\, G(q,Q)} \quad , \tag{1.41}$$

where G is a function of q and Q, eqs. (1.37) and (1.38) read

$$\langle q|\hat{p}|Q\rangle = \frac{\partial G}{\partial q} \langle q|Q\rangle \tag{1.42}$$

$$\langle q|\hat{P}|Q\rangle = -\frac{\partial G}{\partial Q} \langle q|Q\rangle \quad . \tag{1.43}$$

Then if we assume that $\frac{\partial G}{\partial q}$ and $\frac{\partial G}{\partial Q}$ are "well-ordered" functions in the sense of (1.40), these can now become equations between operators

$$\hat{p} = \widehat{\frac{\partial G}{\partial q}} \quad ; \quad \hat{P} = -\widehat{\frac{\partial G}{\partial Q}} \quad . \tag{1.44}$$

Thus we observe that G defined by (1.41) is the quantum equivalent of the generating function. Dirac calls it "analogous to" [P.A.M. Dirac, Phys. Zeit. der Sowjetunion, Band 3, Heft 1 (1933), or as reprinted in "Selected Papers on Quantum Electrodynamics, J. Schwinger, ed., Dover, 1958].

Problems.

A. Show that the Poisson brackets are left invariant by a canonical transformation.

B. Consider an infinitesimal Canonical Transformation

$$\delta f = \{f, \varepsilon_a F_a\} \qquad ,$$

where $\varepsilon_a$ are the parameters of the transformation, and $F_a$ the generators. Show that the operation

$$[\delta_1, \delta_2] f$$

is itself a canonical transformation. The subscripts stand for two different parameters $\varepsilon_{1a}$ and $\varepsilon_{2a}$. What is the generator of this canonical transformation? Do canonical transformations satisfy the group axioms?

C. Take a system with coordinate $q_i$ and momenta $p_i$, $i = 1,2,3$. Find the generators of infinitesimal rotations, and verify the results of problem B.

## 2. The Feynman Path Integral

Dirac now proceeds to apply this analogy to Hamilton's principal function with $q = q'$ at $t$ and $Q = q$ at $T$:

$$<q'_t|q_T> \sim e^{i/\hbar \int_T^t dtL} \quad . \qquad (2.1)$$

Let me emphasize that the $\sim$ sign means just a loose connection, because to arrive at (1.44) Dirac had to make all kinds of assumptions with no way to justify them. In fact, we can see that an equality sign would not be correct for (2.1)* as long as the time interval T-t is finite: split up T-t into N infinitesimal time intervals $t_a = t + a\varepsilon$; $N\varepsilon = T-t$. Let $q_a = q_{ta}$ and use the completeness relation (1.33) for each $t_a$ to write

$$<q'_t|q_T> = \int dq_1 dq_2 .. dq_{N-1} <q'_t|q_1><q_1|q_2> .... <q_{N-1}|q_T> \quad . \qquad (2.2)$$

This is an exact quantum mechanical formula. Now if (2.1) is taken to mean an equality, the integral in the exponent can be split up into many integration regions, yielding an incorrect formula:

$$<q_t|q_T> = <q_t|q_1><q_1|q_2>...<q_{N-1}|q_T> \quad , \qquad (2.3)$$

which differs from the correct one by the absence of the intermediate integrations. Here $q_1, q_2, ...$ are the classical values the trajectory takes at times $t_1, t_2, ...$ .

However, if we assume (2.1) to hold as an equality (up to a constant) only for an <u>infinitesimal</u> time interval, i.e.,

$$<q'_t|q_{t+\delta t}> = A\ e^{-\frac{i}{\hbar}\delta t L(q'_t, q_{t+\delta t})} \quad , \qquad (2.4)$$

where L (in the spirit of the Hamilton-Jacobi theory) is taken to be a function

*Only formulae appearing in previous chapters will carry the chapter label.

of $q'_t$ and $q_{t+\delta t}$, we run into no conflict with the quantum mechanical formula (2.2). This is exactly what Feynman did! [Rev. Mod. Phys. 20, 267 (1948).] This leads to the Feynman Path Integral for the transition amplitude, using (2.4) in (2.2):

$$\langle q'_t | q_T \rangle = \lim_{\substack{N\to\infty \\ N\varepsilon \text{ fixed}}} A^N \int \left( \prod_{i=1}^{N-1} dq_i \right) e^{-\frac{i}{\hbar}\int_T^t dt L(q,\dot{q})} \tag{2.5}$$

$$\equiv \int \mathcal{D}q \; e^{-\frac{i}{\hbar} S(t,T,[q])} \quad , \tag{2.6}$$

where the second expression is just a fancy way of hiding our lack of knowledge about the measure. The boundary conditions are the values the path takes on at the initial and final times. In words, this formula means that if you want to compute the probability amplitude for the particle to be at q' at time t, given that it was at q at time T, you express it as the sum over all possible paths that start at q at T and end at q' at t weighted by the exponential of $-\frac{i}{\hbar}$ times the action evaluated for the particular path. This formulation brings out clearly the distinction between Classical and Quantum Mechanics. In the former the particle takes only one path to go from q to q' while all paths contribute to the latter.

As a by-product, (2.6) gives the clearest relation between Classical and Quantum Mechanics. In the classical limit, $\hbar \to 0$, the integrand will oscillate with greater frequency as q varies and wash itself out unless S is fairly constant; but this happens where S is stationary and q is the classical trajectory [could there be several stationary points?]. Then we see that as $\hbar \to 0$, the classical trajectory is naturally singled out and (2.3) is recovered, but only in the classical limit.

It is this connection, first noted by Dirac, that makes the Action Functional so important. We saw that S, beautiful as it is, was only slightly used in Classical Mechanics where only the knowledge of the location of its extrema was needed. Now Quantum Mechanics uses it everywhere. In retrospect, is it not conceivable to ask whether physicists of earlier centuries asked themselves why so little of S was ever used?

Let us now check explicitly the degree of validity of Feynman's hypothesis (2.4). Let $\hat{H}$ be the time independent Hamiltonian operator for our one-dimensional system. In the Heisenberg picture, the state q at t + $\delta t$ is obtained from the state at t via

$$|q_{t+\delta t}\rangle \simeq |q_t\rangle + \frac{i}{\hbar}\,\delta t\hat{H}|q_t\rangle + O(\delta t)^2 \quad . \tag{2.7}$$

Hence

$$\langle q'_{t+\delta t}|q_t\rangle = \langle q'_t|q_t\rangle - \frac{i}{\hbar}\langle q'_t|\hat{H}|q_t\rangle\delta t + O(\delta t)^2 \quad . \tag{2.8}$$

As a simple example, let

$$\hat{H} = \frac{1}{2}\hat{p}^2 + V(\hat{q}) \tag{2.9}$$

so as to avoid (for the moment) ordering difficulties. Then

$$\langle q'_t|\hat{H}|q_t\rangle = [-\frac{\hbar^2}{2}\frac{\partial^2}{\partial q^2} + V(q)]\langle q'_t|q_t\rangle \tag{2.10}$$

$$= \int_{-\infty}^{+\infty}\frac{d\ell}{2\pi}\,[-\frac{\hbar^2}{2}\frac{\partial^2}{\partial q^2} + V(q)]e^{i\ell(q'-q)} \quad , \tag{2.11}$$

where we have used (1.32) and the integral representation for the $\delta$-function

$$\delta(x-x') = \int_{-\infty}^{+\infty}\frac{d\ell}{2\pi}\,e^{i\ell(x-x')} \quad . \tag{2.12}$$

Putting it all together and performing the q differentiation, we obtain

$$<q'_{t+\delta t}|q_t> = \int_{-\infty}^{+\infty} \frac{d\ell}{2\pi} e^{i\ell(q'-q)}[1 - \frac{i}{\hbar} \delta t H(\ell,q)+O(\delta t')] \quad , \quad (2.13)$$

with

$$H(\ell,q) = \frac{\hbar^2}{2} \ell^2 + V(q) \quad . \quad (2.14)$$

Set

$$q' - q = \frac{dq}{dt} \delta t = \dot{q}\delta t \quad (2.15)$$

and elevate H to an exponent (thus inducing only an error of $O(\delta t)^2$), to obtain

$$<q'_{t+\delta t}|q_t> = \int_{-\infty}^{+\infty} \frac{d\ell}{2\pi} e^{i \frac{\delta t}{\hbar} [\hbar\ell\dot{q} - \frac{1}{2} \hbar^2\ell^2-V(q)]} +O(\delta t)^2 \quad . \quad (2.16)$$

At this stage the manipulations become formal and not really well defined. We want to integrate over $\ell$ and yet the integrand is purely oscillatory. There are two ways out, either we put in by hand a convergence factor of the form $e^{-\varepsilon\ell^2}$ or we take $i\delta t$ formally to be "real", i.e., we continue the expression in Euclidean space by letting $t \to it$. For the moment, keep $i\delta t$ and treat it as a real constant. Then by performing the change of variables

$$\ell \to \ell' = \sqrt{\frac{i\delta t}{\hbar}} (\hbar\ell-\dot{q}) \quad , \quad (2.17)$$

we arrive at

$$<q'_{t+\delta t}|q_t> = \frac{1}{2\pi} e^{\frac{i\delta t}{\hbar} [\frac{1}{2} \dot{q}^2-V(q)]} \sqrt{\hbar} \int_{-\infty}^{+\infty} \frac{d\ell'}{\sqrt{i\delta t}} e^{-\frac{1}{2} \ell'^2} \quad . \quad (2.18)$$

We perform the funny Gaussian integral and obtain

$$<q'_{t+\delta t}|q_t> = \sqrt{\frac{\hbar}{2\pi i\delta t}} e^{\frac{i\delta t}{\hbar}[\frac{1}{2} \dot{q}^2-V(q)]} \quad . \quad (2.19)$$

Remember that $\dot{q}$ is defined by (2.15) and involves both q' and q. The quantity in square brackets is the Lagrangian. Therefore Feynman's hypothesis is correct except that the constant is, to say the least, peculiar. So for a finite time interval

$$\langle q_t^f | q_T^i \rangle = \lim_{\substack{\delta t \to 0 \\ N\delta t \text{ fixed}}} \int \ldots \int \prod_{j=1}^{N-1} \{ dq_j (\frac{\hbar}{2\pi i \delta t})^{1/2} \} e^{\frac{i}{\hbar} \int_T^t L dt} \quad ;$$

$$(q^i \equiv q_0 , \ q^f = q_N) \quad . \tag{2.20}$$

Assuming this limit makes sense, we see that we cannot possibly narrow down the proportionality constant between the transition amplitude and the path integral. This result is strictly true for systems described by Hamiltonians of the form (2.9).

Now suppose we want to calculate the transition amplitude for a system with a classical Hamiltonian of the form

$$H = \frac{1}{2} p^2 v(q) \quad , \tag{2.21}$$

where v(q) is some function of q. The corresponding Hamiltonian operator has to be defined with care since $\hat{p}$ and $\hat{q}$ do not commute. An ordering prescription is in order. We define a symmetric ordering "..." such that

$$\langle q_t' | "\frac{1}{2} \hat{p}^2 v(\hat{q})" | q_t \rangle = \int_{-\infty}^{+\infty} \frac{d\ell}{2\pi} \frac{\hbar^2}{2} \ell^2 v(\frac{q+q'}{2}) e^{i\ell(q'-q)} \quad . \tag{2.22}$$

It is easy to convince oneself that this requirement does define an ordering. In this way q and q' are treated on the same footing.

Then (setting $\hbar = 1$), we obtain in a straightforward way

$$\langle q_{t+\delta t}' | q_t \rangle = \int_{-\infty}^{+\infty} \frac{d\ell}{2\pi} e^{i\ell(q'-q) - i\delta t \frac{1}{2} \ell^2 v(\frac{q+q'}{2})} + O(\delta t^2) \quad , \tag{2.23}$$

so that the integration over $\ell$ becomes more complicated. Making the change of variables

$$\ell' = \sqrt{i\delta t}\,(\ell v^{1/2} - v^{-1/2}\dot{q}) \tag{2.24}$$

with $\dot{q}$ defined in (2.15), we find

$$\langle q'_{t+\delta t}|q_t\rangle = \frac{1}{2\pi}\, e^{-i\delta t \frac{1}{2v}\dot{q}^2} \frac{1}{\sqrt{i\delta t v}} \int_{-\infty}^{+\infty} d\ell'\, e^{-\frac{1}{2}\ell'^2} \tag{2.25}$$

$$= \frac{1}{\sqrt{2\pi i\delta t}}\, e^{-i\delta t \cdot \frac{1}{2} v^{-1}\dot{q}^2} \left(\frac{1}{\sqrt{v}}\right) \quad . \tag{2.26}$$

Now the term in the exponential can be interpreted as the Lagrangian

$$L = \frac{1}{2} v^{-1}\dot{q}^2 \quad , \tag{2.27}$$

as can be seen by forming the Hamiltonian along canonical lines (see Problem), but there is an extra term $v^{-1/2}(\frac{q+q'}{2})$ which in this case contributes to the path integral, giving

$$\langle q'_t|q_T\rangle = \int\ldots\int \prod_{i=1}^{N-1} \frac{dq_i}{\sqrt{2\pi i\delta t}}\, v^{-1/2}(q_i+q_{i-1})\, e^{i\int_T^t dt L} \quad . \tag{2.28}$$

Note that this expression differs from the naive one of weighing a path with $e^{iS}$ only. Thus the magical measure "$\mathcal{D}q$" sometimes contains peculiar surprises, like the factor of $v^{-1/2}$ in this case. The lesson in here is that one arrives at the correct expression for the transition amplitude only by first passing through the Hamiltonian formalism. The more correct expression is

$$\langle q'_t|q_T\rangle = \int\ldots\int \mathcal{D}q\,\mathcal{D}p\; e^{i\int_T^t d\tau [p \frac{dq}{d\tau} - H(p,\langle q\rangle)]} \quad , \tag{2.29}$$

where "$\mathcal{D}p$" stands for $\Pi \frac{dp_i}{2\pi}$ and $\langle q \rangle$ is the average of q in a given interval. Thus the expression for the transition amplitude in terms of the exponential of the Action is to be regarded as a derived expression; the fundamental form is (2.29) given in terms of the Hamiltonian.

Problems.

A. Evaluate (2.11) directly by computing the matrix element of $\hat{H}$ in a momentum basis, and transforming back to the q-basis.

B. Find the exact expression for "$\hat{q}^n\hat{p}$", "$\hat{q}^n\hat{p}^2$" as defined by the ordering equation (2.22).

C. Show that $L = \frac{1}{2} v^{-1}\dot{q}^2$ leads to $H = \frac{1}{2} p^2 v(q)$ by using the usual canonical procedures.

D. Show that if $T < t_1, t_2 < t$

$$\int \mathcal{D}q\mathcal{D}p \; q(t_1)q(t_2)e^{i\int_T^t dt[p\dot{q}-H]} = \langle q_t' | T[\hat{q}(t_1)\hat{q}(t_2)] | q_T \rangle ,$$

where T[..] stands for the time ordered product

$$T[\hat{q}(t_1)\hat{q}(t_2)] = \begin{cases} \hat{q}(t_1)\hat{q}(t_2) \text{ if } t_1 > t_2 \\ \hat{q}(t_2)\hat{q}(t_1) \text{ if } t_2 > t_1 \end{cases} .$$

## 3. The Path Integral and the Forced Harmonic Oscillator

A very good illustration of path integral techniques is obtained by considering the forced harmonic oscillator. We want to calculate the transition amplitude for a real driving force F

$$\langle Q'_t | Q_T \rangle_F = \int \mathcal{D}q \; e^{i\int_T^t dt[\frac{1}{2}\dot{q}^2 - \frac{1}{2}\omega^2 q^2 + F(t)q(t)]} \quad , \qquad (3.1)$$

with the boundary conditions that q at t is Q' and q at T is Q. The integrand as written is purely oscillatory and a particular way to restore convergence is to add a damping term

$$-\frac{1}{2}\varepsilon\int_T^t dt\; q^2(t) \qquad \varepsilon > 0 \quad , \qquad (3.2)$$

and then let $\varepsilon$ go to zero after the end of the calculation. The integrand is now written as

$$\exp\{i\int dt[\frac{1}{2}\dot{q}^2 - \frac{1}{2}(\omega^2 - i\varepsilon)q^2 + Fq]\} \quad .$$

Now suppose we want to compute the transition amplitude for the oscillator, starting in a state Q in the infinite past, to be in a state Q' in the infinite future in the presence of an external driving force. In order to work it out, it is convenient to introduce the Fourier transforms

$$G(t) = \int_{-\infty}^{+\infty} \frac{dE}{\sqrt{2\pi}} e^{iEt} \tilde{G}(E) \qquad (3.3)$$

and its inverse

$$\tilde{G}(E) = \int_{-\infty}^{+\infty} \frac{dt}{\sqrt{2\pi}} e^{-iEt} G(t) \quad , \qquad (3.4)$$

where G is any function of t, and $\tilde{G}$ its transform. We express q(t) and F(t) in (3.1) in terms of their Fourier transforms:

$$\frac{1}{2}[\dot{q}^2-(\omega^2-i\varepsilon)q^2] = \frac{1}{2}\int \frac{dE}{\sqrt{2\pi}}\frac{dE'}{\sqrt{2\pi}} e^{i(E+E')t}[-EE'-\omega^2+i\varepsilon]\tilde{q}(E)\tilde{q}(E) \tag{3.5}$$

$$F_{ext}(t)q(t) = \frac{1}{2}\int \frac{dE}{\sqrt{2\pi}}\frac{dE'}{\sqrt{2\pi}} e^{i(E+E')t}[\tilde{q}(E)\tilde{F}(E')+\tilde{q}(E')\tilde{F}(E)]. \tag{3.6}$$

Integration over t and use of the integral representation (2.12) for the δ-function and then integration over E' yield for the exponent

$$\frac{i}{2}\int_{-\infty}^{+\infty} dE[(E^2-\omega^2+i\varepsilon)\tilde{q}(E)\tilde{q}(-E)+\tilde{q}(E)\tilde{F}(-E)+\tilde{q}(-E)\tilde{F}(E)] \quad . \tag{3.7}$$

By defining the new variables in E-space

$$\tilde{q}'(E) = \tilde{q}(E) + \frac{\tilde{F}(E)}{E^2-\omega^2+i\varepsilon} \quad , \tag{3.8}$$

or in t-space

$$q'(t) = q(t) + \int \frac{dE}{\sqrt{2\pi}} e^{iEt} \frac{\tilde{F}(E)}{E^2-\omega^2+i\varepsilon} \quad , \tag{3.9}$$

the transition amplitude becomes

$$\langle Q'_\infty|Q_{-\infty}\rangle_F = e^{-\frac{i}{2}\int_{-\infty}^{+\infty} dE \frac{\tilde{F}(E)\tilde{F}(-E)}{E^2-\omega^2+i\varepsilon}} \int \mathcal{D}q\, e^{\frac{i}{2}\int_{-\infty}^{+\infty} dE\, \tilde{q}'(E)(E^2-\omega^2+i\varepsilon)\tilde{q}'(-E)} \quad . \tag{3.10}$$

The magic of the path integral now comes in because the Jacobian of the transformation (3.8) is one. Hence

$$\mathcal{D}q' = \mathcal{D}q \quad , \tag{3.11}$$

and we recognize the last term in (3.10) as the transition amplitude with $F = 0$, leading to the result

$$\langle Q'_\infty | Q_{-\infty} \rangle_F = \langle Q'_\infty | Q_{-\infty} \rangle_{F=0} \, e^{-\frac{i}{2} \int_{-\infty}^{+\infty} dE \frac{\tilde{F}(E)\tilde{F}(-E)}{E^2-\omega^2+i\varepsilon}} \quad . \tag{3.12}$$

The F-dependence has then been explicitly worked out. We can massage it further by giving its time form

$$e^{-\frac{i}{2} \int_{-\infty}^{+\infty} dt\, F(t)D(t-t')F(t')dt'} \quad , \tag{3.13}$$

where

$$D(t-t') = \int_{-\infty}^{+\infty} \frac{dE}{2\pi} \frac{e^{-i(t-t')E}}{E^2-\omega^2+i\varepsilon} \quad . \tag{3.14}$$

What is the physical meaning of (3.13)? At infinite times $t = \pm\infty$, assume that there is no driving force. Then the vacuum states at these times will not depend on the existence of F. Let $|\Omega_{\pm\infty}\rangle$ be the vacuum states in the infinite future and past, and express them in terms of the $|Q_{\pm\infty}\rangle$ that appear in (3.12). We have

$$\langle \Omega_{+\infty} | \Omega_{-\infty} \rangle_F = \int dQ' dQ \langle \Omega_{+\infty} | Q'_{+\infty} \rangle \langle Q'_{+\infty} | Q_{-\infty} \rangle_F \langle Q_{-\infty} | \Omega_{-\infty} \rangle \tag{3.15}$$

$$= \int dQ' dQ \langle \Omega_{+\infty} | Q'_{+\infty} \rangle \langle Q'_{+\infty} | Q_{-\infty} \rangle_{F=0} \langle Q_{-\infty} | \Omega_{-\infty} \rangle e^{-\frac{i}{2}\langle FDF \rangle} \quad , \tag{3.16}$$

where we have used (3.12) and the notation <...> to indicate integration over t and t'. Using (3.15) with F = 0 allows us to rewrite (3.16) as

$$\langle \Omega_{+\infty} | \Omega_{-\infty} \rangle_F = \langle \Omega_{+\infty} | \Omega_{-\infty} \rangle_{F=0} \, e^{-\frac{i}{2}\langle FDF \rangle} \quad . \tag{3.17}$$

But $\langle \Omega_{+\infty} | \Omega_{-\infty} \rangle_{F=0}$ is the transition amplitude that the system, starting in its ground state in the infinite past, will be in its ground state in the infinite future, in the absence of any driving force. It has to be equal to 1 (if it is normalizable). Hence we identify (3.13) as the transition amplitude for the system to go from past to future ground state in the presence of an external driving force.

Call (3.13) W[F], and define Z[F] by

$$W[F] \equiv e^{-\frac{i}{2}\langle F_1 D_{12} F_2\rangle_{1,2}} \tag{3.18}$$

$$\equiv e^{iZ[F]} \quad . \tag{3.19}$$

Here again $\langle\ldots\rangle_{12}$ means integration over the dummy variables "1" and "2"; $F_1$ is F(1), etc... . Note that W is normalized so that W[0] = 1.

It is easy to find that (see problem)

$$D(t) = \frac{1}{2i\omega}\,[\theta(t)e^{-i\omega t}+\theta(-t)e^{i\omega t}] \quad , \tag{3.20}$$

where $\theta(x)$ is the step function

$$\theta(x) = \begin{cases} 1 & x > 0 \\ 0 & x < 0 \end{cases} \quad . \tag{3.21}$$

Furthermore, by direct differentiation of (3.14) we see that

$$\left(\frac{d^2}{dt^2} + \omega^2\right)D(t) = -\delta(t) \quad . \tag{3.22}$$

Hence D(t) is the Green's function for the operator $\frac{d^2}{dt^2} + \omega^2$, and the $-i\varepsilon$ prescription enforced by the path integral fixes the boundary conditions. As can be seen from (3.20), D(t) is a mixture of retarded and advanced signals. This is the precursor of the Feynman propagator which describes signal propagation as coming from two sources: positive energy (particle) states moving in positive time and negative energy (antiparticle) states moving backward in time.

At this stage we notice yet another interesting thing: the expression (3.13) can be continued to imaginary E without encountering any singularity when the contour is rotated:

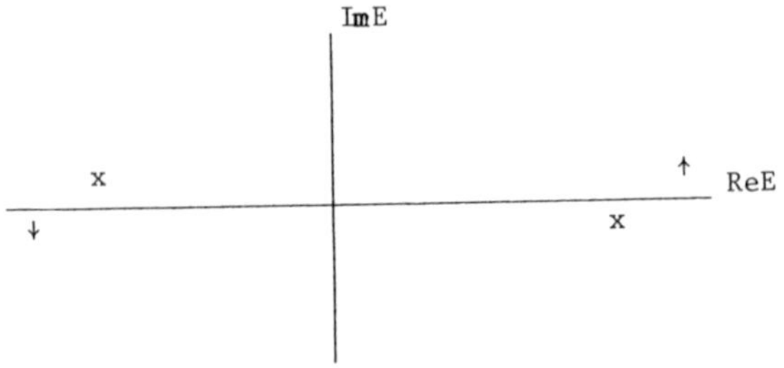


The ability to do this operation, called the Wick rotation, is important because it corresponds to starting from the "Euclidean" ($t \to it$) definition of the path integral. This is the alternative to $\varepsilon$ damping since the Euclidean space integrands are no longer oscillatory. For instance, we could have started from

$$W_E[F] = \int \mathcal{D}q \; e^{-\int_{-\infty}^{+\infty} d\tau \left[\frac{1}{2}\left(\frac{dq}{d\tau}\right)^2 + \frac{1}{2}\omega^2 q^2 - Fq\right]} , \tag{3.23}$$

obtained by letting $\tau = it$ in (3.1). Then the answer in real time is obtained by analytically continuing the expression for $W_E[J]$. This process yields the same pole structure for $D(t)$ as that obtained by the $-i\varepsilon$ prescription (see problem). This procedure will be followed later on while computing Feynman diagrams. However, in Euclidean space, the absence of singularities, although mathematically pleasing, obscures the physical meaning of the field theoretic amplitudes. In practice, one takes for granted the right to continue and computes wherever it is easier. Still on formal matters, the difference between Euclidean and Minkowski descriptions can be formidable (meaning of $-i\varepsilon$ for fermions, "light-cone" gauge, ...).

Problems.

A. Show that, using the techniques of appendix A,

$$\int \prod_{1}^{N} dz_i dz_i^* \, e^{-z^+ C z} = \frac{\pi^N}{\det C} .$$

What conditions must be satisfied by C (assume C is hermitian). Optional: prove the above when C is not hermitian.

B. Consider the Green's functions

$$D_a = \int \frac{dE}{2\pi} \frac{e^{-iEt}}{E^2-\omega^2-i\varepsilon E} , \qquad D_r = \int \frac{dE}{2\pi} \frac{e^{-iEt}}{E^2-\omega^2+i\varepsilon E} \qquad \varepsilon > 0 .$$

Evaluate these using the method of residues. Give the physical meaning of their boundary conditions. Can you come up with any way to obtain them from a path integral formulation? Express D of the text in terms of $D_a$ and $D_r$.

*C. Starting from

$$W_E[F] \sim \int \mathcal{L} q \, e^{-\int d\tau [\frac{1}{2}(\frac{dq}{d\tau})^2 + \frac{1}{2} \omega^2 q^2 - Fq]} ,$$

show that

$$W_E[F] = W_E[0] e^{-\frac{1}{2} \int d\tau d\sigma F(\tau) D_E(\tau-\sigma) F(\sigma)}$$

where $D_E(\tau)$ is to be worked out. Then use analytic continuation to obtain the D of the text.

D. Show that

$$D(t) = \int \frac{dE}{2\pi} \frac{e^{-iEt}}{E^2-\omega^2+i\varepsilon} \qquad \varepsilon > 0$$

$$= \frac{1}{2i\omega} [\theta(t) e^{-i\omega t} + \theta(-t) e^{i\omega t}] .$$

## III - THE FEYNMAN PATH INTEGRAL IN FIELD THEORY.

### 1. The Generating Functional

We now want to generalize the path integral to field theory. Reasoning by analogy with quantum mechanics, and taking for convenience a real scalar field as an example, we can describe the states of the system at a given time t by the ket$|\phi(\vec{x})\rangle_t$. We call it a shape state. We can compute the transition amplitude between a shape at $t_0$ and a new shape at a later time t, but we are immediately faced with the need to resolve the shape ket in terms of the possible physical states of the system. The way one identifies the states of the system usually relies very much on the success of the perturbation theory: one starts with a zero$^{th}$ order approximation to the theory in which the states are easily recognized. The full theory is recovered by adding a small perturbation to the idealized $0^{th}$ order theory. Then the effects of this small perturbation on the idealized $0^{th}$ order states are computed. These steps are possible only when one can recognize the full theory in terms of a small perturbation on a simple system and, having done that, calculate corrections to the idealized $0^{th}$ order states. As an example of this procedure, consider quantum electrodynamics QED where a small parameter $\alpha \approx (137)^{-1}$ is easily recognized. Then the $0^{th}$ order theory is that for which $\alpha = 0$. It is easily identified in terms of idealized photon and electron (or muon, tau, quark, ...) states. The effect of the interaction is then computed on these states and their interactions order by order in $\alpha$. After some trickery (i.e., renormalization theory), these corrections are found to yield physical electron and photon states and their interactions. The point of this discussion is to emphasize the reliance of the success of QED on our ability to recognize the emergence of idealized electron- and photon-like states in a $0^{th}$ order theory. This procedure was made possible because $\alpha$ is a small number, otherwise such a recognition could not have been made. An example of a yet unresolved theory is QCD, quantum-chromodynamics, which is thought to describe the interactions of quarks and gluons (the QCD equivalent of photons). It is believed that quarks are not physical particles but that bound states of quarks such as protons, $\pi$-mesons, etc., are physical. Yet there is nothing to tell us *a priori* that quarks are *not* physical states. Operationally then we have to decide the size of the quark couplings among themselves. If they are small, then quarks could serve as physical states (here physical applies to states that survive in isolation); if they are big, it is not good form to talk of quarks because quarks would tend to bind among themselves, and not appear as asymptotic states.

Thus the knowledge of the physical states of a field theory depends very much on the solution. But this is what we are trying to find! We have to obtain a path integral formulation that does not rely on the knowledge of its physical states (we'll derive that). The trick that gets us off the hook is very simple. Whatever the states, everyone agrees there must be a state of least energy, call it the vacuum state. It may be a very complicated structure (e.g., superconductivity), and may be inhabited by all kinds of strange objects, but nevertheless it is thought to exist. Suppose now we ask for the transition amplitude of the system from the vacuum state at $t = -\infty$ to the vacuum state at $+\infty$ *in the presence of an arbitrary driving force*. This means that at any time we reserve the right to drive the system in any way we please and watch it respond. This ought to tell us all we want to know provided we are sufficiently clever to apply probes that will give recognizable responses. This will then be the strategy: a) work out the amplitude $\langle\Omega|\Omega\rangle_J$ for an arbitrary source J(x), b) interpret (more exactly recognize) the results in terms of scattering amplitudes, c) use these amplitudes to calculate the physical consequences of the theory.

Traditionally the source will be attached to a local field, the rationale being that this provides a generic driving term since all possible sources can be built in terms of it. When perturbation theory is applicable, the local fields will naturally be interpreted in terms of particles.

We start with the simplest field theory: a self-interacting scalar field, described by the action

$$S = \int d^4x[\frac{1}{2}\partial_\mu\phi\partial^\mu\phi - \frac{1}{2}m^2\phi^2 - V(\phi)] \quad . \tag{1.1}$$

$$= \int d^4x\mathcal{L}(\phi,\partial_\mu\phi) \quad . \tag{1.2}$$

To construct the Hamiltonian density $\mathcal{H}$, define the canonical momentum

$$\pi(x) = \frac{\partial\mathcal{L}}{\partial[\partial_0\phi]} = \partial_0\phi = \dot{\phi} \quad , \tag{1.3}$$

and then perform a Legendre transformation

$$\mathcal{H}(\pi,\phi,\vec{\nabla}\phi) = \pi\dot{\phi} - \mathcal{L} \tag{1.4}$$

$$= \frac{1}{2}(\pi^2+\vec{\nabla}\phi\cdot\vec{\nabla}\phi+m^2\phi^2)+V(\phi) \quad . \tag{1.5}$$

$\mathcal{H}$ is positive definite if $m^2 > 0$ and $V > 0$.

The vacuum to vacuum amplitude is defined to be

$$\langle\Omega|\Omega\rangle_J \equiv W[J] = N\int\mathcal{D}\phi\mathcal{D}\pi e^{i\langle\pi\dot{\phi}-\mathcal{H}+J\phi\rangle} \quad , \tag{1.6}$$

where N is a constant (usually ill-defined), the <...> now means integration over spacetime, and J(x) is an arbitrary source. We use the techniques of the previous chapter to integrate over $\pi$ to obtain

$$W[J] = N'\int \mathcal{D}\phi\, e^{i<\frac{1}{2}\partial_\mu\phi\partial^\mu\phi - \frac{1}{2}m^2\phi^2 - V(\phi) + J\phi>} \quad . \tag{1.7}$$

In this case $\mathcal{D}\phi$ (or $\mathcal{D}\pi$) stands for the product of all the $d\phi_k$ where $\phi_k$ is the value of $\phi$ at $x = x_k$.

The integrand in (1.7) is oscillatory and even path integrals are not well defined. There are two ways to remedy this problem:

a) put in a convergence factor $e^{-\frac{1}{2}\varepsilon<\phi^2>}$ with $\varepsilon > 0$,

or b) define W in Euclidean space by setting

$$x_0 = -i\bar{x}_0 \qquad d^4x = -id^4\bar{x}$$

$$\partial_\mu\phi\partial^\mu\phi = -\bar{\partial}_\mu\phi\bar{\partial}_\mu\phi \quad ,$$

where the bar denotes Euclidean space variables, $\bar{\partial}_\mu = \frac{\partial}{\partial\bar{x}^\mu}$. Then eq. (1.7) becomes

$$W_E[J] = N_E\int \mathcal{D}\phi\, e^{-<\frac{1}{2}\bar{\partial}_\mu\phi\bar{\partial}_\mu\phi + \frac{1}{2}m^2\phi^2 + V(\phi) - J\phi>} \quad . \tag{1.8}$$

The exponent of the integrand is now negative definite for positive $m^2$ and V.

In either case, the generating functional is used to manufacture the Green's functions which are the coefficients of the functional expansion

$$W[J] = \sum_{N=0}^{\infty} \frac{(i)^N}{N!} <J_1J_2..J_NG^{(N)}(1,2,...,N)>_{1,..N} \quad , \tag{1.9}$$

or

$$G^{(N)}(1,2,..,N) = \frac{1}{(i)^N}\frac{\delta}{\delta J_1}\frac{\delta}{\delta J_2}\cdot\cdot\frac{\delta}{\delta J_N}W[J]\Big|_{J=0} \tag{1.10}$$

where as usual $J_i = J(x_i)$, etc., ..., $<..>_{1,..N}$ means integration over $d^4x_1..d^4x_N$. The task at hand is to compute the functions $G^{(N)}(x_1,.....,x_N)$, perturbatively or otherwise. In p-space they will be identified as transition amplitudes. This is not trivial since transition amplitudes must satisfy unitarity and completeness criteria. The Euclidean space functional $W_E[J]$ is used to construct these functions, $G_E^{(N)}(\bar{x}_1,..,\bar{x}_N)$; they are related to the $G^{(N)}$ by analytic continuation (Wick rotation), which presupposes that no singularities are encountered in the process of contour rotation. This is sufficient to determine the singularity structure of $G^{(N)}$, but to show that it is consistent with unitarity is not a trivial matter. These rather obscure remarks will hopefully become transparent in the light of explicit calculations.

## 2. The Feynman Propagator

In this section we evaluate W[J] when V = 0. We choose to do it in Minkowski space with the ε-procedure. Let

$$W_0[J] \equiv N\int \mathcal{D}\phi\, e^{i<\frac{1}{2}\partial_\mu\phi\partial^\mu\phi - \frac{1}{2}(m^2-i\varepsilon)\phi^2+J\phi>} \quad . \tag{2.1}$$

It is most easily evaluated in Fourier transform (momentum) space, following the same techniques used for the driven harmonic oscillator. We introduce the four-dimensional Fourier transform

$$\widetilde{F}(p) = \int_{-\infty}^{+\infty}\frac{d^4x}{(2\pi)^2}\, e^{-ip\cdot x}\, F(x) \quad , \tag{2.2}$$

$$F(x) = \int_{-\infty}^{+\infty} \frac{d^4p}{(2\pi)^2} e^{ip\cdot x} \tilde{F}(p) \quad , \quad (2.3)$$

and

$$\delta^{(4)}(x-x') = \delta(x^0-x^{0\prime})\delta(\vec{x}-\vec{x}') \quad ,$$

$$= \int_{-\infty}^{+\infty} \frac{d^4p}{(2\pi)^4} e^{i(x-x')\cdot p} \quad , \quad (2.4)$$

where $x\cdot p = x^0p^0 - \vec{x}\cdot\vec{p}$, and F is any sufficiently well-behaved function. The exponent of the integrand is easily expressed in terms of the Fourier transforms of $\phi$ and J; it reads

$$\frac{i}{2} \int d^4p[\tilde{\phi}'(p)[p^2-m^2+i\varepsilon]\tilde{\phi}'(-p)-\tilde{J}(p)[p^2-m^2+i\varepsilon]^{-1}\tilde{J}(-p)] \quad , \quad (2.5)$$

where

$$\tilde{\phi}'(p) = \tilde{\phi}(p) + [p^2-m^2+i\varepsilon]^{-1}\tilde{J}(p) \quad . \quad (2.6)$$

The new variable $\phi'$ differs from $\phi$ in function space by a constant, so that

$$\mathcal{D}\phi = \mathcal{D}\phi' \quad . \quad (2.7)$$

Putting it all together, we find

$$W_0[J] = Ne^{-\frac{i}{2}\int d^4p \frac{|\tilde{J}(p)|^2}{p^2-m^2+i\varepsilon}} \int \mathcal{D}\phi' e^{i\langle \frac{1}{2}\partial_\mu\phi'\partial^\mu\phi' - \frac{1}{2}(m^2-i\varepsilon)\phi'^2\rangle} \quad , \quad (2.8)$$

where we observed that the $\phi'$-dependent term was just the same as the $\phi$ term in (2.1) with J = 0. Hence

$$W_0[J] = W_0[0]e^{-\frac{i}{2}\int d^4p \frac{\tilde{J}(p)\tilde{J}(-p)}{p^2-m^2+i\varepsilon}} \quad . \quad (2.9)$$

By adjusting N, we can take $W_0[0] = 1$. Note that $W_0[0]$ can be formally calculated using formulae of Appendix A. The important thing is that we have succeeded in finding the explicit dependence of $W_0[J]$ on J. The use of the inverse Fourier transform yields

$$W_0[J] = W_0[0]e^{-\frac{i}{2}\langle J_1\Delta_{F12}J_2\rangle_{12}} \quad , \qquad (2.10)$$

where $\Delta_{F12}$ stands for $\Delta_F(x_1-x_2)$:

$$\Delta_F(x-y) = \int \frac{d^4p}{(2\pi)^4} \frac{e^{-ip\cdot(x-y)}}{p^2-m^2+i\varepsilon} \quad . \qquad (2.11)$$

It is the Feynman propagator. We now interpret the Green's functions obtained from $W_0$. From (1.10) we find

$$G_0^{(2)}(x_1,x_2) = \Delta_F(x_1-x_2) \qquad (2.12)$$

$$G_0^{(4)}(x_1,x_2,x_3,x_4) = -[\Delta_F(x_1-x_2)\Delta_F(x_3-x_4)+\Delta_F(x_1-x_3)\Delta_F(x_2-x_4) +\Delta_F(x_1-x_4)\Delta_F(x_2-x_3)] \quad , \qquad (2.13)$$

etc. ...

together with the vanishing of the G's with odd number of variables. This fact is easy to understand since $W_0[J]$ depends only on $J^2$. In passing, note that all G's are functions of only the difference of coordinates, reflecting the translation invariance of the theory. Another lesson is that the higher Green's functions can all be understood in terms of $G_0^{(2)}$. Hence it would appear more convenient to set

$$W[J] = e^{iZ[J]} \quad , \qquad (2.14)$$

and define new Green's functions in terms of Z[J]

$$iZ[J] = \sum_N \frac{(i)^N}{N!} \langle G_c^{(N)}(1,\ldots,N)J_1 \ldots J_N \rangle_{1..N} \qquad . \qquad (2.15)$$

Then, at least in the case of $W_0$, we see that $G_c$ is much simpler than G.

We now turn to the physical meaning of the Green's functions generated by $W_0$. By direct computation, we find

$$(\partial_\mu \partial^\mu + m^2)\Delta_F(x) = -\delta^{(4)}(x) \qquad , \qquad (2.16)$$

thus identifying $\Delta_F$ with the Green's function of the operator $\Box + m^2$. Its boundary conditions are determined from the $-i\varepsilon$ procedure dictated by the path integral. We can therefore identify $\Delta_F(x-y)$ with the propagator of a signal from x to y. The signals it propagates are single particle and antiparticle states, since those are solutions of the Klein-Gordon equation

$$(\Box + m^2)\phi = 0 \qquad . \qquad (2.17)$$

The $-i\varepsilon$ procedure tells us which of the solutions are propagated. One finds that positive energy solutions of the Klein-Gordon equation are propagated forward in time while negative energy solutions are propagated backwards in time (see problem).

Since these solutions are to be identified with particle (antiparticle) states of energy $E = p^0 = \sqrt{\vec{p}^2 + m^2}$ $(-\sqrt{\vec{p}^2 + m^2})$, we arrive at the very nice physical picture that information propagates forward in time with particle and backward in time with antiparticle. As an example, ask how many ways a quantum number can be carried from x to y given that we have a particle which carries one unit and an antiparticle with minus one unit. The quantum number could be the electric charge and the particle a $\pi^+$ meson. There are two ways:

by propagating a $\pi^+$ meson from x to y thus destroying a charge $+1$ at x and depositing it at y or by using a $\pi^-$, the antiparticle of $\pi^+$, to carry a negative charge from y to x.

The lesson is: 1) we have recognized the Green's function as the propagator of certain signals, and 2) we know which signals it propagates. Then it naturally follows that the states are, in our example, going to be particles of mass $m^2$, and we interpret $G_0^{(2)}(x-y)$ as the amplitude for this particle to go from x to y. We can invent a diagrammatic representation in x-space, associating with $\Delta_F(x-y)$ a line connecting the two space-time points x and y:

$$G_0^{(2)}(x,y): \qquad \underset{x}{\bullet} \!\!\!\!\rule[0.5ex]{3em}{0.4pt}\!\!\!\! \underset{y}{\bullet}$$

For higher Green's functions we just diagrammatically add the contributions of, say (2.13),

$$G_0^{(4)}(x_1,x_2,x_3,x_4): \left(\begin{matrix} \underset{x_1}{\bullet}\!\rule[0.5ex]{3em}{0.4pt}\!\underset{x_2}{\bullet} \\[1em] \underset{x_3}{\bullet}\!\rule[0.5ex]{3em}{0.4pt}\!\underset{x_4}{\bullet} \end{matrix}\right) + \left(\begin{matrix} \underset{x_1}{\bullet}\!\rule[0.5ex]{3em}{0.4pt}\!\underset{x_3}{\bullet} \\[1em] \underset{x_2}{\bullet}\!\rule[0.5ex]{3em}{0.4pt}\!\underset{x_4}{\bullet} \end{matrix}\right) + \left(\begin{matrix} \underset{x_1}{\bullet}\!\rule[0.5ex]{3em}{0.4pt}\!\underset{x_4}{\bullet} \\[1em] \underset{x_2}{\bullet}\!\rule[0.5ex]{3em}{0.4pt}\!\underset{x_3}{\bullet} \end{matrix}\right).$$

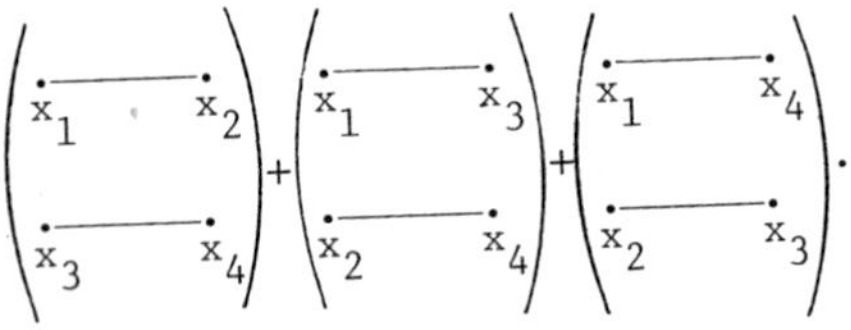

It is clear that $G_0^{(4)}$ is a rather disconnected object. It can be interpreted as the amplitude for, say, a transition from $x_1,x_2$ to $x_3,x_4$. In this approximation there are only so many ways for signal propagation, all expressed diagrammatically in the above picture.

A much more transparent interpretation is obtained in the Fourier transformed space. We have seen that the nature of $\Delta_F$ impels us to interpret $p_\mu$ as the four-momentum of a particle state. This is consistent with translation invariance, leading to p-conservation. Indeed, since G depends only on differences of x's, the naive Fourier transform

$$\int d^4x_1..d^4x_N \, e^{-i(p_1x_1+...+p_Nx_N)} G^{(N)}(x_1,..x_N)$$

necessarily contains a $\delta$-function of $(p_1+...+p_N)$. So, instead we set

$$\tilde{G}^{(N)}(p_1,..,p_N)(2\pi)^4\delta^{(4)}(p_1+..+p_N) =$$

$$\int d^4x_1..d^4x_N e^{-i(p_1x_1+..+p_Nx_N)} G^{(N)}(x_1,..,x_N), \qquad (2.18)$$

with $\tilde{G}^{(N)}(p_1,..,p_N)$ defined only when $p_1+...+p_N = 0$. For example,

$$\tilde{G}_0^{(2)}(p,-p) = \frac{1}{p^2-m^2+i\varepsilon} \qquad (2.19)$$

gives the amplitude that a particle of momentum p and mass $m^2$ propagates. We can represent this diagrammatically as well

$\tilde{G}_0^{(2)}(p) =$ 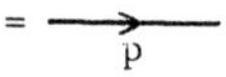

. (2.20)

In general, however, we will represent the Green's function $\tilde{G}^{(N)}(p_1,...p_N)$ as a blob with N lines, labeled by $p_1,p_2,..,p_N$, entering it and with $p_1+p_2+..+p_N = 0$, which reflects the conservation of momentum:

$\tilde{G}^{(N)}(p_1,..,p_N) =$ 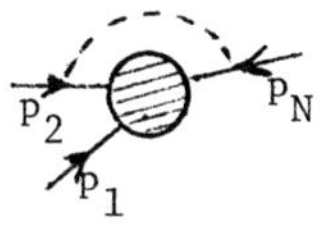

. (2.21)

It will be interpreted, say, as the scattering amplitude of states of momenta $p_1,..p_j$ into states of momenta $p_{j+1},..p_N$, if we take the lines j+1, ...,N to be outgoing. Again note that the form of $\tilde{G}_0^{(2)}$ is what suggests the nature of the external states. We will later discuss the unitarity constraints imposed on $\tilde{G}$.

Problems.

A. Starting from the Euclidean formulation for $W_E[J]$, work out the corresponding expression for the Feynman propagator and show that by analytic continuation in Minkowski space it reduces to the usual one.

B. Given $\Delta_F(x)$, show that it propagates positive energy signals forward in time, and negative energy signals backward in time.

C. Find the real and imaginary parts of $\Delta_F(x)$; interpret physically. Can you express $\Delta_F$ in terms of $\mathrm{Im}\Delta_F$?

3. The Effective Action

Out of the generating functional we can construct local quantities which lend themselves to familiar interpretations. For instance,

$$\frac{\delta W_0}{\delta J(x)} = -i\langle \Delta_c(x-1)J_1\rangle_1 W_0[J] \quad , \qquad (3.1)$$

so that

$$\phi_{c\ell}^{(0)}(x) \equiv -i \frac{\delta \, \ell n \, W_0}{\delta J(x)} = \frac{\delta Z_0}{\delta J(x)} \qquad (3.2)$$

satisfies the classical equation of motion [using (2.16)]

$$(\Box+m^2)\phi_{c\ell}^{(0)}(x) = J(x) \quad . \qquad (3.3)$$

In fact, we can use (3.3) to replace J(x) in terms of $\phi_{c\ell}^{(0)}(x)$. Formally it comes down to performing a functional Legendre transformation; introducing

$$\Gamma_0[\phi_{c\ell}^{(0)}] = Z_0[J] - \langle J\phi_{c\ell}^{(0)}\rangle \quad , \qquad (3.4)$$

we see by using (3.2) that $\Gamma_0$ is independent of J. In this case it is easy to find the explicit form of $\Gamma_0$ by replacing J in terms of $\phi_{c\ell}^{(0)}$. We find (integrating by parts as we go along)

$$\Gamma_0[\phi_{c\ell}^{(0)}] = -\frac{1}{2}\langle[(\Box+m^2)\phi_{c\ell}^{(0)}]_1\Delta_{F12}[(\Box+m^2)\phi_{c\ell}^{(0)}]_2\rangle - \langle\phi_{c\ell}^{(0)}(\Box+m^2)\phi_{c\ell}^{(0)}\rangle$$

$$= -\frac{1}{2}\langle\phi_{c\ell}^{(0)}(\Box+m^2)\phi_{c\ell}^{(0)}\rangle \quad , \qquad (3.5)$$

using (2.16). Integration by parts yields now the final form

$$\Gamma_0[\phi_{c\ell}^{(0)}] = \frac{1}{2}\int d^4x[\partial_\mu\phi_{c\ell}^{(0)}\partial^\mu\phi_{c\ell}^{(0)} - m^2\phi_{c\ell}^{(0)\,2}] \quad , \qquad (3.6)$$

which is the free action we had started from.

A similar procedure can be carried out in the general case $V \neq 0$. We form

$$\phi_{c\ell}(x) \equiv -i\frac{\delta \, \ell n \, W}{\delta J} = \frac{\delta Z[J]}{\delta J} \quad , \qquad (3.7)$$

and try to compute the effective action

$$\Gamma[\phi_{c\ell}] = Z[J] - \langle J\phi_{c\ell}\rangle \quad , \tag{3.8}$$

with now

$$J(x) = -\frac{\delta\Gamma[\phi_{c\ell}]}{\delta\phi_{c\ell}(x)} \quad , \tag{3.9}$$

as seen by differentiating (3.8) with respect to $\phi_{c\ell}$. (Of course, $\Gamma[\phi_{c\ell}]$ depends only on $\phi_{c\ell}$ and Z[J] only on J.) By the way, we observe that since $\Gamma$ is an effective action, (3.9) is proportional to its equation of motion coming from extremizing $\Gamma$. In the V = 0 case this is obvious from (3.3).

In order to derive an equation of motion for $\phi_{c\ell}(x)$, we have to write W[J] in a manageable form. We write

$$W[J] \equiv N\int\mathcal{D}\phi e^{i\langle \frac{1}{2}\partial_\mu\phi\partial^\mu\phi - \frac{1}{2}(m^2-i\varepsilon)\phi^2 - V(\phi)+J\phi\rangle} \quad , \tag{3.10}$$

$$= N\int\mathcal{D}\phi e^{-i\langle V(\phi)\rangle} e^{i\langle \frac{1}{2}\partial_\mu\phi\partial^\mu\phi - \frac{1}{2}(m^2-i\varepsilon)\phi^2+J\phi\rangle} \quad . \tag{3.11}$$

Now comes the trick: observe that

$$\frac{1}{i}\frac{\delta}{\delta J(x)} e^{i\langle J\phi\rangle} = \phi(x)e^{i\langle J\phi\rangle} \quad , \tag{3.12}$$

and since J and $\phi$ are independent variables, the same will be true for any function of $\phi$. In particular

$$e^{-i\langle V(\phi)\rangle}e^{i\langle J\phi\rangle} = e^{-i\langle V(\frac{1}{i}\frac{\delta}{\delta J})\rangle}e^{i\langle J\phi\rangle} \quad . \tag{3.13}$$

This allows us to take the V dependent term out of the integral

$$W[J] = e^{-i\langle V(\frac{1}{i}\frac{\delta}{\delta J})\rangle} N\int\mathcal{D}\phi e^{i\langle \frac{1}{2}\partial_\mu\phi\partial^\mu\phi - \frac{1}{2}(m^2-i\varepsilon)\phi^2+J\phi\rangle} \tag{3.14}$$

$$= e^{-i\langle V(\frac{1}{i}\frac{\delta}{\delta J})\rangle} W_0[J] \quad , \tag{3.15}$$

or

$$e^{iZ[J]} = W[J] = N\, e^{-i\langle V(\frac{1}{i}\frac{\delta}{\delta J})\rangle}\, e^{-\frac{i}{2}\langle J_1 \Delta_{F12} J_2\rangle} \quad . \qquad (3.16)$$

This equation will be the starting point of the perturbative evaluation of W[J]. For the moment, we use it to derive an equation for $\phi_{c\ell}$. From (3.16)

$$\frac{\delta W}{\delta J_x} = -i\, e^{-i\langle V(-i\frac{\delta}{\delta J})\rangle} \langle \Delta_{Fx1} J_1\rangle_1\, W_0[J]$$

$$= -i\, e^{-i\langle V(-i\frac{\delta}{\delta J})\rangle} \langle \Delta_{Fx1} J_1\rangle_1\, e^{i\langle V(-i\frac{\delta}{\delta J})\rangle}\, W[J] \quad . \qquad (3.17)$$

It follows that

$$(\Box_x + m^2)\,\frac{\delta W}{\delta J_x} = i\, \mathcal{O}_x W[J] \quad , \qquad (3.18)$$

where

$$\mathcal{O}_x = e^{-i\langle V(-i\frac{\delta}{\delta J})\rangle}\, J_x\, e^{i\langle V(-i\frac{\delta}{\delta J})\rangle} \quad . \qquad (3.19)$$

We can evaluate $\mathcal{O}_x$ by means of yet another trick. Set

$$\mathcal{O}_x(\lambda) = e^{-i\lambda\langle V(-i\frac{\delta}{\delta J})\rangle}\, J_x\, e^{i\lambda\langle V(-i\frac{\delta}{\delta J})\rangle} \quad , \qquad (3.20)$$

where $\lambda$ is a parameter. Clearly

$$\frac{d\mathcal{O}_x(\lambda)}{d\lambda} = e^{-i\lambda\langle V(-i\frac{\delta}{\delta J})\rangle} [-i\langle V(-i\frac{\delta}{\delta J})\rangle, J_x]\, e^{i\lambda\langle V(-i\frac{\delta}{\delta J})\rangle} \quad . \qquad (3.21)$$

But

$$[V(-i\frac{\delta}{\delta J_y}), J_x] = -i\, V'(-i\frac{\delta}{\delta J_y})\,\delta^{(4)}(x-y) \quad , \qquad (3.22)$$

where V' is the derivative of V with respect to its argument. Integrating over y, we find

$$\frac{d\mathcal{O}_x(\lambda)}{d\lambda} = -V'(-i\frac{\delta}{\delta J_x}) \quad . \qquad (3.23)$$

This equation is now integrated over $\lambda$ to yield

$$\mathcal{O}_x = \mathcal{O}_x(\lambda=1) = J(x) - V'(-i\frac{\delta}{\delta J_x}) \quad . \tag{3.24}$$

Hence

$$(\Box_x+m^2)\frac{\delta W}{\delta J} = i(J(x) - V'(-i\frac{\delta}{\delta J_x}))W[J] \quad , \tag{3.25}$$

or

$$(\Box_x+m^2)\phi_{c\ell}(x) = J(x) - \frac{1}{W[J]}V'(-i\frac{\delta}{\delta J(x)})W[J] \quad . \tag{3.26}$$

The last term clearly resembles a force. For example, take

$$V = \frac{\lambda}{4!}\phi^4 \quad , \quad \lambda \text{ dimensionless} \quad . \tag{3.27}$$

Then

$$\frac{1}{W[J]}V'(-i\frac{\delta}{\delta J_x})W[J] = \frac{\lambda}{3!}(-i)^3\frac{1}{W}\frac{\delta^3}{\delta J_x^3}W[J]$$

$$= \frac{\lambda}{3!}[\phi_{c\ell}^3(x) - \frac{\delta^2\phi_{c\ell}}{\delta J_x^2} - 3i\phi_{c\ell}\frac{\delta\phi_{c\ell}}{\delta J_x} \quad , \tag{3.28}$$

and finally

$$(\Box_x+m^2)\phi_{c\ell}(x) = J(x) - \frac{\lambda}{3!}\phi_{c\ell}^3(x) + \frac{\lambda}{3!}\frac{\delta^2\phi_{c\ell}(x)}{\delta J^2(x)} + \frac{i\lambda}{4}\frac{\delta\phi_{c\ell}^2(x)}{\delta J(x)} \quad . \tag{3.29}$$

The first two terms on the right hand side give the classical equation of motion modified by the last two terms, which must amount to corrections from the quantum theory (see problem).

In the case $V \neq 0$, the explicit form of the effective action is, of course, not known. We can expand it functionally in terms of $\phi_{c\ell}$ as

$$\Gamma[\phi_{c\ell}] = \int d^4x[-V^e(\phi_{c\ell}) + \frac{1}{2}F(\phi_{c\ell})\partial_\mu\phi_{c\ell}\partial^\mu\phi_{c\ell}$$

$$+ \text{ higher order derivatives}] \quad , \tag{3.30}$$

where we take into account now local effects by including arbitrarily high derivatives of $\phi_{c\ell}$. We have arbitrary functions $V^e(\phi_{c\ell})$, $F(\phi_{c\ell})$, etc., to be determined. $V^e$ is clearly an effective potential. By expressing J in terms of $\phi_{c\ell}$ using (3.29) and integrating (3.9), we see that

$$V^e(\phi_{c\ell}) = \frac{\lambda}{4!}\,\phi_{c\ell}^4 + \frac{m^2}{2}\,\phi_{c\ell}^2 + \mathcal{O}(\hbar) \tag{3.31}$$

and

$$F(\phi_{c\ell}) = 1 + \text{corrections} \quad . \tag{3.32}$$

Alternatively, we can expand the effective action in terms of $\phi_c$ nonlocal way:

$$\Gamma[\phi_{c\ell}] = \sum_N \frac{1}{N!} \langle \Gamma^{(N)}(1,\ldots\ldots,N)\phi_{c\ell}(1)\ldots\ldots\phi_{c\ell}(N)\rangle_{1,\ldots N} \quad . \tag{3.33}$$

The coefficients $\Gamma^{(N)}(x_1,\ldots,x_N)$ are called the proper vertices. They depend on the differences $x_i - x_j$ because of translation invariance so that their Fourier transforms are introduced via

$$\tilde{\Gamma}^{(N)}(p_1,\ldots,p_N)(2\pi)^4\delta(p_1+\ldots+p_N) =$$

$$\int d^4x_1\ldots d^4x_N\; e^{-i(p_1x_1+\ldots+p_Nx_N)}\,\Gamma^{(N)}(x_1,\ldots,x_N) \quad , \tag{3.34}$$

with $\tilde{\Gamma}^{(N)}$ being defined only when the sum of its arguments vanishes.

Problems.

A. By a judicious set of insertions of $\hbar$ when needed, show that the non-classical terms in the equation for $\phi_{c\ell}$ do indeed vanish as $\hbar \to 0$.

B. Suppose that $V = \frac{1}{2}\,\delta m^2 \phi^2$ in the scalar field action. Find the equation obeyed by $\phi_{c\ell}$

## 4. Saddle-Point Evaluation of the Path Integral

Integrals of the form

$$I \equiv \int dx\ e^{-a(x)} \quad , \qquad (4.1)$$

where $a(x)$ is a function of $x$, can be approximated by expanding $a(x)$ around $x_0$ where $a(x)$ is stationary:

$$a(x) \simeq a(x_0) + \frac{1}{2}(x-x_0)^2 a''(x_0) + .. \quad . \qquad (4.2)$$

Then

$$I \simeq e^{-a(x_0)} \int dx\ e^{-\frac{1}{2}(x-x_0)^2 a''(x_0)} \quad , \qquad (4.3)$$

and the integral is easily performed if $a''(x_0) > 0$ (it's a Gaussian), neglecting the higher derivatives. The success of this approximation rests on the fact that the integrand is largest when $a(x)$ is smallest and that the points away from the minimum do not significantly contribute, as in the figure

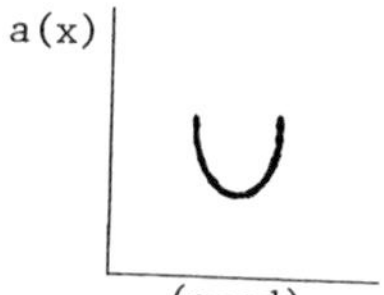

(good)

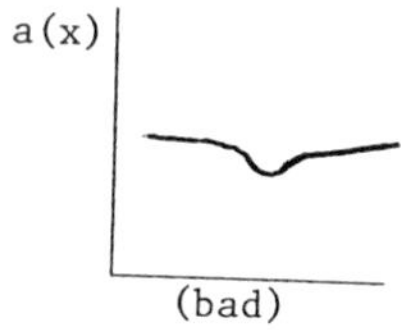

(bad)

In this section we apply this technique to the Euclidean space generating functional.

We start from the Euclidean space definition of the generating functional

$$W_E[J] = N_E \int \mathcal{D}\phi\ e^{-S_E[\phi,J]} \quad , \qquad (4.4)$$

where

$$S_E[\phi,J] = \int d^4\bar{x}\left[\frac{1}{2}\bar{\partial}_\mu\phi\bar{\partial}_\mu\phi + \frac{1}{2}m^2\phi^2 + V(\phi) - J\phi\right] \quad . \qquad (4.5)$$

We then expand the action around a field configuration $\phi_0$

$$S_E[\phi,J] = S_E[\phi_0,J] + \langle\frac{\delta S_E}{\delta\phi}(\phi-\phi_0)\rangle$$

$$+ \frac{1}{2}\langle\frac{\delta^2 S_E}{\delta\phi_1\delta\phi_2}(\phi-\phi_0)_1(\phi-\phi_0)_2\rangle_{1,2}+\ldots \quad , \tag{4.6}$$

with the functional derivative evaluated at $\phi_0$. We take $S_E$ to be stationary at $\phi_0$, which means that $\phi_0$ obeys the classical equations of motion with the source term

$$\left.\frac{\delta S_E}{\delta\phi}\right|_0 = -\bar{\partial}_\mu\bar{\partial}_\mu\phi_0 + m^2\phi_0 + V'(\phi_0) - J = 0 \quad . \tag{4.7}$$

It follows that (after integration by parts)

$$S_E[\phi_0,J] = \frac{1}{2}\int d^4\bar{x}(2-\phi_0\frac{d}{d\phi_0})(-J\phi_0+V(\phi_0)) \quad , \tag{4.8}$$

while

$$\frac{\delta^2 S}{\delta\phi_1\delta\phi_2} = [-\bar{\partial}_\mu\bar{\partial}_\mu+m^2+V''(\phi)]_1\delta(\bar{x}_1-\bar{x}_2) \tag{4.9}$$

is an operator. In the spirit of the saddle point evaluation, the generating functional now becomes

$$W_E[J] \simeq N_E\, e^{-S_E[\phi_0,J]}\int\mathcal{D}\phi\; e^{-\frac{1}{2}\langle\phi_1\frac{\delta^2 S_E}{\delta\phi_1\delta\phi_2}\phi_2\rangle_{1,2}} \quad . \tag{4.10}$$

The Gaussian integral can be done (see Appendix A), with the formal result

$$W_E[J] \simeq N_E'\, e^{-S_E[\phi_0,J]}\{\det([-\bar{\partial}_\mu\bar{\partial}_\mu+m^2+V''(\phi_0)]\delta_{12})\}^{-1/2} . \tag{4.11}$$

Clearly this expression needs some getting used to. We can rewrite it in a slightly more suggestive form by using the identity

$$\det M = e^{\mathrm{Tr}\,\ell n\, M} \tag{4.12}$$

as

$$W_E[J] = N_E' \, e^{-S_E[\phi_0,J] - \frac{1}{2} \mathrm{Tr}\, \ell n\, [\{-\bar{\partial}_\mu \bar{\partial}_\mu + m^2 + V''(\phi_0)\}\delta_{12}]}, \quad (4.13)$$

which clearly indicates we are computing corrections to Z[J]. The physical meaning of this approximation can be understood by carefully putting back all the $\hbar$ factors. Then it is seen that it corresponds to an asymptotic series in $\hbar$ (see problem). The first term $S_E[\phi_0,J]$ gives the classical contribution to the Green's functions (remember Dirac's identification). The next term, of $\mathcal{O}(\hbar)$, gives the first quantum correction to the Green's functions. [The determinant of an operator is understood to mean the product of its eigenvalues.] We start by computing the classical contributions to W[J]. It must be remembered that $\phi_0$, being the solution of (4.7), is a functional of J. The procedure is therefore very simple: a) calculate the functional dependence of $\phi_0$ on J, b) insert it in (4.8) and, c) by comparing the resulting expression with the expansion (2.15), extract the Green's functions $G_c^{(N)}(1,..,N)$. Alas there are grave theoretical difficulties in carrying out step a). The equation obeyed by $\phi_0$ is a nonlinear differential equation (for $V' \neq 0$), and no one has succeeded in solving it in closed form. The best one can do is to solve it in perturbation theory. Specifically, take the $\phi^4$ potential and expand around $\lambda = 0$.

We write

$$\phi_0 = \phi^{(0)} + \lambda\phi^{(1)} + \lambda^2\phi^{(2)} + \ldots \quad (4.14)$$

so that

$$S_E = -\frac{1}{2}\int d^4\bar{x}[J(\phi^{(0)}+\lambda\phi^{(1)}+\ldots) + \frac{\lambda}{12}(\phi^{(0)}+\lambda\phi^{(1)}+..)^4] \quad (4.15)$$

$$= -\frac{1}{2}\int d^4\bar{x}\, J\phi^{(0)} - \frac{\lambda}{2}\int d^4\bar{x}[J\phi^{(1)} + \frac{1}{12}\phi^{(0)4}] + \mathcal{O}(\lambda^2). \quad (4.16)$$

If we define the Euclidean Green's function (in an obvious notation)

$$(\bar{\partial}_\mu \bar{\partial}_\mu - m^2) G_{xy} = -\delta_{xy} \quad , \tag{4.17}$$

it follows that

$$\phi^{(0)}(x) = \langle G_{xa} J_a \rangle_a$$

$$\phi^{(1)}(x) = -\frac{1}{6} \langle G_{xy} G_{ya} G_{yb} G_{yc} J_a J_b J_c \rangle_{abcy}, \text{ etc...} \quad . \tag{4.18}$$

Thus,

$$S_E[J] = -\frac{1}{2} \langle J_a G_{ab} J_b \rangle_{ab} + \frac{\lambda}{4!} \langle G_{xa} G_{xb} G_{xc} G_{xd} J_a J_b J_c J_d \rangle_{abcdx}$$

$$- \frac{\lambda^2}{3.4!} \langle G_{xa} G_{xb} G_{xc} G_{xy} G_{yd} G_{ye} G_{yf} J_a J_b J_c J_d J_e J_f \rangle_{abcdefxy} + \mathcal{O}(\lambda^3) . \tag{4.19}$$

Correspondingly, the (connected) Euclidean Green's functions are given by

$$G_E^{(N)}(\bar{x}_1, \ldots, \bar{x}_N) = - \frac{\delta^N Z_E}{\delta J_1 \ldots \delta J_N} \quad , \tag{4.20}$$

where

$$W_E[J] = N_E e^{-Z_E[J]} \quad . \tag{4.21}$$

In this classical approximation we find the connected Green's functions to be

$$G_E^{(2)}(\bar{x}_1, \bar{x}_2) = G(\bar{x}_1, \bar{x}_2) = \int \frac{d^4\bar{p}}{(2\pi)^4} \frac{e^{i\bar{p}\cdot(\bar{x}_1 - \bar{x}_2)}}{\bar{p}^2 + m^2} \tag{4.22}$$

$$G_E^{(4)}(\bar{x}_1, \bar{x}_2, \bar{x}_3, \bar{x}_4) = -\lambda \int d^4\bar{y}\, G(\bar{x}_1, y) G(\bar{x}_2, y) G(\bar{x}_3, y) G(\bar{x}_4, y) \tag{4.23}$$

$$G_E^{(6)}(\bar{x}_1, \bar{x}_2, \bar{x}_3, \bar{x}_4, \bar{x}_5, \bar{x}_6) = \lambda^2 \int d^4\bar{x} d^4\bar{y}\, G(\bar{x}, \bar{y})$$

$$P(\bar{x}, \bar{y}, \bar{x}_1, \bar{x}_2, \bar{x}_3, \bar{x}_4, \bar{x}_5, \bar{x}_6) \quad , \tag{4.24}$$

where

$$P(\bar{x},\bar{y},\{\bar{x}_i\}) = \sum_{(ijk)} G(\bar{x},\bar{x}_i)G(\bar{x},\bar{x}_j)G(\bar{x},\bar{x}_k)G(\bar{y},\bar{x}_\ell)G(\bar{y},\bar{x}_m)G(\bar{y},\bar{x}_n), \tag{4.25}$$

where the sum runs over all the following values of the triples, (ijk) = (123), (124), (125), (126), (134), (135), (136), (145), (146), (156), with ($\ell$mn) assuming the complementary value [e.g., ($\ell$mn) = (456) when (ijk) = (123)]. Note that ijk runs only over half of the possible values. This is because the expression for P is symmetric under the interchange $\bar{x} \to \bar{y}$.

In this classical approximation and to order $\lambda^2$ these are the only non-zero Green's functions.

The momentum space Green's functions, defined by

$$\tilde{G}_E^{(N)}(\bar{p}_1,..,\bar{p}_N)(2\pi)^4\delta(\bar{p}_1+..+\bar{p}_N) = \int d^4\bar{x}_1..d^4\bar{x}_N\, e^{i\bar{p}_1\bar{x}_1+..+i\bar{p}_N x_N}\, G_E^{(N)}(\bar{x}_1,..,\bar{x}_N) \tag{4.26}$$

are easily seen to be given by

$$\tilde{G}_E^{(2)}(\bar{p}_1,\bar{p}_2=-\bar{p}_1) = \frac{1}{\bar{p}_1^2+m^2} + \mathcal{O}(\hbar) \tag{4.27}$$

$$\tilde{G}_E^{(4)}(\bar{p}_1,\bar{p}_2,\bar{p}_3,\bar{p}_4) = \left(\frac{1}{\bar{p}_1^2+m^2}\,\frac{1}{\bar{p}_2^2+m^2}\,\frac{1}{\bar{p}_3^2+m^1}\,\frac{1}{\bar{p}_4^2+m^2}\right)\lambda + \mathcal{O}(\hbar) \tag{4.28}$$

$$\tilde{G}_E^{(6)}(\bar{p}_1,\ldots,\bar{p}_6) = \left[\prod_{i=1}^{6}\frac{1}{(\bar{p}_i^2+m^2)}\right]\sum_{(ijk)}\frac{\lambda^2}{(\bar{p}_i+\bar{p}_j+\bar{p}_k)^2+m^2} + \mathcal{O}(\hbar) \tag{4.29}$$

where again the sum is over the same triples as in (4.25). In the above expressions the $\tilde{G}$'s are to be evaluated only when the sum of their arguments vanishes.

In the perturbative evaluation of $\phi_0$, we note that $\phi^{(k)}$ always depends on $J^{1+2k}$. Hence the $\lambda^k$ order contributes solely to $G^{2(k+1)}$. This is an artifact of the approximation, which neglects contributions of order $\hbar$.

Following Feynman, we develop a pictogram for these Green's functions. We represent $\tilde{G}_E^{(N)}(\bar{p}_1,\ldots,\bar{p}_N)$ by a blob with N external lines

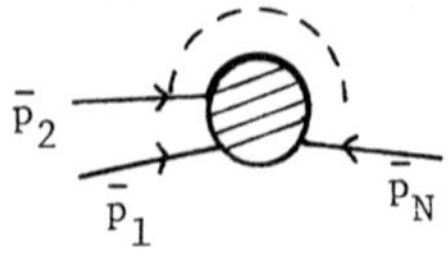


each line carrying a $\bar{p}$ label, and with all arrows pointing into the blob. This blob is then represented diagrammatically by the following rules:

$\longrightarrow$ means $\dfrac{1}{\bar{p}^2+m^2}$, the propagator factor

 means $-\lambda$, the vertex.

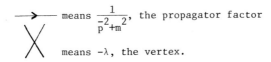

It is understood that the net amount of $\bar{p}$ flowing through the vertex is conserved. The vertex X has no arrows on its lines, thus indicating that the propagator factor for the lines is not included. Comparison with (4.27-4.29) leads to

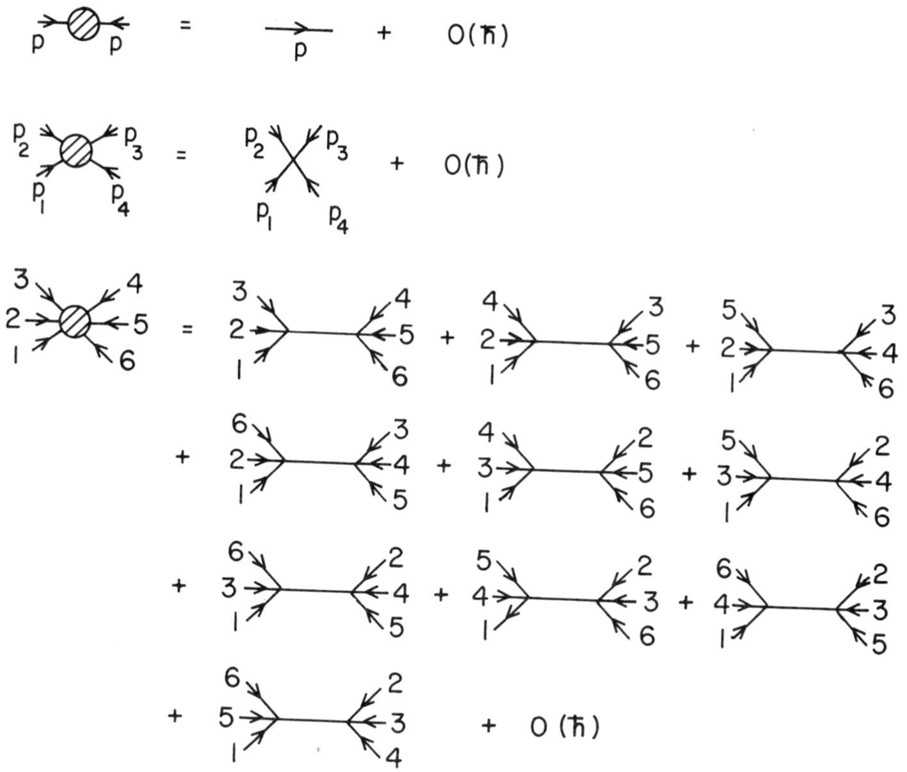


It appears then that the Feynman rules, to this order in $\hbar$, are to draw all possible arrangements using —— and X as basic building blocks with no closed circuits (also called loops). Such diagrams are called tree diagrams. It is not hard to see that this approximation represents the Green's functions by their tree diagrams.

Consider a diagram with E external lines, I internal lines and $V_n$ vertices, each with n lines (in our case n = 4). Each internal line hooks on to two vertex lines. Hence the number of external lines is just equal to the number of vertex lines minus twice the number of internal lines

$$E = nV_n - 2I \qquad .$$

To $\mathcal{O}(\hbar^\circ)$, we have $E = 2k+2$, $n = 4$, $V_4 = k$ hence

$$I = V_4 - 1 \qquad \mathcal{O}(\hbar^\circ) \text{ only} \qquad ,$$

so that there cannot be any closed loop in the Feynman diagram: so they are the tree diagrams. The rule for an arbitrary (tree) diagram is to draw all topologically inequivalent diagrams with the external lines identified. For example, in $\tilde{G}^{(6)}$ when the same lines emanate from the vertex, the diagram enters only once because their rearrangements would give topologically equivalent diagrams [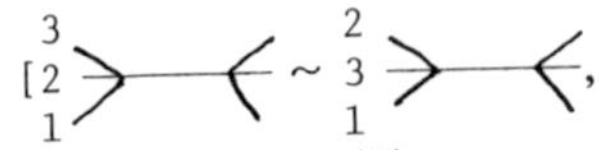

, etc.]. These rules make it easy to write the expression for $\tilde{G}^{(8)}$, ...: we first draw all possible inequivalent tree arrangements with the external lines identified, and then use the rules in reverse to get the analytical expression. This use of the pictorial representation has proved to be an essential tool in the perturbative evaluation of the Green's functions.

It is interesting to rewrite these results in terms of the classical field $\phi_{c\ell}(x)$ and to see the form of the resulting effective action. In Euclidean space we define the classical field as

$$\phi_{c\ell}(\bar{x}) = -\frac{\delta Z_E}{\delta J(\bar{x})} \simeq -\frac{\delta S_E}{\delta J(\bar{x})} + \mathcal{O}(\hbar) \qquad , \qquad (4.30)$$

which, using (4.19) gives us $\phi_{c\ell}(\bar{x})$ as a functional of J, order by order in $\lambda$. Then, we invert the equation in perturbation theory, and find $J(\bar{x})$ as a functional of $\phi_{c\ell}$. The result is

$$J(\bar{x}) = (\bar{\partial}^2 - m^2)\phi_{c\ell}(\bar{x}) - \frac{\lambda}{3!}\phi_{c\ell}^3(\bar{x}) \qquad . \qquad (4.31)$$

The remarkable thing is that there are no terms of higher order in $\lambda$ in this equation. By comparing with (4.7), we conclude that

$$\phi_{c\ell}(\bar{x}) = \phi_0(\bar{x}) + \mathcal{O}(\hbar) \quad . \tag{4.32}$$

Integration of (3.9) gives us immediately the effective action to this order

$$\Gamma_E[\phi_{c\ell}] = -\int d^4\bar{x}[\frac{1}{2}\,\phi_{c\ell}(\bar{\partial}^2 - m^2)\phi_{c\ell} - \frac{\lambda}{4!}\,\phi^4_{c\ell}(\bar{x})] \quad . \tag{4.33}$$

Thus, $\Gamma_E$ is the classical action. We can therefore derive the expression for the proper vertices. In this approximation, we see that $\tilde{\Gamma}^{(2)}$ is minus the inverse propagator and that $\tilde{\Gamma}^{(n)}$ for $n > 4$ vanish. The higher order diagrams do not appear. The reason is that the $\tilde{\Gamma}^{(N)}$ generate only Feynman graphs that cannot become disconnected by cutting off one of their internal lines. Such graphs are called one-particle irreducible. As we have seen, all the tree graphs are one-particle reducible except for the lowest one.

Problems.

A. Show that the saddle point evaluation of the path integral corresponds to an asymptotic expansion in $\hbar$.

B. Solve the equation

$$(\partial^2 - m^2 - \frac{\lambda}{3!}\phi^2)\phi = -J$$

order by order about $\lambda = 0$. If we set $\phi = \phi^{(0)} + \lambda\phi^{(1)} + \lambda^2\phi^{(2)} + \ldots$, derive the explicit expressions for $\phi^{(2)}$ and $\phi^{(3)}$.

C. For $\lambda\phi^4$ theory, find the effective classical action to order $\lambda^3$, and derive the classical Euclidean Green's functions to order $\lambda^3$, both in $\bar{x}$- and $\bar{p}$-space.

## 5. First Quantum Corrections - ζ-function Evaluation of Determinants

The $\mathcal{O}(\hbar)$ correction to the effective action is computed by evaluating the determinant of (4.11). This determinant is to be interpreted as the product of the eigenvalues of the operator. In one possible procedure, the space is truncated (by, say, a box), resulting in discrete eigenvalues. Their product is computed and then the size of the box is let go to infinity. In the following we want to make use of a powerful formal technique for computing the determinant of operators.

Consider an operator A with positive and real discrete eigenvalues $a_1, \ldots a_n, \ldots$; call its eigenfunctions $f_n(x)$

$$Af_n(x) = a_n f_n(x) \quad . \tag{5.1}$$

We form the construct

$$\zeta_A(s) = \sum_n \frac{1}{a_n^s} \quad , \tag{5.2}$$

called the ζ-function associated to A. [If A is the harmonic oscillator transformation, then ζ is indeed Riemann's ζ-function.] Then the sum extends over all the eigenvalues and A is a real variable. We note that

$$\left.\frac{d\zeta_A(s)}{ds}\right|_{s=0} = -\sum_n \ln a_n e^{-sa_n}\Big|_{s=0} = -\ln\,(\prod_n a_n) \quad , \tag{5.3}$$

leading to

$$\det A \equiv \prod_n a_n = e^{-\zeta_A'(0)} \quad . \tag{5.4}$$

The advantage of this representation for det A is that for many operators of physical interest $\zeta_A$ is not singular at A = 0. In fact, introduce the "Heat Function"

$$G(x,y,\tau) \equiv \sum_n e^{-a_n \tau} f_n(x) f_n^*(y) \quad , \qquad (5.5)$$

which obeys the differential equation (heat equation)

$$A_x G(x,y,\tau) = -\frac{\partial}{\partial\tau} G(x,y,\tau) \quad , \qquad (5.6)$$

as can be seen by inspection. The $\zeta$-function can now be expressed in terms of this "Heat Function" very easily:

$$\zeta_A(s) = \frac{1}{\Gamma(s)} \int_0^\infty d\tau \; \tau^{s-1} \int dx G(x,x,\tau) \quad , \qquad (5.7)$$

using the orthogonality of the eigenfunctions and the well-known representation of the $\Gamma$-function. This equation is the desired analytic representation of $\zeta_A(s)$. Note that

$$G(x,y,\tau=0) = \delta(x-y) \quad , \qquad (5.8)$$

using the orthonormality of the eigenfunctions. Thus a possible way of computing det A emerges: 1) find the solution of eq. (5.6) subject to the initial condition (5.8), 2) insert the solution into (5.7), to compute $\zeta_A(s)$, and use (5.4) to obtain det A.

This procedure can be generalized to our problem. The operator is now $[-\bar{\partial}^2+m^2 + \frac{\lambda}{2}\phi_0^2(\bar{x})]$, where $\phi_0(\bar{x})$ is a solution of the classical equations with a source J.

It is easy to check that the solution of the equation

$$-\bar{\partial}_x^2 G_0(\bar{x},\bar{y},\tau) = -\frac{\partial G_0}{\partial\tau} \quad , \qquad (5.9)$$

with the boundary condition (5.8), is (in four dimensions only!)

$$G_0(\bar{x},\bar{y},\tau) = \frac{1}{16\pi^2\tau^2} e^{-\frac{1}{4\tau}(\bar{x}-\bar{y})^2} \quad . \qquad (5.10)$$

This does not yet solve our problem. In particular the resulting $\zeta_{-\bar{\partial}^2}(s)$ computed from (5.9) does not exist. We want to find $G(\bar{x},\bar{y},\tau)$ subject to (5.8) which obeys

$$[-\bar{\partial}_x^2+m^2 + \frac{\lambda}{2}\phi_0^2(\bar{x})]G(\bar{x},\bar{y},\tau) = -\frac{\partial G(\bar{x},\bar{y},\tau)}{\partial\tau} \quad . \quad (5.11)$$

It is clear that for an arbitrary $\phi_0(\bar{x})$ this equation is very hard to solve. Still, let us see what we can do. If we write the effective action in the form

$$\Gamma_E[\phi_{c\ell}] = \Gamma_E\phi[\phi_{c\ell}] + \hbar\Gamma_E^{(1)}[\phi_{c\ell}] + .. \quad , \quad (5.12)$$

we see that

$$\Gamma_E^{(1)}[\phi_{c\ell}] = -\frac{1}{2}\zeta'_{[-\bar{\partial}^2+m^2 + \frac{\lambda}{2}\phi_{c\ell}^2(\bar{x})]}(0) \quad , \quad (5.13)$$

where we have replaced $\phi_0$ by $\phi_{c\ell}$ which does not induce any error up to $\mathcal{O}(\hbar)$, and used (5.4) and (4.11).

On the other hand, we can set

$$\Gamma_E[\phi_{c\ell}] = \int d^4\bar{x}[V(\phi_{c\ell}(\bar{x}))+F(\phi_{c\ell})\partial_\mu\phi_{c\ell}(\bar{x})\partial_\mu\phi_{c\ell}(\bar{x})+..] \quad . \quad (5.14)$$

Hence, if we want to calculate the $\mathcal{O}(\hbar)$ contribution to $V(\phi_{c\ell})$, it suffices to consider a constant field configuration: suppose we set

$$\phi_{c\ell}(\bar{x}) = v \quad , \quad (5.15)$$

where v is a constant independent of $\bar{x}$. Then

$$\Gamma_E[\phi_{c\ell}] = \int d^4\bar{x}\; V(v) \quad , \quad (5.16)$$

and it is proportional to $\int d^4\bar{x}$, the infinite volume element, because the Euclidean space $R_4$ is not bounded. However, if we make believe we are on $S_4$, the surface of a sphere in five dimensions, we get a finite volume element (the surface of the sphere). This procedure avoids this infrared divergence. Later we can let the radius of the sphere go to infinity.

It follows that the $\mathcal{O}(\hbar)$ contribution to the potential is given by

$$V(v)\int d^4\bar{x} = -\frac{1}{2}\,\zeta'_{[-\bar{\partial}^2+m^2+\frac{\lambda}{2}v^2]}(0) \quad . \tag{5.17}$$

With a constant v, (5.11) can be integrated very easily. We find

$$G(\bar{x},\bar{y},\tau) = \frac{\mu^4}{16\pi^2\tau^2}\, e^{\mu^2(\bar{x}-\bar{y})^2/4\tau}\, e^{-(m^2+\frac{\lambda}{2}v^2)\frac{\tau}{\mu^2}} \quad , \tag{5.18}$$

where we have inserted an arbitrary factor μ with dimensions of mass to make τ dimensionless. Then, by using (5.7), we arrive at

$$\zeta(s) = \frac{1}{\Gamma(s)}\int_0^\infty d\tau\, \tau^{s-1}\int d^4\bar{x}\, \frac{\mu^4}{16\pi^2\tau^2}\, e^{-(m^2+\frac{\lambda}{2}v^2)\tau/\mu^2} \tag{5.19}$$

$$= \frac{\mu^4}{16\pi^2}\left(\frac{m^2+\frac{\lambda}{2}v^2}{\mu^2}\right)^{2-s}\frac{\Gamma(s-2)}{\Gamma(s)}\int d^4\bar{x} \tag{5.20}$$

where we have rescaled τ [the integration over τ is strictly valid only when $s-2 > 0$, but we define ζ(s) everywhere by analytic continuation]. Note the appearance of the volume factor $\int d^4\bar{x}$ which accounts for the one in (5.17). Comparison yields

$$V(v) = -\frac{\mu^4}{32\pi^2}\frac{d}{ds}\left\{\frac{1}{(s-2)(s-1)}\left(\frac{m^2+\frac{\lambda}{2}v^2}{\mu^2}\right)^{2-s}\right\}\Bigg|_{s=0} \quad , \tag{5.21}$$

$$= \frac{1}{64\pi^2}\,[m^2+\frac{\lambda}{2}v^2]^2\left(-\frac{3}{2}+\ell n\,\frac{m^2+\frac{\lambda}{2}v^2}{\mu^2}\right) \quad . \tag{5.22}$$

Now that we have the functional form of V, we can state that the effective potential of the theory is given by

$$V[\phi_{c\ell}] = \frac{1}{2} m^2 \phi_{c\ell}^2(\bar{x}) + \frac{\lambda}{4!} \phi_{c\ell}^4(\bar{x}) + \frac{\hbar}{64\pi^2} (m^2 + \frac{\lambda}{2} \phi_{c\ell}^2)^2$$

$$\cdot \left[- \frac{3}{2} + \ln \frac{m^2 + \frac{\lambda}{2} \phi_{c\ell}^2}{\mu^2}\right] + \mathcal{O}(\hbar^2) \quad . \qquad (5.23)$$

This result is quite peculiar, because it seems to depend on the unknown scale $\mu^2$, which was introduced arbitrarily. Does it mean that the potential thus obtained is arbitrary? Observe that V depends on the parameters $m^2$ and $\lambda$. These have not really been defined except as input parameters in the classical Lagrangian. For simplicity, take $m^2 = 0$ to start with. Then it is easy to see that automatically

$$\frac{d^2V}{d\phi^2} = 0 \qquad \text{at } \phi = 0 \quad . \qquad (5.24)$$

We define the mass squared as the coefficient of the $\phi^2$ term in $\mathcal{L}$ evaluated at $\phi = 0$; it is seen to be zero to $\mathcal{O}(\hbar)$, if it is classically zero. Next, what about $\lambda$? Let us define it to be the coefficient of the fourth derivative of V evaluated at some constant point $\phi = M$

$$\lambda \equiv \frac{d^4V}{d\phi^4} \qquad \text{at } \phi = M \quad . \qquad (5.25)$$

Note that we cannot take $\phi = 0$ as in the previous case because of the divergence coming from the logarithm [infrared divergence]. This is typical of theories where $m^2 = 0$ classically.

The condition (5.25) requires

$$\ln \frac{\lambda M^2}{2\mu^2} = - \frac{8}{3} \quad , \qquad (5.26)$$

as seen by differentiating (5.23), setting $m^2 = 0$ and using (5.25). Thus we can eliminate $2\mu^2/\lambda$ in favor of $M^2$ and express the result as

$$V(\phi_{c\ell}) = \frac{\lambda}{4!}\phi_{c\ell}^4 + \frac{\lambda^2\phi_{c\ell}^4}{256\pi^2}\left[\ell n\frac{\phi_{c\ell}^2}{M^2} - \frac{25}{6}\right] \quad , \tag{5.27}$$

in accordance with the result of S. Coleman and E. Weinberg, Phys. Rev. D7, 1888 (1973). This little exercise shows that we must carefully define the input parameters in the Lagrangian in order to handle the quantum corrections. The result (5.27) still seems to depend on one arbitrary scale $M^2$ but it really does not, because, given the normalization condition, if we change the scale from $M^2$ to $M'^2$, we have to change at the same time $\lambda$ to $\lambda'$, where

$$\lambda' = \lambda + \frac{3\lambda^2}{16\pi^2}\ell n\frac{M'}{M} \tag{5.28}$$

(using 5.25). We see that the potential

$$V(\phi_{c\ell}) = \frac{\lambda'}{4!}\phi_{c\ell}^4 + \frac{\lambda'^2\phi_{c\ell}^4}{256\pi^2}\left[\ell n\frac{\phi_{c\ell}^2}{M'^2} - \frac{25}{6}\right] + \mathcal{O}(\lambda^3) \tag{5.29}$$

is form invariant under this reparametrization:

$$V(\lambda', M') = V(\lambda, M) \quad . \tag{5.30}$$

This shows that the physics does not change, only our way of interpreting the constants.

Problems.

A. Suppose that the classical potential is given by $V_{c\ell} = \frac{1}{6} f\phi^3$, where f has dimension of mass. Find by the steepest descent method the first quantum correction to this potential. Interpret the resulting potential physically.

B. Repeat problem A. for $V_{c\ell} = \frac{1}{6} f\phi^3 + \frac{\lambda}{4!}\phi^4$. Interpret physically.

**C. Find the solution of the heat equation

$$(\bar{\partial}^2 - m^2) G(\bar{x},\bar{y},\tau) = \frac{\partial G}{\partial \tau}$$

$$G(\bar{x},\bar{y},0) = \delta(\bar{x}-\bar{y})$$

in d-dimensions. Use your result to compute the effective potential for the theory defined by

$$\int d^6\bar{x}[\frac{1}{2}\bar{\partial}_\mu\phi\bar{\partial}_\mu\phi + m^2\phi^2 + \frac{\lambda}{3!}\phi^3]$$

in six-dimensions. In particular find the λ-rescaling necessary to provide invariance of the result under a scale transformation. Interpret the sign and plot the variation of λ with scale.

## 6. Scaling of Determinants. The Scale Dependent Coupling Constant

The ζ-function technique for evaluating determinant of operators makes it particularly simple to derive the scaling properties of these determinants. Under a scale change

$$A \to A' = e^{ad} A \quad , \tag{6.1}$$

where d is the (natural) dimension of A. The definition of the ζ-function leads to

$$\zeta_{A'}(s) = e^{-sad} \zeta_A(s) \quad , \tag{6.2}$$

from which

$$\det(e^{ad} A) = e^{ad\zeta_A(0)} \det(A) \quad . \tag{6.3}$$

An illustrative application of this formula is obtained as follows: Under a dilatation

$$x_\mu \to x'_\mu = e^a x_\mu \quad , \qquad \phi_{c\ell} \to \phi'_{c\ell} = e^{-a} \phi_{c\ell} \quad , \tag{6.4}$$

the classical action with $m^2 = 0$

$$S_E[\phi_{c\ell}] = \int d^4\bar{x} [\frac{1}{2} \phi_{c\ell} \bar{\partial}^2 \phi_{c\ell} - \frac{\lambda}{4!} \phi_{c\ell}^4] \tag{6.5}$$

suffers no change. On the other hand, the path integral for this action is not scale invariant. Indeed, in the steepest descent approximation, we find that the change in the effective action is to $\mathcal{O}(\hbar)$,

$$S_E^{eff}[\phi_{c\ell}] \to S_E'^{eff}[\phi_{c\ell}] = S_E^{eff}[\phi_{c\ell}] - \hbar a \zeta_{[-\bar{\partial}^2 + \frac{\lambda}{2}\phi_{c\ell}^2]}(0) \quad . \tag{6.6}$$

The ζ-function for the operator $-\bar{\partial}^2 + \frac{\lambda}{2} \phi_{c\ell}^2$ is calculated by assuming for $G(\bar{x},\bar{y},\tau)$ the asymptotic expansion (setting $\mu^2 = 1$)

$$G(\bar{x},\bar{y},\tau) = \frac{e^{-(\bar{x}-\bar{y})^2/4\tau}}{16\pi^2\tau^2} e^{-\varepsilon\tau} \sum_{n=0}^{\infty} a_n(\bar{x},\bar{y}) \tau^n \quad , \tag{6.7}$$

where we have inserted an artificial convergence factor with $\varepsilon > 0$. For the reader unhappy at this procedure, imagine that $m^2 \neq 0$ to start with. The boundary condition (5.8) requires that

$$a_0(\bar{x},\bar{x}) = 1 \quad . \tag{6.8}$$

Furthermore, the differential equation (5.11) applied to the form (6.7) yields recursion relations for the $a_n(\bar{x},\bar{y})$ coefficients

$$(\bar{x}-\bar{y})_\mu \frac{\partial}{\partial x_\mu} a_0(\bar{x},\bar{y}) = 0 \tag{6.9}$$

$$[(n+1)+(\bar{x}-\bar{y})_\mu \frac{\partial}{\partial \bar{x}_\mu}]a_{n+1}(\bar{x},\bar{y}) = (\bar{\partial}_x^2 - \frac{\lambda}{2}\phi_{c\ell}^2(\bar{x})+\varepsilon)a_n(\bar{x},\bar{y})$$

$$n = 0,\ 1,\ .. \quad . \tag{6.10}$$

They can be solved, giving

$$a_1(\bar{x},\bar{x}) = -\frac{\lambda}{2}\phi_{c\ell}^2(\bar{x})+\varepsilon, \qquad a_2(\bar{x},\bar{x}) = \frac{\lambda^2}{8}\phi_{c\ell}^4(\bar{x}) - \frac{\lambda}{6}\bar{\partial}^2\phi_{c\ell}(\bar{x})+\varepsilon\lambda\phi_{c\ell}^2 \quad . \tag{6.11}$$

The resulting $\zeta$-function, evaluated at $s = 0$, is now given by

$$\zeta(0) = \frac{1}{16\pi^2}[\frac{\varepsilon^4}{2}\int d^4\bar{x} + \frac{\varepsilon^2\lambda}{2}\int d^4\bar{x}\phi_{c\ell}^2(\bar{x})+\int d^4\bar{x}\ \frac{\lambda^2}{8}\ \phi_{c\ell}^4(\bar{x})] \quad , \tag{6.12}$$

where we have used the definition (5.7), (6.7) and (6.11). The $\bar{\partial}^2$ term in $a_2(\bar{x},\bar{x})$ has been integrated out. As we take $\varepsilon$ to zero, we obtain the final result

$$S_E^{\prime eff} = S_E^{eff} - \hbar a \frac{\lambda^2}{8\cdot 16\pi^2}\int d^4\bar{x}\phi_{c\ell}^4(\bar{x}) \quad . \tag{6.13}$$

Thus we see that the sole effect of the dilatation (to this order in $\hbar$) is to change the coupling constant $\lambda$ by

$$\frac{\lambda}{4!} \to \frac{\lambda'}{4!} = \frac{\lambda}{4!} - \hbar a \frac{\lambda^2}{8\cdot 16\pi^2} \quad , \tag{6.14}$$

i.e.,

$$\lambda \to \lambda' = \lambda - \frac{3\lambda^2}{16\pi^2}\,\hbar a \quad . \qquad (6.15)$$

This very important formula tells us that the coupling constant, which is classically a dimensionless parameter, develops as a result of quantum effects a scale dependence. In this particular case, it tells us that at large scales the coupling constant decreases, which means that the non-interacting theory is in some sense a good approximation for asymptotic states. As the scale decreases, the coupling starts increasing, and even though we may have started from a small value of $\lambda$ at an initial scale, $\lambda$ may increase invalidating results obtained on the basis of perturbation in $\lambda$. Note that this scaling law is exactly the same as that obtained in the previous paragraph [recall that $a = -\ln\frac{M'}{M}$]. This result is exact to $\mathcal{O}(\hbar)$. It is customary to define the $\beta$-function

$$\beta = \frac{d\lambda(M^2)}{d\ln M^2} = \frac{3\lambda^2}{32\pi^2}\,\hbar + \ldots \quad , \qquad (6.16)$$

which in this case is positive.

Thus we have learned from a different point of view that in Quantum Field Theories, the coupling constants have to be defined at some scale because even though they may be classically scale independent, they develop quantum scale dependences.

Problems.

**A. When $m^2 \neq 0$, the classical action with $V_{c\ell} = \frac{1}{2} m^2 \phi_{c\ell}^2 + \frac{\lambda}{4!} \phi_{c\ell}^4$ is no longer dilatation invariant. Find the changes in the effective action stemming from a dilatation. In particular, find the change in $m^2$, both classical and quantum (to $\mathcal{O}(\hbar)$).

**B. Introduce the new asymptotic expansion for $G(x,y,\tau)$

$$G(\bar{x},\bar{y},\tau) = \frac{e^{-(\bar{x}-\bar{y})^2/4\tau}}{16\pi^2\tau^2} e^{-\frac{\lambda}{2}\phi^2(\bar{x})\tau} \sum_{n=0}^{\infty} b_n(\bar{x},\bar{y})\tau^n ,$$

corresponding to the operator $-\partial^2 + \frac{\lambda}{2}\phi^2(\bar{x})$. Find the recursion relations for the $b_n$ coefficients, and work out the form of $b_n(\bar{x},\bar{x})$ for $n = 0, 1, 2, 3$.

## IV - PERTURBATIVE EVALUATION OF THE FPI: $\phi^4$ THEORY.

### 1. Feynman Rules for $\lambda\phi^4$ Theory

In the following, we proceed with the conventional (perturbative) evaluation of the Green's functions in Euclidean space. We start from

$$W_E[J] = e^{-Z_E[J]} = N\int \mathcal{D}\phi \; e^{-\int d^4\bar{x}[\frac{1}{2}\bar{\partial}_\mu\phi\bar{\partial}_\mu\phi + \frac{1}{2}m^2\phi^2 + V(\phi) - J\phi]} , \tag{1.1}$$

where N is an arbitrary (infinite) normalization constant. The connected Green's functions are given by

$$G_E^{(N)}(\bar{x}_1,\ldots,\bar{x}_N) = -\left.\frac{\delta^N Z_E[J]}{\delta J_1 \ldots \delta J_N}\right|_{J=0} . \tag{1.2}$$

They will be calculated by perturbing in the potential V. For simplicity in the following, we neglect the subscript E and the bar over x, which indicate Euclidean space. Later when confusion with Minkowski space might occur, they will be reinstated. Using the trick of Section 3 of Chapter III, we obtain

$$W[J] = N\, e^{-\langle V(\frac{\delta}{\delta J})\rangle} e^{-Z^0[J]} , \tag{1.3}$$

where

$$Z^0[J] = -\frac{1}{2}\langle J(x)\Delta_F(x-y)J(y)\rangle_{xy} \tag{1.4}$$

and

$$\Delta_F(x-y) = \int \frac{d^4p}{(2\pi)^4}\frac{e^{ipx}}{p^2+m^2} . \tag{1.5}$$

A little algebraic rearrangement yields

$$Z[J] = -\ell n\, N + Z^0[J] - \ell n\,(1+e^{Z^0}(e^{-\langle V(\frac{\delta}{\delta J})\rangle}-1)e^{-Z^0}) , \tag{1.6}$$

which is ready for a perturbative expansion in the potential V. If we let

$$\delta = e^{Z^0}\left(e^{-\langle V(\frac{\delta}{\delta J})\rangle}-1\right)e^{-Z^0} \quad , \quad (1.7)$$

we arrive at

$$Z[J] = -\ln N + Z^0[J] - \delta[J] + \frac{1}{2}\,\delta^2[J] - \frac{1}{3}\,\delta^3[J] + \ldots \quad (1.8)$$

In particular for $V = \frac{\lambda}{4!}\,\phi^4$, we can expand in powers of the dimensionless (in four dimensions) coupling constant $\lambda$. Setting

$$\delta = \delta_1\lambda + \delta_2\lambda^2 + \ldots \quad , \quad (1.9)$$

we find

$$Z[J] = -\ln N + Z^0[J] - \lambda\delta_1[J] - \lambda^2(\delta_2[J] - \frac{1}{2}\,\delta_1^2[J])$$

$$- \lambda^3(\delta_3[J] - \delta_1[J]\delta_2[J] + \frac{1}{3}\,\delta_1^3[J]) + \ldots .$$

(1.10)

From expanding the exponential in (1.7) we find

$$\delta_1[J] = -\frac{1}{4!}\,e^{Z^0[J]}\langle\frac{\delta^4}{\delta J^4}\rangle e^{-Z^0[J]} \quad (1.11)$$

$$\delta_2[J] = \frac{1}{2(4!)^2}\,e^{Z^0[J]}\langle\frac{\delta^4}{\delta J_1^4}\rangle_1\langle\frac{\delta^4}{\delta J_2^4}\rangle_2\,e^{-Z^0[J]}, \text{ etc. } \ldots . \quad (1.12)$$

Using the explicit form (1.4) for $Z^0$, we arrive at

$$\delta_1[J] = -\frac{1}{4!}\,[\langle\Delta_{xa}\Delta_{xb}\Delta_{xc}\Delta_{xd}J_aJ_bJ_cJ_d\rangle + 6\langle\Delta_{xx}\Delta_{xa}\Delta_{xb}J_aJ_b\rangle$$

$$+ 3\langle\Delta_{xx}^2\rangle] \quad (1.13)$$

where all variables x, a, b, c, d, are integrated over in the relevant $\langle\ldots\rangle$. Similarly, we evaluate $\delta_2$ in a slightly trickier fashion: we note that

$$\delta_2[J] = -\frac{1}{2(4!)} e^{Z^0} \langle \frac{\delta^4}{\delta J_1^4} \rangle_1 e^{-Z^0} \delta_1[J] \qquad , \qquad (1.14)$$

by inserting $e^{-Z^0} e^{Z^0}$ in the middle of (1.12). Next the expansion

$$\frac{\delta^4}{\delta J^4} e^{-Z^0} = \frac{\delta^4 e^{-Z^0}}{\delta J^4} + 4 \frac{\delta^3 e^{-Z^0}}{\delta J^3} \frac{\delta}{\delta J} + 6 \frac{\delta^2 e^{-Z^0}}{\delta J^2} \frac{\delta^2}{\delta J^2}$$

$$+ 4 \frac{\delta e^{-Z^0}}{\delta J} \frac{\delta^3}{\delta J^3} + e^{-Z^0} \frac{\delta^4}{\delta J^4} \qquad (1.15)$$

allows us to write

$$\delta_2 = \frac{1}{2} \delta_1^2 - \frac{1}{2(4!)} e^{Z^0} \langle (4 \frac{\delta^3 e^{-Z^0}}{\delta J_1^3} \frac{\delta}{\delta J_1} + 6 \frac{\delta^2 e^{-Z^0}}{\delta J_1^2} \frac{\delta^2}{\delta J_1^2}$$

$$+ 4 \frac{\delta e^{-Z^0}}{\delta J_1} \frac{\delta^3}{\delta J_1^3} + e^{-Z^0} \frac{\delta^4}{\delta J_1^4}) \rangle_1 \delta_1[J] \qquad . \qquad (1.16)$$

Comparison with the expansion (1.10) for Z[J] shows that the "disconnected" part $\frac{1}{2} \delta_1^2$ drops out. By disconnected we mean a contribution which can be written as the product of two or more functionals of J. This concept will become obvious in the diagrammatic representation. The fact that Z generates only connected pieces is true to all orders (see problem). For example, the order $\lambda^3$ contribution in (1.10) is connected: write

$$\delta_3 = -\frac{1}{3!} \langle e^{Z^0} V_x V_y V_z e^{-Z^0} \rangle_{xyz}$$

$$= -\frac{1}{3!} \langle (e^{Z^0} V_x e^{-Z^0})(e^{Z^0} V_y e^{-Z^0})(e^{Z^0} V_z e^{-Z^0}) \rangle_{xyz}$$

$$- \frac{1}{2} \langle (e^{Z^0} V_x e^{-Z^0})(e^{Z^0} V_y V_z e^{-Z^0}) \rangle_{xyz} + \delta_3^c \qquad (1.17)$$

$$= \frac{1}{3!}\,\delta_1^3[J] + \delta_1[J]\delta_2^c[J] + \delta_3^c[J] \qquad . \qquad (1.18)$$

In the above $\delta_2^c$, $\delta_3^c$ stand for the connected pieces. To arrive at this form, we have used the fact that there are only two types of "disconnectedness": all three x, y, z disconnected, and only one disconnected from the other two; and there are three ways to obtain the latter possibility. The parentheses in (1.17) serve to shield other terms from the action of the derivative operators within them. It follows that the term appearing in the expansion of Z can be rewritten, using (1.18):

$$\delta_3 - \delta_1\delta_2 + \frac{1}{3}\delta_1^3 = \delta_3^c + \frac{1}{3!}\delta_1^3 + \delta_1\delta_2^c - \delta_1(\delta_2^c + \frac{1}{2}\delta_1^2) + \frac{1}{3}\delta_1^3$$

$$= \delta_3^c \qquad . \qquad (1.19)$$

Now, the explicit evaluation of the connected part of $\delta_2$ yields, save for the J-independent part,

$$\delta_2^c[J] = +\frac{1}{2}\langle J_a\Delta_{ax}(\frac{1}{6}\Delta_{xy}^3 + \frac{1}{4}\Delta_{xx}\Delta_{yy}\Delta_{xy})\Delta_{yb}J_b\rangle_{xyab}$$

$$+ \frac{1}{8}\langle J_a\Delta_{ax}\Delta_{yy}\Delta_{xy}^2\Delta_{xb}J_b\rangle_{xyab}$$

$$+ \frac{2}{4!}\langle J_a\Delta_{ax}\Delta_{xx}\Delta_{xy}\Delta_{yb}\Delta_{yc}\Delta_{yd}J_bJ_cJ_d\rangle_{xyabcd}$$

$$+ \frac{3}{2(4!)}\langle J_aJ_b\Delta_{ax}\Delta_{bx}\Delta_{xy}^2\Delta_{yc}\Delta_{yd}J_cJ_d\rangle_{xyabcd}$$

$$+ \frac{1}{2(3!)^2}\langle J_aJ_bJ_c\Delta_{ax}\Delta_{bx}\Delta_{cx}\Delta_{xy}\Delta_{yd}\Delta_{ye}\Delta_{yf}J_dJ_eJ_f\rangle_{xyabcdef} \, . \qquad (1.20)$$

The resulting connected Green's functions follow from (1.2):

$$G^{(2)}(x_1,x_2) = \Delta(x_1-x_2) - \frac{\lambda}{2}\int d^4y\ \Delta(x_1-y)\Delta(y-y)\Delta(y-x_2)$$

$$+ \frac{\lambda^2}{6}\int d^4x d^4y\ \Delta(x_1-x)\Delta^3(x-y)\Delta(y-x_2)$$

$$+ \frac{\lambda^2}{4}\int d^4x d^4y\ \Delta(x_1-x)\Delta^2(x-y)\Delta(y-y)\Delta(x-x_2)$$

$$+ \frac{\lambda^2}{4}\int d^4x d^4y\ \Delta(x_1-x)\Delta(x-x)\Delta(x-y)\Delta(y-y)\Delta(y-x_2)$$

$$+ \mathcal{O}(\lambda^3) \qquad , \qquad (1.21)$$

$$G^{(4)}(x_1,x_2,x_3,x_4) = -\lambda\int d^4x\ \Delta(x_1-x)\Delta(x_2-x)\Delta(x_3-x)\Delta(x_4-x)$$

$$+ \frac{\lambda^2}{2}\int d^4x d^4y\ \Delta^2(x-y)[\Delta(x_1-x)\Delta(x_2-x)\Delta(x_3-y)\Delta(x_4-y)$$

$$+ \Delta(x_1-x)\Delta(x_3-x)\Delta(x_2-y)\Delta(x_4-y)$$

$$+ \Delta(x_1-x)\Delta(x_4-x)\Delta(x_2-y)\Delta(x_3-y)]$$

$$+ \frac{\lambda^2}{2}\int d^4x d^4y\ \Delta(y-y)\Delta(x-y)[\Delta(x_1-x)\Delta(x_2-x)\Delta(x_3-x)$$

$$\Delta(x_4-y) + \text{cyclic permutations}]$$

$$+ \mathcal{O}(\lambda^3) \qquad , \qquad (1.22)$$

and finally

$$G^{(6)}(x_1,\ldots,x_6) = \lambda^2\int d^4x d^4y\ \Delta(x-y) \sum_{(ijk)} \Delta(x_i-x)\Delta(x_j-x)\Delta(x_k-x)$$

$$\Delta(x_\ell-y)\Delta(x_m-y)\Delta(x_n-y) + \mathcal{O}(\lambda^3) \qquad , \qquad (1.23)$$

where the sum in the last expression runs over the triples (ijk) = (123), (124), (125), (126), (134), (135), (136), (145), (146), (156), with ($\ell$mn) assuming the complementary value, i.e., ($\ell$mn) = (456) when (ijk) = (123), etc. The remaining Green's functions get no contribution to this order in $\lambda$. Note that the $\lambda^0$ contribution to $G^{(2)}$, the $\lambda$ contribution to $G^{(4)}$ and the $\lambda^2$ contribution to $G^{(6)}$ were all previously obtained in the classical approximation of the last chapter.

It is straightforward to derive the p-space Green's functions, using (4.26) of Chapter III. We find

$$\begin{aligned}
\tilde{G}^{(2)}(p,-p) &= \frac{1}{p^2+m^2} - \frac{\lambda}{2}\frac{1}{(p^2+m^2)^2}\int\frac{d^4q}{(2\pi)^4}\frac{1}{q^2+m^2} \\
&+ \frac{\lambda^2}{6}\frac{1}{(p^2+m^2)^2}\int\frac{d^4q_1}{(2\pi)^4}\frac{d^4q_2}{(2\pi)^4}\frac{d^4q_3}{(2\pi)^4}\frac{\delta(p-q_1-q_2-q_3)(2\pi)^4}{(q_1^2+m^2)(q_2^2+m^2)(q_3^2+m^2)} \\
&+ \frac{\lambda^2}{4}\frac{1}{(p^2+m^2)^2}\int\frac{d^4q}{(2\pi)^4}\frac{1}{q^2+m^2}\int\frac{d^4\ell_1}{(2\pi)^4}\frac{d^4\ell_2}{(2\pi)^4}\frac{\delta(p-\ell_1-\ell_2)(2\pi)^4}{(\ell_1^2+m^2)(\ell_2^2+m^2)} \\
&+ \frac{\lambda^2}{4}\frac{1}{(p^2+m^2)^2}\int\frac{d^4q}{(2\pi)^4}\frac{1}{q^2+m^2}\frac{1}{p^2+m^2}\int\frac{d^4\ell}{(2\pi)^4}\frac{1}{\ell^2+m^2} \\
&+ \mathcal{O}(\lambda^3)
\end{aligned} \tag{1.24}$$

$$\tilde{G}^{(4)}(p_1,p_2,p_3,p_4) = \prod_{i=1}^{4} \frac{1}{(p_i^2+m^2)} \{-\lambda + \frac{1}{2}\lambda^2 \int \frac{d^4q}{(2\pi)^4} \frac{1}{q^2+m^2} \sum_{i=1}^{4} \frac{1}{p_i^2+m^2}$$

$$+ \frac{\lambda^2}{2} \int \frac{d^4q_1}{(2\pi)^4} \frac{d^4q_2}{(2\pi)^4} \frac{1}{(q_1^2+m^2)(q_2^2+m^2)} \sum_{(ij)} \delta(q_1+q_2-p_i-p_j)(2\pi)^4$$

$$+ \mathcal{O}(\lambda^3) \quad . \qquad (1.25)$$

In the last expression, the sum ij runs over (ij) = (12), (13), (14) only. Finally $\tilde{G}^{(6)}$ is given by (4.29) of Chapter III. These expressions are clearly unwieldy. One needs to devise a clever way of remembering how to generate them. This is exactly what the Feynman rules achieve. We now proceed to state them:

1. For each factor $\frac{1}{p^2+m^2}$ draw a line with momentum p flowing through it:

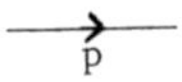


$$: \quad \frac{1}{p^2+m^2} \quad .$$

2. For each factor of $-\lambda/4!$ draw a four-point vertex with the understanding that the net momentum flowing into the vertex is zero:

$$-\frac{\lambda}{4!}$$

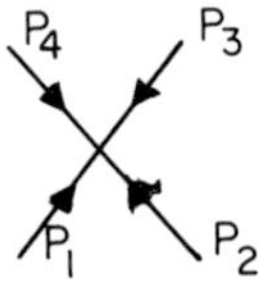


$$(p_1+p_2+p_3+p_4 = 0) \quad .$$

3. In order to get the contribution to $\tilde{G}^{(N)}(p_1,\ldots,p_N)$, draw all possible arrangements which are topologically inequivalent after having identified the external legs. The number of ways a given diagram can be drawn is the topological weight of the diagram.

4. After having conserved momentum at every vertex, integrate over the internal loop momenta with $\int \frac{d^4q}{(2\pi)^4}$.

The result gives the desired Green's function. Perhaps a more systematic way to describe these rules is to attach to each vertex the factor $-\frac{\lambda}{4!}\,\delta(\Sigma p)(2\pi)^4$, where $\Sigma p$ is the net incoming momentum at that vertex. Then one integrates over all internal momenta. In this manner one obtains an overall $(2\pi)^4\delta(\Sigma p)$ where $\Sigma p$ is the net momentum flow into the Green's function. For instance, the expression (1.24) is diagrammatically rewritten as

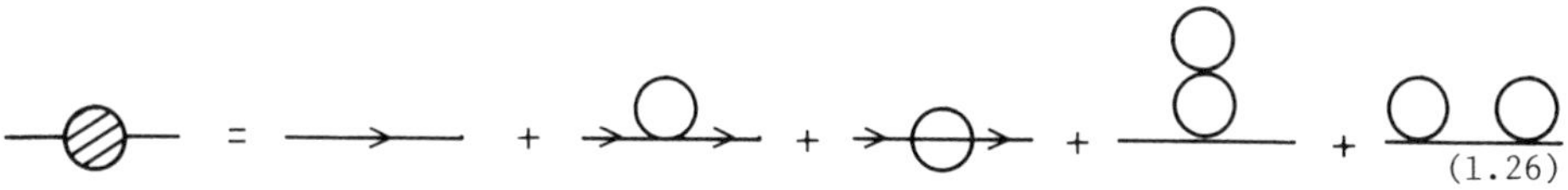

(1.26)

From the rules we can easily obtain the analytical expressions corresponding to these diagrams:

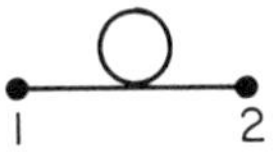


a) We need one vertex and three propagators. There are four ways to attach the first leg of the vertex to 1, three ways to attach the second leg to 2. Hence the weight $\frac{1}{4!}\,4\cdot 3 = \frac{1}{2}$. The vertex counts for $-\lambda$. Let q be the momentum circulating around the loop. The rules then give

$$-\frac{\lambda}{2}\int \frac{d^4q}{(2\pi)^4}\,(2\pi)^4\delta(p_1+p_2+q-q)\,\frac{1}{q^2+m^2}\,\frac{1}{(p^2+m^2)^2}.$$

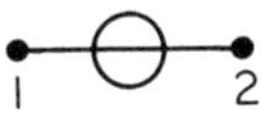


b) We need two vertices. There are four ways to attach the first leg of the first vertex to 1, four ways to attach the first leg of the second vertex to 2, three ways to sew the second leg of the first vertex to the second, and two ways to sew the third leg of the first vertex to the second. Hence the weight $\frac{1}{4!} \cdot \frac{1}{4!} 4 \cdot 4 \cdot 3 \cdot 2 = \frac{1}{6}$. Note that we did not count that we initially had two vertices to play with. This is because the diagrams are the same irrespective of which vertex was used. The strength of this diagram is $(-\lambda)^2 = \lambda^2$. If we call the momenta flowing in the internal legs $q_1$, $q_2$, $q_3$, the Feynman rules give

$$\frac{\lambda^2}{6} \frac{1}{(p^2+m^2)^2} \int \frac{d^4q_1}{(2\pi)^4} \frac{d^4q_2}{(2\pi)^4} \frac{d^4q_3}{(2\pi)^4} (2\pi)^8 \frac{\delta(p_1-q_1-q_2-q_3)\delta(p_2+q_1+q_2+q_3)}{(q_1^2+m^2)(q_2^2+m^2)(q_3^2+m^2)} .$$

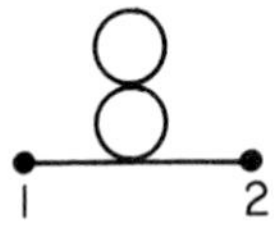


c) We need two vertices; four ways to attach the first leg to 1, four ways to attach the third leg of the tied vertex to the other, three ways to tie the fourth leg of the first vertex to the second one. Hence $\frac{1}{4!} \frac{1}{4!} 4 \cdot 3 \cdot 4 \cdot 3 = \frac{1}{4}$. Strength $(-\lambda)^2$.

d) We need two vertices; four ways to attach one vertex to 1, four ways to attach the other vertex to 2. This leaves three legs from each vertex free to be tied together one way. For each vertex there are three ways to close the buckle. Hence $\frac{1}{4!} \frac{1}{4!} 4 \cdot 4 \cdot 3 \cdot 3 = \frac{1}{4}$. Strength $(-\lambda)^2$.

Thus we see that the Feynman rules have reduced the problem to that faced by a child assembling a "Leggo" set. The basic tools are the propagator (line) and the vertex. With a bit of skill one can read those directly from the Lagrangian. The same applies for the four-point function:

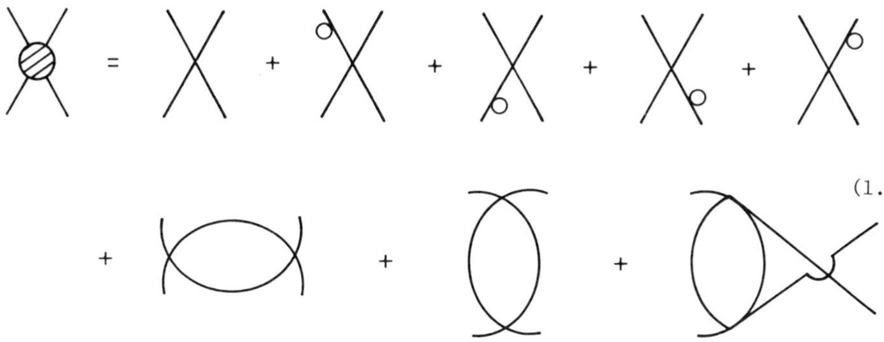

(1.27)

corresponding to the analytical expression (1.25). We leave it to the reader to verify the correctness of the numerical factors in (1.25).

Let us remark that in the expression for $\tilde{G}^{(N)}(p_1,\ldots,p_N)$, we will have the multiplicative factor $\prod_{i=1}^{N}(p_i^2+m^2)^{-1}$ corresponding to the propagation of the external legs.

It is much simpler to deal with the Green's functions generated by the effective Action. Their relation to $\tilde{G}^{(n)}$ is very simple: while $\tilde{G}^{(n)}$ are connected, the $\tilde{\Gamma}^{(n)}$ are one particle irreducible. In particular $\tilde{\Gamma}^{(2)}(p)$ is minus the inverse propagator. We have already come in contact with this result in conjunction with the tree diagrams, but the result holds true in all orders of perturbation theory. As a result $\tilde{\Gamma}^{(4)}$ contains only the diagrams

$$\tilde{\Gamma}^{(4)} =$$

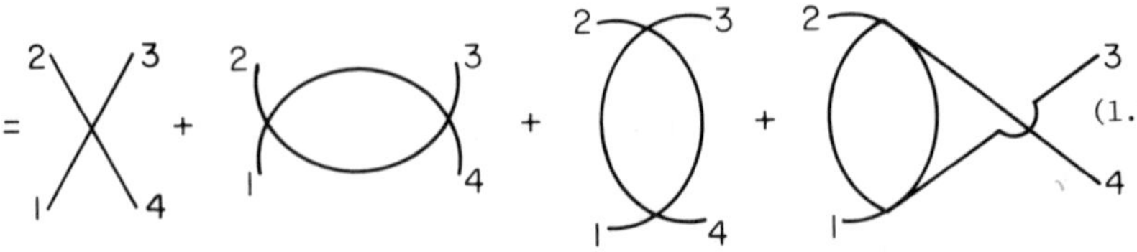

(1.28)

with no propagators for the external legs. In order to show that the $\tilde{\Gamma}^{(n)}$ are one particle irreducible, it is convenient to use as a starting point the defining equations:

$$\frac{\delta\Gamma[\phi]}{\delta\phi} = -J \qquad ; \qquad \frac{\delta Z[J]}{\delta J} = \phi \qquad , \qquad (1.29)$$

and keep on differentiating them.

After all this work, let us go back to the Lagrangian where all the ingredients of the Feynman rules are easily identified: the four-particle vertex coming from the $\lambda\phi^4$ term, and the propagator coming from the kinetic and mass terms. Hence, after a little bit of practice, one can just read off the Feynman rules from $\mathcal{L}$. The difficult part is to get the right signs and weight factors in front of the diagrams.

Problems.

A. Draw the contributions to the two point function $\tilde{G}^{(2)}$ that are of order $\lambda^3$.

B. Draw the contributions to $\tilde{G}^{(4)}$ of order $\lambda^3$.

C. Write the analytical expressions for the diagrams of problem A, including the weights.

*D. Derive the Feynman rules for $V = \frac{\lambda}{4!}\phi^4 + \frac{\mu}{3!}\phi^3$.

**E. Show that Z[J] generates only connected Feynman diagrams.

**F. Derive the form of the effective Action $S^{eff}[\phi_{c\ell}]$ to order $\lambda^2$, and show that the one particle reducible diagrams do not appear in the Green's functions it generates.

## 2. Divergences of Feynman Diagrams

No sooner has the beauty of the Feynman rules sunk in than one realizes that most of the loop integrations diverge! For instance the $\mathcal{O}(\lambda)$ contribution to $\tilde{G}^{(2)}$ involves the integral

$$\int \frac{d^4q}{(2\pi)^4} \frac{1}{q^2+m^2} \qquad ;$$

it clearly diverges when $q \to \infty$ since the integrand does not have enough juice to make up for the measure. Such a divergence is called an ultraviolet divergence. It occurs for large momenta or, equivalently, small distances [in x-space it comes from $\Delta(0)$], and clearly has to do with the fact that one is taking too many derivatives with respect to J at the same point. Another example occurs in the "fish" diagram

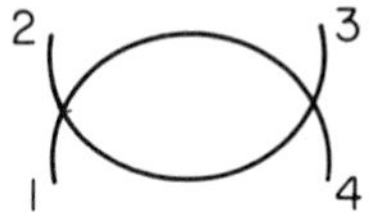


$$\sim \frac{\lambda^2}{2} \int \frac{d^4q_1}{(2\pi)^4} \frac{d^4q_2}{(2\pi)^4} \frac{\delta(q_1+q_2-p_1-p_2)}{(q_1^2+m^2)(q_2^2+m^2)}$$

$$= \frac{1}{(2\pi)^4} \frac{\lambda}{2} \int \frac{d^4q}{(2\pi)^4} \frac{1}{(q^2+m^2)((q-p_1-p_2)^2+m^2)} \quad .$$

Here as $q \to \infty$, the integral behaves as $\frac{d^4q}{q^4}$, which is a logarithm: log q. It also diverges! In passing we note that when $m^2 = 0$, $(p_1+p_2)^2 = 0$ it also diverges for low q. Such a divergence is called an infrared divergence. Typically such divergences occur only in the massless case and for special value of the external momenta [in this case $p_1+p_2 = 0$ because we are in Euclidean space]. For the moment we take $m^2 \neq 0$ and concentrate on the ultraviolet divergences.

On the face of it, it is catastrophic to our program to find that our carefully constructed Green's functions diverge. Still, with the hindsight of History we do not get discouraged, but try to learn more about these divergences. We will find that they appear in a very traceable way, and that by a suitable redefinition of the fields and coupling constants, they will disappear! This is the miracle of renormalization which, as we shall see, occurs only for certain theories.

By a mixture of topology and power counting we now show how to tell where the divergences are. Consider a Feynman diagram with V vertices, E external lines and I internal lines. For a start we assume only scalar particles are involved.

The number of independent internal momenta is the number of loops, L, in the diagram. The I internal momenta satisfy V-1 relations among themselves (the -1 appears here because of overall momentum conservation), so that

$$L = I - V + 1 \quad . \qquad (2.1)$$

This relation enables us to compute the naive count of powers of momenta for the diagram. This yields the superficial (apparent) degree of divergence of the diagram D. To compute it, we note that there are

- L independent loop integrations, each providing in d dimensions d powers of momenta
- I internal momenta, each providing a propagator with two inverse powers of momenta. Hence

$$D = dL - 2I \quad . \qquad (2.2)$$

We need one more relation among V, E and I. Suppose $V_N$ stands for the number of vertices with N legs. In a diagram with $V_N$ such vertices, we have $NV_N$ lines which are either external or internal. An internal line counts twice because it originates and terminates at a vertex, so that

$$NV_N = E + 2I \quad . \qquad (2.3)$$

These relations allow us to express D in terms of the number of external lines and vertices

$$D = d - \frac{1}{2}(d-2)E + V_N(\frac{N-2}{2} d-N) \quad . \quad (2.4)$$

In four dimensions, this reduces to

$$D = 4 - E + (N-4)V_N \qquad [\text{four-dimensions}] \quad . \quad (2.5)$$

Furthermore, in the theory of interest to us, N = 4. Hence

$$D = 4 - E \qquad [\text{for } \lambda\phi^4 \text{ theory in four dimensions}] \quad . \quad (2.6)$$

The important result here is that the superficial degree of divergence does not depend on the number of vertices, but only on the number of external legs! Thus we have only two candidates with $D \geq 0$

- $\tilde{G}^{(2)}$ with D = 2 superficial quadratic divergence
- $\tilde{G}^{(4)}$ with D = 0 superficial logarithmic divergence.

Note that these two- and four-point interactions are already present in the Lagrangian, a fact which will prove crucial for renormalization. Also D = 0 does not necessarily mean a logarithmic divergence: the fundamental vertex has D = 0.

This analysis does not prove that $\tilde{G}^{(6)}$, $\tilde{G}^{(8)}$, ... which have a negative D, converge. This is why D is called superficial. Consider a "n-particle reducible" diagram with E external lines; it is a diagram which can be disconnected by cutting at least n internal lines. In general, if $D_1$ and $D_2$ are the superficial degrees of divergence of the two blobs shown below, then the whole n-particle reducible diagram has

$$D = D_1 + D_2 + 4(n-1) - 2n \quad , \quad (2.7)$$

since the two blobs are connected by n propagators and n-1 loops:

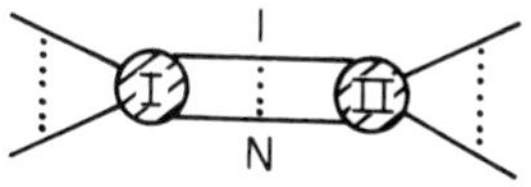


Note that by definition I and II are at least n-particle reducible themselves.

In our case, when n = 1, we could have $D_1 = D_2 = 0$ and yet we would obtain $D = -2$, making the diagram apparently convergent. An example of this situation is the "dinosaur" diagram

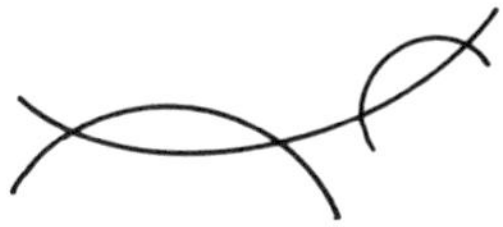

which clearly diverges because of the two divergent loop integrations. Another example of a one particle reducible graph is

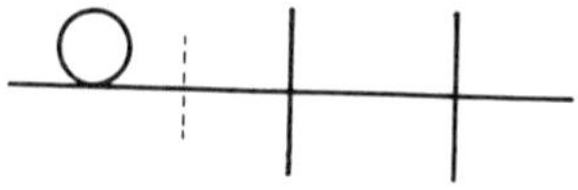

in this case $D_1 = 2$, $D_2 = -2$; it is apparently convergent, but is not because of the "wart" on one of the legs. This should also serve to show that a four point function may be more divergent than it would naively seem to be: witness this quadratically divergent four-point function

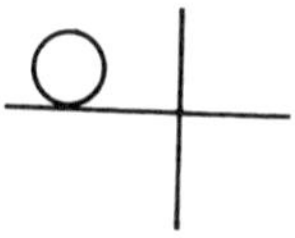

Let us remark that by considering $\tilde{\Gamma}^{(n)}$ we avoid such one particle reducible

diagrams. When n = 2, $D = D_1 + D_2$ and we can have $D_1$ (or $D_2$) sufficiently negative to offset $D_2$ (or $D_1$) and yield a negative D. An example is pictured by the "lobster" diagram

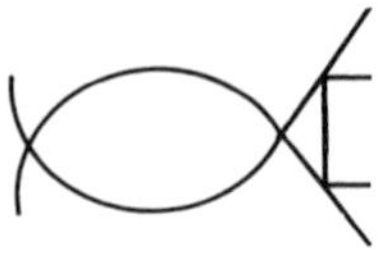

Similarly, when n = 3, we can have diagrams like

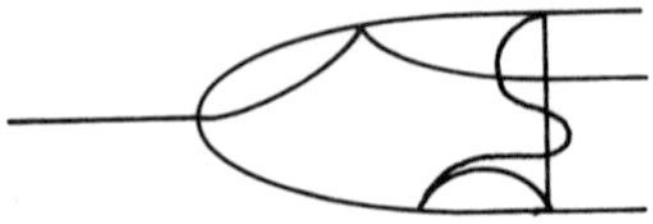

[in this complicated diagram, each vertex has at most one external line emanating from it]. In $\lambda\phi^4$ theory diagrams can be at most three-particle irreducible because any vertex attached to an external line can be disconnected from the diagram by cutting its remaining three legs.

The procedure of hunting for diagrams which have a negative D and yet are divergent is now clear. Take any Feynman diagram and catalog it in terms of its reducibility; in our case it is either 1-, 2- or 3-particle reducible. If it is three-particle reducible, decompose it into units, and look for breakups of the form

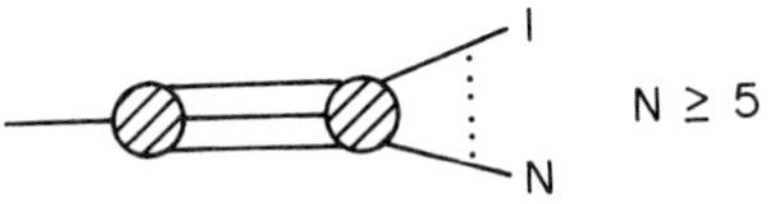


Then in this type of diagram, we have a primitively divergent four-point function, where the second blob itself can be decomposed in the same way.

Similarly, hidden divergences in two-particle reducible diagrams will arise in diagrams of the form

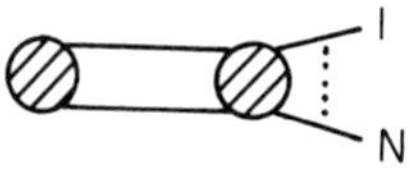


$N > 4$ ;

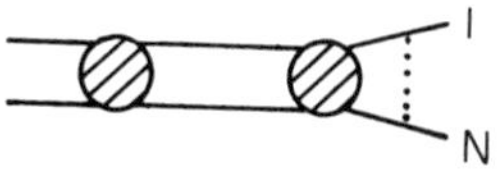


$N > 2$

with the same breakup to be repeated in the second blob. Finally, one-particle reducible diagrams which can be decomposed in the form

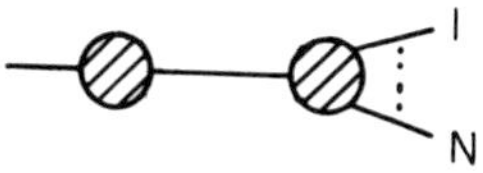


$N > 5$ ;

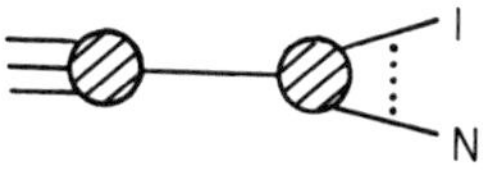


$N > 3$

will have hidden divergences. The same decomposition can be carried out for the second blobs until all such structures have been uncovered.

This exhaustive catalog shows that truly convergent diagrams do not contain hidden two and four-point functions.

One can understand the origin of hidden ultraviolet divergences in any diagram in a more pedestrian fashion. Consider any loop residing inside a diagram. Integration over the loop momentum in four dimensions will lead to a UV divergence if the loop is bounded by one or two propagators (internal lines) at most. Any more will give UV convergence. A loop bounded by one propagator involves only one vertex,

leaving two free legs, which in turn may be attached to the rest of the diagram (or one external, one attached). In this case one can isolate this two point function from the insides of the diagram. A loop bounded by two propagators involves two vertices and therefore four free legs

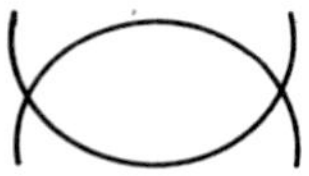

which can be attached to the rest of the diagram, or else up to three can serve as external lines. In all these cases one is led to isolate from the diagram a four-point function, or a two point function if two of the four legs are attached together (in this case the divergence becomes quadratic). In this way, one sees that UV divergences inside a diagram originate from such loops and that such loops appear in two- and four-point functions nested inside the diagram.

Thus, for the $\lambda\phi^4$ theory in four dimensions, a Feynman diagram is truly convergent if its superficial degree of convergence D is positive <u>and</u> if it cannot be split up into three-, two- or one-particle reducible parts of the kind just described which can contain isolated two- and four-point function blobs. Stated more elegantly, a Feynman diagram is convergent if its superficial degree of convergence and that of all its subgraphs are positive. This is known as Weinberg's theorem, and it holds irrespective of the field theory.

This means that the generic sources of the divergences are the two- and four-point functions and nothing else. They are the culprits! So, if we control them in $\tilde{G}^{(2)}$ and $\tilde{G}^{(4)}$, we have the possibility of controlling the divergences of all the other $\tilde{G}^{(N)}$'s.

The graphs which contain the generic divergences are said to be <u>primitively</u> divergent. The fact in $\lambda\phi^4$ theory that the primitively divergent interactions are finite in number (two- and four-point interactions) and are of the type that appears in the Lagrangian, is a necessary ingredient for the successful removal of the ultraviolet divergences by clever redefinitions. A theory for which this is possible is said to be renormalizable. We can see from (2.4) that very few theories of interacting scalars satisfy these requirements (see problem).

- When $d = 4$, we see that $D$ grows with the number of vertices for which $N > 4$. Hence $\phi^5$, $\phi^6$, ... theories in four-dimensions, although perfectly reasonable classically, lead to an infinite number of primitively divergent diagrams (the more vertices the more divergent!). In this case the situation quickly gets out of hand and the hope of tagging the divergences disappears, and hence the renormalizability.
- When $d = 2$ (one space and one time dimension) the situation is reversed. There

$$D = 2 - 2V_N \qquad \text{(two-dimensions)} \qquad , \qquad (2.8)$$

  and $D$ does not depend on $N$, which labels the type of interaction! It depends only on the number of vertices, and the more vertices, the more convergent the Feynman diagram!! Also $D$ does not depend on the number of external legs! So the only primitively divergent diagrams have one or no vertex. Since divergences occur because of loop integrations, this means that divergences occur only when a leg from one vertex is connected to the same vertex, and not from the interaction between two or more vertices. Such self-inflicted divergences are called "normal ordering" divergences. In two dimensions, the only ultraviolet divergences come from "normal ordering," and not from the type of interaction.

Finally we note that when $d \geq 7$, there are no theories with a finite number of primitively divergent graphs. The last theory in higher dimensions is $\lambda\phi^3$ in six dimensions, where $\lambda$ is dimensionless, since $\phi$ now has dimension -2. There the primitively divergent diagrams are few since V does not appear in the expression for D

$$D = 6 - 2E \qquad (\phi^3 \text{ in six dimensions}) \qquad , \qquad (2.9)$$

so that the one-, two- and three-point functions are primitively divergent (quartic, quadratic and logarithmic, respectively). This theory, although having an unsatisfactory potential unbounded from below, is interesting in that it shares with the more complicated gauge theories the property of asymptotic freedom.

To summarize this section, we have noticed the appearance of ultraviolet divergences in Feynman diagrams with loops (the bad news), but we have seen that we can, at least in our theory, narrow them down as coming only from two primitively divergent Green's function (the good news). Hence, if we can arrange to stop divergences from appearing in $\tilde{G}^{(2)}$ and $\tilde{G}^{(4)}$, we have a hope of stemming the flood and obtaining convergent answers!

Problems.

A. In four dimensions, find all primitively divergent diagrams for the $\phi^3$ theory. For each give examples in lowest order of perturbation theory.

B. In three dimensions (d = 3), list all theories of interacting scalars with a finite number of primitively divergent graphs. Give graphical examples for each.

C. Repeat B. when d = 5, and show that when $d \geq 7$ there are no theories with a finite number of primitively divergent graphs.

D. For d = 2, 3, 5, 6, find the dimensions of the various couplings constants in the theories where there is a finite number of primitively divergent graphs.

## 3. Dimensional Regularization of Feynman Integrals

In the following, we proceed to evaluate the Feynman diagrams. The procedure is straightforward for the UV convergent ones, while special measures have to be taken to evaluate the divergent ones. In those we are confronted with integrals of the form

$$I_4(k) = \int_{-\infty}^{+\infty} d^4\ell \; F(\ell,k) \qquad , \qquad (3.1)$$

where for large $\ell$, F behaves either as $\ell^2$ or $\ell^4$. The basic idea behind the technique of dimensional regularization is that by lowering the number of dimensions over which one integrates, the divergences trivially disappear. For instance, if $F \to \ell^4$, then in two dimensions the integral (3.1) converges at the UV end.

Mathematically, we can introduce the function

$$I(\omega,k) = \int d^{2\omega}\ell \; F(\ell,k) \qquad , \qquad (3.2)$$

as a function of the (complex) variable $\omega$. Evaluate it in a domain where I has no singularities in the $\omega$ plane. Then invent a function $I'(\omega,k)$ which coincides with I in the domain of convergence of (3.2) in the $\omega$-plane, and has well-defined singularities outside of the domain of convergence. We say by analytic continuation that I and I' are the same function.

A nice example, which is the basis for the method of analytic continuation, is the difference between the Euler and Weierstrass representations of the $\Gamma$-function. For Re z > 0, the Euler representation is

$$\Gamma(z) = \int_0^\infty dt \; e^{-t} t^{z-1} \qquad . \qquad (3.3)$$

As such it diverges when Re z < 0, because as t approaches zero, the integral behaves as $dt/t^{1+|\mathrm{Re}\, z|}$, which leads to an infinity. Starting from (3.3) we can split up the troublesome integration limit

$$\Gamma(z) = \sum_{n=0}^{\infty} \frac{(-1)^n}{n!} \int_0^{\alpha} dt\, t^{n+z-1} + \int_{\alpha}^{\infty} dt\, e^{-t} t^{z-1} \quad , \tag{3.4}$$

where $\alpha$ is totally arbitrary. The second integral is well defined even when Re $z < 0$ as long as $\alpha > 0$. The first integral has simple poles whenever z is a negative integer or zero. We find

$$\Gamma(z) = \sum_{n=0}^{\infty} \frac{(-1)^n}{n!} \frac{\alpha^{n+z}}{(z+n)} + \int_{\alpha}^{\infty} dt\, e^{-t} t^{z-1} \quad . \tag{3.5}$$

This form is valid everywhere in the z-plane. Furthermore it should not depend on the arbitrary coefficient $\alpha$ (you can check that $\frac{d\Gamma}{d\alpha} = 0$). When $\alpha = 1$, it is the Weierstrass representation of the $\Gamma$-function. Still, to isolate the singularities we did introduce an arbitrary scale in the process, although the end result is independent of it.

Our problem is that integral expressions like (3.2) are like Euler's. We want to find the equivalent of Weierstrass' representation. Our procedure will be as follows: 1) establish a finite domain of convergence for the loop integral in the $\omega$-plane. For divergent integrals, it will typically lie to the left of the $\omega = 2$ line; 2) construct a new function which overlaps with the loop integral in its domain of convergence, but is defined in a larger domain which encloses the point $\omega = 2$; 3) take the limit $\omega \to 2$.

We now show how this is done in the case of one loop diagrams, following the procedures of 't Hooft and Veltman, Nucl. Phys. 44B, 189 (1972). Let us take as one example the quadratically divergent tadpole diagram. We first split up the domain of integration as

$$d^{2\omega}\ell \to d^4\ell\; d^{2\omega-4}\ell \quad .$$

Next in the $2\omega-4$ space, introduce polar coordinates, and call L the length of the $2\omega-4$ dimensional $\ell$-vector. The integral now reads

$$I = \int d^4\ell \int d\Omega_{2\omega-4} \int_0^\infty dL\ L^{2\omega-5} \frac{1}{(L^2+\ell^2+m^2)} \quad . \qquad (3.6)$$

Integration over the angles can be performed (see Appendix B), with the result

$$I = \frac{2\pi^{\omega-2}}{\Gamma(\omega-2)} \int d^4\ell \int_0^\infty dL\ L^{2\omega-5} \frac{1}{(L^2+\ell^2+m^2)} \quad . \qquad (3.7)$$

This expression is not well defined because it is UV divergent for $\omega \geq 1$, and the integration over L diverges at the lower end ("infrared") whenever $\omega \leq 2$. Thus there is no overlapping region in the $\omega$ plane when I is well defined. The IR divergence is, however, an artifact of the break up of the measure. Observe that by writing

$$L^{2\omega-6} = \frac{1}{\omega-2} \frac{d}{dL^2} (L^2)^{\omega-2} \quad , \qquad (3.8)$$

and integrating by parts over $L^2$, and throwing away the surface term, we obtain

$$I = \frac{\pi^{\omega-2}}{\Gamma(\omega-1)} \int d^4\ell \int_0^\infty dL^2\ (L^2)^{\omega-2} \left(-\frac{d}{dL^2}\right) \frac{1}{L^2+\ell^2+m^2} \quad , \qquad (3.9)$$

where we have used $\Gamma(\omega-1) = (\omega-2)\Gamma(\omega-2)$. Now the representation (3.9) has an infrared divergence for $\omega \leq 1$ and the same UV divergence for $\omega \geq 1$, that is, still no overlapping region of convergence. So we do it again, and obtain

$$I = \frac{\pi^{\omega-2}}{\Gamma(\omega)} \int d^4\ell \int_0^\infty dL^2\ (L^2)^{\omega-1} \left(-\frac{d}{dL^2}\right)^2 \frac{1}{L^2+\ell^2+m^2} \quad , \qquad (3.10)$$

an expression which is well defined for $0 < \omega < 1$. Note that we had to move the IR convergence region two units to obtain a nonzero region of convergence. Had the loop integral been logarithmically divergent, one such step would have sufficed.

Having obtained an expression for I convergent in a finite domain (in this case $0 < \omega < 1$), we want to continue I of (3.10) to the physical point $\omega = 2$. It goes as follows: insert in the integrand the clever expression

$$1 = \frac{1}{5}\left(\frac{\partial L}{\partial L} + \frac{\partial \ell_\mu}{\partial \ell_\mu}\right) \quad , \qquad (3.11)$$

and integrate by parts in the region of convergence. We obtain

$$I = -\frac{1}{5}\frac{2\pi^{\omega-2}}{\Gamma(\omega)} \int d^4\ell \int_0^\infty dL^2 \left[\ell_\mu \frac{\partial}{\partial \ell_\mu} + L^2 \frac{\partial}{\partial L^2}\right] \frac{(L^2)^{\omega-1}}{(L^2+\ell^2+m^2)^3} \, . \qquad (3.12)$$

A little bit of algebra gives, after reexpressing the RHS in terms of I,

$$I = -\frac{3m^2}{\omega-1}\frac{2\pi^{\omega-2}}{\Gamma(\omega)} \int d^4\ell \int_0^\infty dL^2 \frac{(L^2)^{\omega-1}}{(L^2+\ell^2+m^2)^4} \, . \qquad (3.13)$$

This expression displays explicitly a pole at $\omega = 1$. The integral now diverges at the upper end when $\omega \geq 2$. So we repeat the process and reinsert (3.11) in (3.13). The result is predictable:

$$I = \frac{2\cdot 3\cdot 4m^4}{(\omega-1)(\omega-2)}\frac{\pi^{\omega-2}}{\Gamma(\omega)} \int d^4\ell \int_0^\infty dL^2 \frac{(L^2)^{\omega-1}}{(L^2+\ell^2+m^2)^5} \, . \qquad (3.14)$$

This is the desired result. The only hint of a divergence comes from the simple pole at $\omega = 2$, since the integral now converges.

Let us summarize. We first define a finite integral in the $\omega$-plane to be what we mean by $\int d^{2\omega}\ell\, F(\ell,k)$; in this case it is the expression given by (3.10), and constitutes our starting point. Then if the region of convergence does not include $\omega = 2$, we continue analytically by means of the trick (3.11).

It would be nice to show that for a convergent integral, the procedure that leads to (3.10) indeed gives the right answer. Take as an example the convergent integral

$$I = \int d^{2\omega}\ell \, \frac{1}{(\ell^2+m^2)^6} \quad .$$

It is easy to see that the same procedure leads to the expression

$$I(\omega) = \frac{\pi^{\omega-2}}{\Gamma(\omega-1)} \int d^4\ell \int_0^\infty dL^2 \, (L^2)^{\omega-2} \left(-\frac{d}{dL^2}\right) \frac{1}{(L^2+\ell^2+m^2)^6} \quad , \tag{3.15}$$

which is perfectly finite at $\omega = 2$. We find

$$I(2) = \int d^4\ell \, \frac{(-1)}{(L^2+\ell^2+m^2)^6}\Bigg|_0^\infty = \int d^4\ell \, \frac{1}{(L^2+\ell^2+m^2)^6} \quad ,$$

as desired. Therefore, the procedure is entirely consistent.

After all this ponderous work, let us turn to a more cavalier interpretation of integrals in $2\omega$-dimensions, as derived in Appendix B.

If we were to blindly plug in the formulae of Appendix B, we would obtain

$$\int \frac{d^{2\omega}\ell}{(\ell^2+m^2)} = \pi^\omega \, \frac{\Gamma(1-\omega)}{\Gamma(1)} \, \frac{1}{(m^2)^{1-\omega}} \quad . \tag{3.16}$$

We expand around $\omega = 2$, using for $n = 0, 1, 2, \ldots$ and $\varepsilon \to 0$ the formula

$$\Gamma(-n+\varepsilon) = \frac{(-1)^n}{n!} \left[\frac{1}{\varepsilon} + \psi(n+1) + \frac{1}{2}\varepsilon\left[\frac{\pi^2}{3} + \psi^2(n+1) - \psi'(n+1)\right] + \mathcal{O}(\varepsilon^2)\right] \tag{3.17}$$

where

$$\psi(n+1) = 1 + \frac{1}{2} + \ldots + \frac{1}{n} - \gamma, \quad \left[\psi(s) = \frac{d \ln \Gamma(s)}{ds}\right] \quad , \tag{3.18}$$

$$\psi'(n+1) = \frac{\pi^2}{6} + \sum_{k=1}^{n} \frac{1}{k^2} \, , \quad \psi'(1) = \frac{\pi^2}{6} \tag{3.19}$$

$\gamma$ being the Euler-Mascheroni constant

$$\psi(1) = -\gamma = -0.5772\ldots \quad . \tag{3.20}$$

The result is

$$\lim_{\omega\to 2} \int \frac{d^{2\omega}\ell}{(\ell^2+m^2)} = -\pi^2 m^2 [\frac{1}{\omega-2} + \psi(2)] \quad . \tag{3.21}$$

It can be shown that this is the same result as that obtained by integrating (3.14).

From now on, we are going to use the naive formulae of Appendix B, and not concern ourselves with their justification since we know they are legitimate, thanks to 't Hooft and Veltman.

Problems.

A. Show that the naive procedure of Appendix B and the more careful one outlined in this section lead to the same results for the integrals

$$I = \int d^{2\omega}\ell \; \frac{1}{(\ell^2+m^2)} \quad , \qquad I = \int d^{2\omega}\ell \; \frac{1}{(\ell^2+m^2)^2} \quad .$$

B. Prove the last four formulae of Appendix B.

## 4. Evaluation of Feynman Integrals

Using the techniques of the previous section and the formulae of Appendix B, we proceed to evaluate the low order Feynman diagrams for $\lambda\phi^4$ theory.

We have seen that the technique of dimensional regularization is predicated on the evaluation of Feynman integrals in $2\omega$ dimensions, where the coupling constant $\lambda$ is no longer dimensionless. We find it convenient to redefine it in terms of a dimensionless coupling constant by the artifact

$$\lambda_{old} = \lambda_{new}(\mu^2)^{2-\omega} \qquad , \qquad (4.1)$$

where $\lambda_{new}$ is dimensionless and $\mu^2$ is an arbitrary constant with the dimension of mass. Alternatively, we can say that we evaluate the Green's functions for the theory defined by the action

$$S_\omega[\phi] = \int d^{2\omega}x[\frac{1}{2}\,\partial_\mu\phi\partial_\mu\phi + \frac{1}{2}\,m^2\phi^2 + \frac{\lambda}{4!}\,(\mu^2)^{2-\omega}\phi^4] \qquad . \qquad (4.2)$$

The Feynman rules for this theory are the same as the one for the theory in four-dimensions with three exceptions: 1) the scalar product between vectors is summed over their $2\omega$ components, 2) the loop integrals appear with $\int \frac{d^{2\omega}\ell}{(2\pi)^{2\omega}}$ and 3) the vertex strength $-\lambda$ is replaced by $(-\lambda)(\mu^2)^{2-\omega}$.

Let us evaluate the lowest order diagrams for this theory. We start with the "tadpole" diagram (in conventional terms such a diagram comes from not normal ordering the interaction term). It is given by

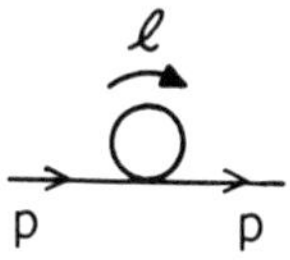


$$\equiv T = \frac{1}{2}\,(-\lambda)(\mu^2)^{2-\omega} \int \frac{d^{2\omega}\ell}{(2\pi)^{2\omega}}\,\frac{1}{(\ell^2+m^2)} \qquad (4.3)$$

$$= -\,\frac{\lambda m^2}{2(4\pi)^2}\,\left(\frac{4\pi\mu^2}{m^2}\right)^{2-\omega}\,\Gamma(1-\omega) \qquad , \qquad (4.4)$$

where we have used (B-16), and have kept $m^2$ in front because the diagram has dimension of mass squared. By expanding around $\omega = 2$, we obtain

$$T = -\frac{\lambda m^2}{32\pi^2}\left[1 + (2-\omega)\ln\frac{4\pi\mu^2}{m^2} + \ldots\right]\left[-\frac{1}{2-\omega} - \psi(2) + ..\right] \tag{4.5}$$

$$= \frac{\lambda m^2}{32\pi^2}\left\{\frac{1}{2-\omega} + \psi(2) - \ln\frac{m^2}{4\pi\mu^2} + \mathcal{O}(2-\omega)\right\} \quad . \tag{4.6}$$

Observe that in this formula, the judicious introduction of the arbitrary scale $\mu^2$ allows us to keep track of dimensions, and that the pole from the expansion of $\Gamma$ cancels against the first order term in the expansion of $\left(\frac{4\pi\mu^2}{m^2}\right)^{2-\omega}$, leaving us with a finite contribution as $\omega \to 2$. This feature will be present in the evaluation of all divergent graphs. We conclude that the divergence in T appears as a simple pole, and that the finite part of T, which in this case does not depend on the external momenta, is totally arbitrary because a change in $\mu^2$ affects it.

The next diagram is the "fish"

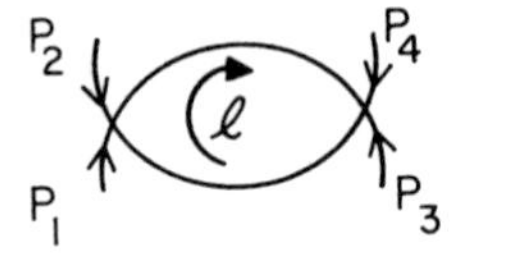


$$P_1 + P_2 = P_3 + P_4$$

The Feynman rules give

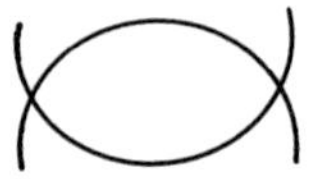

$$= \frac{1}{2}(-\lambda)^2(\mu^2)^{4-2\omega}\int\frac{d^{2\omega}\ell}{(2\pi)^{2\omega}}\frac{1}{(\ell^2+m^2)}\frac{1}{[(\ell-p)^2+m^2]} \quad . \tag{4.7}$$

When there are more than one propagator taking part in a loop integration, it is convenient to introduce the Feynman parametrization based on the formula

$$\frac{1}{D_1^{a_1}D_2^{a_2}\cdots D_k^{a_k}} = \frac{\Gamma(a_1+a_2+\ldots+a_k)}{\Gamma(a_1)\Gamma(a_2)\ldots\Gamma(a_k)} \int_0^1\cdots\int_0^1 dx_1\ldots dx_k \; \frac{\delta(1-x_1-\ldots-x_k)\,x_1^{a_1-1}\cdots x_k^{a_k-1}}{(D_1x_1+\ldots+D_ka_k)^{a_1+\ldots+a_k}} \; . \tag{4.8}$$

It allows for a convenient rearrangement of the loop momenta. In this case we use it in the form

$$\frac{1}{[\ell^2+m^2][(\ell-p)^2+m^2]} = \int_0^1 dx \, \frac{1}{[\ell^2+m^2-2\ell\cdot p(1-x)+p^2(1-x)]^2} \; . \tag{4.9}$$

The denominator can be rewritten in the form

$$\ell'^2 + m^2 + p^2x(1-x) \qquad ,$$

where

$$\ell' = \ell - p(1-x) \qquad . \tag{4.10}$$

Since we are dealing with convergent integrals, we use $d^{2\omega}\ell' = d^{2\omega}\ell$ and relabel the loop integral from $\ell$ to $\ell'$. The result of these manipulations yields

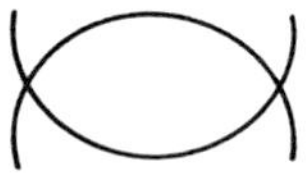

$$= \frac{\lambda^2}{2}(\mu^2)^{4-2\omega}\int_0^1 dx \int \frac{d^{2\omega}\ell}{(2\pi)^{2\omega}} \frac{1}{[\ell^2+m^2+p^2x(1-x)]^2} \qquad . \tag{4.11}$$

Thanks to this trick, we can now use (B-16) and integrate, obtaining

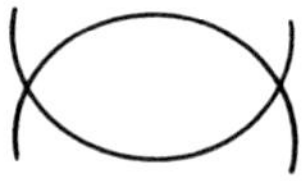

$$= \frac{\lambda^2}{2}(\mu^2)^{4-2\omega}\int_0^1 dx \, \frac{\Gamma(2-\omega)}{(4\pi)^{\omega}} \frac{1}{[m^2+p^2x(1-x)]^{2-\omega}} \qquad . \tag{4.12}$$

Before expanding, remember that this diagram has dimensions $(\mu^2)^{2-\omega}$, which we show explicitly. After expansion we obtain

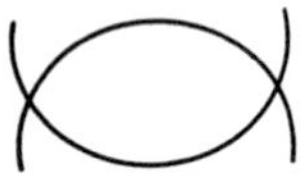

$$= (\mu^2)^{2-\omega} \frac{\lambda^2}{32\pi^2} \int_0^1 dx \{ \frac{1}{2-\omega} + \psi(1) - \ell n\, (\frac{m^2+p^2x(1-x)}{4\pi\mu^2}) + \mathcal{O}(2-\omega)\} \qquad (4.13)$$

$$= (\mu^2)^{2-\omega} \frac{\lambda^2}{32\pi^2} [\frac{1}{2-\omega} + \psi(1) - \int_0^1 dx\, \ell n\, (\frac{m^2+p^2x(1-x)}{4\pi\mu^2}) + \mathcal{O}(2-\omega)] \; . \qquad (4.14)$$

Again, observe that the finite part is arbitrary depending on $\mu^2$, although this time it depends also on the external momenta. Let us emphasize that this arbitrariness in the finite part is generic to the method because of the separation of a divergent expression into a divergence plus a finite part [after all $\infty + 5 = \infty + 6$!]

There remains to integrate over the Feynman parameter x. Since x(1-x) is always positive [$0 < x(1-x) < 1/4$] over the range of integration, the argument of the logarithm is always positive, making the integral easy to evaluate. We use the formula

$$\int_0^1 dx\, \ell n\, [1 + \frac{4}{a} x(1-x)] = -2 + \sqrt{1+a}\, \ell n\, (\frac{\sqrt{1+a}+1}{\sqrt{1+a}-1}), \quad a > 0 \quad . \qquad (4.15)$$

The result is then

$$= (\mu^2)^{2-\omega} \frac{\lambda^2}{32\pi^2} [\frac{1}{2-\omega} + \psi(1) + 2 + \ell n \frac{4\pi\mu^2}{m^2} - \sqrt{1+\frac{4m^2}{p^2}}\, \ell n\, \{\frac{\sqrt{1+\frac{4m^2}{p^2}}+1}{\sqrt{1+\frac{4m^2}{p^2}}-1}\} + \mathcal{O}(2-\omega)] \quad . \qquad (4.16)$$

In the evaluation of the four-point function, there will be three such contributions with $p = p_1 + p_2$, $p = p_1 + p_3$ and $p = p_1 + p_4$, corresponding to the s-, t- and u-channel contributions [caution: here all momenta are incoming]. This diagram is computed in the Euclidean domain; continuation to Minkowski space will entail changing the sign of $p^2$ and carefully interpreting the result. As it stands, however, the finite part has no interesting analytical structure as long as $p^2 > 0$. We will return to this point when the interpretation of the result is discussed in Minkowski space. Using the same techniques, we compute the "double scoop" diagram:

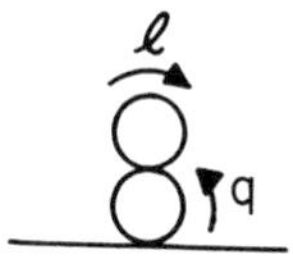


$$= \frac{1}{4} \lambda^2 (\mu^2)^{4-2\omega} \int \frac{d^{2\omega}\ell}{(2\pi)^{2\omega}} \frac{1}{\ell^2+m^2} \int \frac{d^{2\omega}q}{(2\pi)^{2\omega}} \frac{1}{(q^2+m^2)^2} \tag{4.17}$$

$$= -\frac{\lambda^2 m^2}{1024\pi^4} \Big[\frac{1}{(2-\omega)^2} + \frac{1}{(2-\omega)} \Big[2 \ln \frac{4\pi\mu^2}{m^2} + \psi(2) + \psi(1)\Big]$$

$$+ 2 \ln^2 \frac{4\pi\mu^2}{m^2} + 2 \ln \frac{4\pi\mu^2}{m^2} (\psi(2)+\psi(1)) + \frac{1}{2} ([\psi(2)+\psi(1)]^2$$

$$+ \frac{2\pi^2}{3} - \psi'(2)-\psi'(1)) + \mathcal{O}(\omega-2)\Big] \quad . \tag{4.18}$$

Note the appearance of a double pole and the arbitrariness of the residue of the simple pole and of the finite part.

Finally, we calculate in detail the "setting sun" diagram

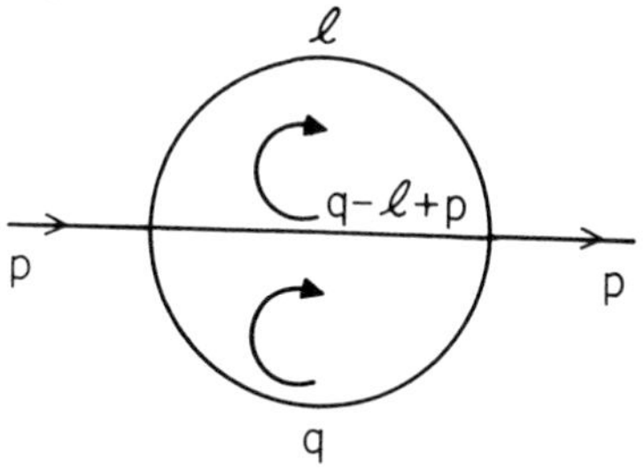

It is worth doing in some detail as it is a two-loop diagram. Call the diagram $\Sigma(p)$. The Feynman rules give

$$\Sigma(p) = \frac{\lambda^2}{6}(\mu^2)^{4-2\omega}\int\frac{d^{2\omega}\ell}{(2\pi)^{2\omega}}\int\frac{d^{2\omega}q}{(2\pi)^{2\omega}}\frac{1}{(\ell^2+m^2)(q^2+m^2)([q+p-\ell]^2+m^2)} \quad . \tag{4.19}$$

In diagrams involving several loops, ultraviolet divergences will find their way into parametric integrals. For convenience one wants to have as few divergences as possible in these integrals. This means that special techniques have to be applied on (4.19) before introducing the Feynman parameters, unless the integral is at worse logarithmically divergent. Using the trick

$$1 = \frac{1}{4\omega}\left[\frac{\partial\ell_\mu}{\partial\ell_\mu} + \frac{\partial q_\mu}{\partial q_\mu}\right] \tag{4.20}$$

in the integrand of (4.19) and integrating by parts, we find

$$\Sigma(p) = -\frac{1}{4\omega}\frac{\lambda^2}{6}(\mu^2)^{4-2\omega}\int\frac{d^{2\omega}\ell}{(2\pi)^{2\omega}}\int\frac{d^{2\omega}q}{(2\pi)^{2\omega}}\left(\ell_\mu\frac{\partial}{\partial\ell_\mu} + q_\mu\frac{\partial}{\partial q_\mu}\right)\cdot$$

$$\cdot\frac{1}{(\ell^2+m^2)(q^2+m^2)([q+p-\ell]^2+m^2)} \quad , \tag{4.21}$$

where we have discarded the surface terms, in keeping with the philosophy of analytical continuation we discussed in the one loop diagram [see 't Hooft and Veltman, Nucl. Phys. B44, 189 (1972), and also Curtright and Ghandour, Annals of Phys. 106, 209 (1977)].

Explicit differentiation gives

$$\Sigma(p) = -\frac{1}{(2\omega-3)}\frac{\lambda^2}{6}(\mu^2)^{4-2\omega}\int\frac{d^{2\omega}\ell}{(2\pi)^{2\omega}}\frac{d^{2\omega}q}{(2\pi)^{2\omega}}$$

$$\frac{3m^2+p\cdot(p+q-\ell)}{(q^2+m^2)(\ell^2+m^2)([q-\ell+p]^2+m^2)^2} \tag{4.22}$$

$$= -\frac{1}{2\omega-3}\frac{\lambda^2}{6}(\mu^2)^{4-2\omega}[3m^2K(p)+p^\mu K_\mu(p)] \quad , \tag{4.23}$$

and

$$K(p) = \int\frac{d^{2\omega}\ell}{(2\pi)^{2\omega}}\frac{d^{2\omega}q}{(2\pi)^{2\omega}}\frac{1}{(q^2+m^2)^2(\ell^2+m^2)[q-\ell+p)^2+m^2]} \tag{4.24}$$

$$K_\mu(p) = \int\frac{d^{2\omega}\ell}{(2\pi)^{2\omega}}\int\frac{d^{2\omega}q}{(2\pi)^{2\omega}}\frac{(p+q-\ell)_\mu}{(q^2+m^2)(\ell^2+m^2)[(q-\ell+p)^2+m^2]^2} , \tag{4.25}$$

where we have freely made several linear changes of variables over loop momenta. Now K(p) is log divergent and $K_\mu(p)$ is linearly divergent. We first evaluate K(p).

It is prudent to introduce Feynman parameters one loop at a time, starting from the most divergent one. So doing we obtain

$$K(p) = \int\frac{d^{2\omega}\ell}{(2\pi)^{2\omega}}\int\frac{d^{2\omega}q}{(2\pi)^{2\omega}}\frac{1}{(q^2+m^2)^2}\int_0^1 dx\frac{1}{[\ell^2+m^2+(p+q)^2x(1-x)]^2} . \tag{4.26}$$

Integrate over ℓ, using (B-16), to obtain

$$K(p) = \frac{\Gamma(2-\omega)}{(4\pi)^\omega}\int_0^1 dx\int\frac{d^{2\omega}q}{(2\pi)^{2\omega}}\frac{1}{(q^2+m^2)^2}\frac{1}{[m^2+(p+q)^2x(1-x)]^{2-\omega}} . \tag{4.27}$$

Proceed to use (4.8) to rewrite

$$K(p) = \frac{\Gamma(4-\omega)}{(4\pi)^\omega} \int_0^1 dx[x(1-x)]^{\omega-2} \int_0^1 dy\ y^{1-\omega}(1-y) \int \frac{d^{2\omega}q}{(2\pi)^{2\omega}}$$

$$[q^2+p^2y(1-y)+m^2(1-y + \frac{y}{x(1-x)})]^{\omega-4} \quad . \tag{4.28}$$

Integration over q finally gives

$$K(p) = \frac{\Gamma(4-2\omega)}{(4\pi)^{2\omega}} \int_0^1 dx[x(1-x)]^{\omega-2} \int_0^1 dy\ y^{1-\omega}(1-y)$$

$$[p^2y(1-y)+m^2(1-y + \frac{y}{x(1-x)})]^{2\omega-4} \quad . \tag{4.29}$$

It is convenient to introduce

$$2-\omega \equiv \varepsilon(> 0 \text{ because of the analytic continuation}) \tag{4.30}$$

and expand (4.29) around $\varepsilon = 0$. The parametric integral has a pole at $\varepsilon = 0$ coming from $y = 0$. Set then

$$K(p) = \frac{\Gamma(2\varepsilon)}{(4\pi)^{4-2\varepsilon}} \int_0^1 dx[x(1-x)]^{-\varepsilon} \int_0^1 dy\ y^{-1+\varepsilon}(1-y)$$

$$[p^2y(1-y)+m^2(1-y + \frac{y}{x(1-x)})]^{-2\varepsilon} \quad . \tag{4.31}$$

Using

$$y^{-1+\varepsilon} = \frac{1}{\varepsilon}\frac{d}{dy}\, y^\varepsilon \quad , \tag{4.32}$$

and integrating by parts, we find

$$K(p) = \frac{\Gamma(2\varepsilon)}{(4\pi)^{4-2\varepsilon}} \frac{1}{\varepsilon} \int_0^1 dx(x(1-x))^{-\varepsilon} \int_0^1 dy\ y^\varepsilon \cdot$$

$$\cdot\ \{1+2\varepsilon(1-y)\frac{d}{dy}\ln[p^2y(1-y)+m^2(1-y + \frac{y}{x(1-x)})]\} \cdot$$

$$\cdot\quad [p^2y(1-y)+m^2(1-y + \frac{y}{x(1-x)})]^{-2\varepsilon} \quad . \tag{4.33}$$

Now the parametric integral can be done by expanding around $\varepsilon = 0$.

The evaluation of $K_\mu$ proceeds in a similar way, giving

$$p_\mu K_\mu = p^2 \frac{\Gamma(2\varepsilon)}{(4\pi)^{4-2\varepsilon}} \int_0^1 dx[x(1-x)]^{-\varepsilon} \int_0^1 dy\ y^\varepsilon (1-y)[p^2 y(1-y)+m^2(1-y + \frac{y}{x(1-x)})]^{-2\varepsilon} . \tag{4.34}$$

Here the parametric integral converges so that the $p^2$ singularity is only at the simple pole.

Expanding around $\varepsilon = 0$ gives

$$K(p) = \frac{\Gamma(2\varepsilon)}{(4\pi)^{4-2\varepsilon}} \frac{1}{\varepsilon} \{1+\varepsilon-2\varepsilon \ln m^2 + \mathcal{O}(\varepsilon^2)\} \tag{4.35}$$

$$p_\mu K_\mu(p) = p^2 \frac{\Gamma(2\varepsilon)}{(4\pi)^{4-2\varepsilon}} [\frac{1}{2} + \mathcal{O}(\varepsilon)] \quad . \tag{4.36}$$

The $\mathcal{O}(\varepsilon^2)[\mathcal{O}(\varepsilon)]$ terms in the parametric integral for $K(p)[p_\mu K_\mu(p)]$ are very complicated, and give contributions to the finite part of $\Sigma(p)$.

Putting it all together we find

$$\Sigma(p) = -\frac{\lambda^2}{6(16\pi^2)^2} [\frac{3m^2}{2\varepsilon^2} + \frac{3m^2}{\varepsilon} [\frac{3}{2} + \psi(1) + \ln \frac{4\pi\mu^2}{m^2}] + \frac{1}{4\varepsilon} p^2 + \text{finite}] \quad . \tag{4.37}$$

Observe that we now have arbitrariness at the level of the simple pole (as well as at the level of the finite part).

The finite part of $\Sigma(p)$ is very difficult to obtain. It cannot be done in closed form and necessitates the introduction of the "dilogarithm" (or Spence) function defined by

$$L_{i2}(x) \equiv -\int_0^1 \frac{dt}{t} \log(1-xt) \quad . \tag{4.38}$$

Experience shows that when masses are present, the finite parts of two-loop diagrams are quite lengthy to evaluate.

Problems.

A. Show that for $a > 0$

$$I(a) = \int_0^1 dx\, \ell n\, [1 + \frac{4}{a}\, x(1-x)] = -2 + \sqrt{1+a}\, \ell n \frac{\sqrt{1+a} + 1}{\sqrt{1+a} - 1} .$$

*B. Now let $z = \frac{4}{a}$. Find the singularity structure of $I(z)$ in the complex z-plane and specify $I(z)$ when z is real.

C. Derive the form of the finite part of $\Sigma(p)$ when $m^2 = 0$.

D. Derive the expression for $p_\mu K_\mu$ [eq. 4.34].

**E. Express $K(p)$ to order $\varepsilon^2$ and express the resulting integrals in terms of dilogarithms.

## 5. Renormalization

In the previous sections we have shown how to evaluate Feynman diagrams. In $\lambda\phi^4$ theory we found that some diagrams are ultraviolet divergent, with the divergences showing up only in two- and four-point Green functions (the primitively divergent diagrams). When dimensional regularization was used to regularize these diagrams, the infinities resulted in poles in the dimension plane in $\varepsilon = 2 - n/2 > 0$, $n$ being the number of space-time dimensions; in addition the finite part of these diagrams was arbitrary, depending in our scheme on a mysterious mass parameter $\mu$.

We are now going to show how to eliminate these poles order by order in $\lambda$. The technique is very simple: alter the Feynman rules at each order so as to obtain a finite result as $\varepsilon \to 0$. As a first example, consider the tadpole diagram

$$= m^2 \frac{\hat{\lambda}}{2} \left[\frac{1}{\varepsilon} + \psi(2) - \ln \hat{m}^2 + \mathcal{O}(\varepsilon)\right] \quad , \qquad (5.1)$$

where

$$\hat{\lambda} \equiv \frac{\lambda}{16\pi^2} \ , \qquad \hat{m}^2 \equiv \frac{m^2}{4\pi\mu^2} \quad . \qquad (5.2)$$

This infinity can be hidden away by adding to $\mathcal{L}$ the additional term

$$\frac{m^2}{4} \hat{\lambda}\left[\frac{1}{\varepsilon} + F_1(\varepsilon, \hat{m}^2)\right]\phi^2 \quad , \qquad (5.3)$$

which we consider to be an extra interaction term. Here $F_1$ is an arbitrary dimensionless function, analytic as $\varepsilon \to 0$; its presence reflects the arbitrariness of the procedure. This extra term results in a new Feynman rule indicated by

$$= -\frac{1}{2} m^2 \hat{\lambda}\left[\frac{1}{\varepsilon} + F_1\right] \quad . \qquad (5.4)$$

Thus if we compute the inverse propagator to $\mathcal{O}(\lambda)$, we find (remember that the correction to the inverse propagator picks up a minus sign for inverting)

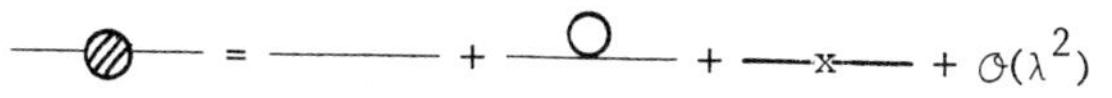

$$\tilde{\Gamma}^{(2)}_{new}(p) = p^2 + m^2[1 - \frac{1}{2}\hat{\lambda}(\psi(2) - \ell n\, \hat{m}^2 - F_1)] + \mathcal{O}(\lambda^2) \quad . \qquad (5.5)$$

This very naive procedure makes the theory finite to order $\lambda$. The extra term (5.3) is called a <u>counterterm</u>. It is crucial to note that its dependence on the field (and its derivatives) is the same as that of a term already appearing in $\mathcal{L}$ (in this case the mass term).

Now we proceed to order $\lambda^2$. The one particle irreducible (1PI) four-point function is given by

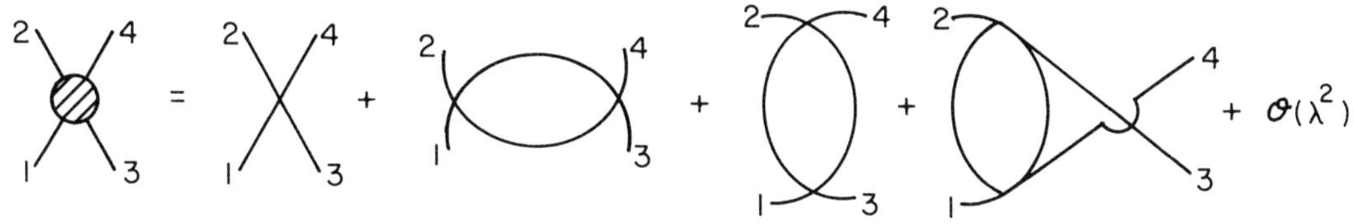

$$\tilde{\Gamma}^{(4)}_{(p_1,p_2,p_3,p_4)} = -\mu^{2\varepsilon}\lambda[1 - \frac{3}{2}\hat{\lambda}(\frac{1}{2} + \psi(1) + 2 - \ell n\, \hat{m}^2 - \frac{1}{3}A(s,t,u) + \mathcal{O}(\varepsilon))] + \mathcal{O}(\lambda^3) \quad , \qquad (5.6)$$

where

$$A(s,t,u) = \sum_{z=s,t,u} (1 + \frac{4m^2}{z})^{1/2} \ell n \frac{\sqrt{1 + \frac{4m^2}{z}} + 1}{\sqrt{1 + \frac{4m^2}{z}} - 1} \quad , \qquad (5.7)$$

and where s, t and u are the Mandelstam variables

$$s = (p_1+p_2)^2 \qquad t = (p_1+p_3)^2 \qquad u = (p_1+p_4)^2 \quad . \qquad (5.8)$$

As it stands $\tilde{\Gamma}^{(4)}$ is divergent as $\varepsilon \to 0$. To remedy this situation, we add yet another term to $\mathcal{L}$

$$\frac{1}{4!}\,\mu^{2\varepsilon}\lambda \cdot \frac{3\hat{\lambda}}{2}\left(\frac{1}{\varepsilon} + G_1(\varepsilon,\hat{m}^2)\right)\phi^4 \qquad , \qquad (5.9)$$

where $G_1$ is an arbitrary dimensionless function of $\varepsilon$, analytic as $\varepsilon \to 0$. This new counterterm results in an additional Feynman rule denoted by

$$= -\frac{3}{2}\,\mu^{2\varepsilon}\lambda\,\hat{\lambda}\left[\frac{1}{\varepsilon} + G_1\right] \qquad . \qquad (5.10)$$

It is added in the calculation of the new $\tilde{\Gamma}^{(4)}$, yielding a finite result as $\varepsilon \to 0$:

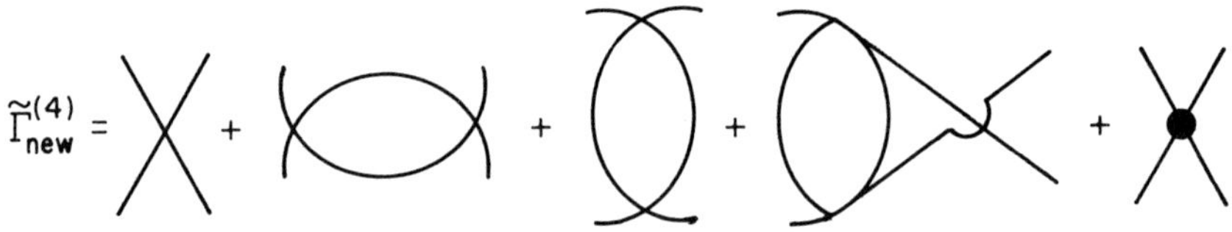

$$= -\mu^{2\varepsilon}\lambda\left[1 - \frac{3}{2}\,\hat{\lambda}\left(-G_1+\psi(1)+2-\ln\hat{m}^2 - \frac{1}{3}A(s,t,u)\right)\right]+\mathcal{O}(\lambda^3) \; . \qquad (5.11)$$

The contribution to $\tilde{\Gamma}^{(2)}$ in $\mathcal{O}(\lambda^2)$ can be handled in a similar way, but there the additional Feynman rules (5.4) and (5.10) must be used. Hence we have diagrammatically for the inverse propagator

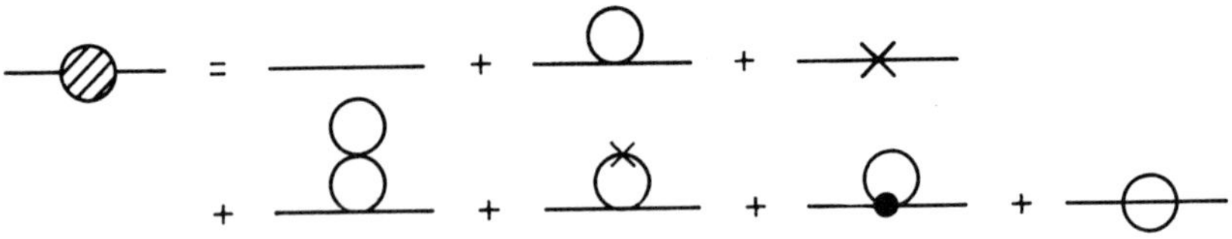

(5.12)

where the extra Feynman rules have induced to this order two new diagrams. These are easily calculated to be (see problem)

 $$\frac{m^2}{4}\hat{\lambda}^2\left[\frac{1}{\varepsilon^2}+\frac{1}{\varepsilon}(\psi(1)+F_1-\ell n\,\hat{m}^2)+\ldots.\right] \quad , \tag{5.13}$$

where we show only the poles in $\varepsilon$. We also have

 $$\frac{3m^2}{4}\hat{\lambda}^2\left[\frac{1}{\varepsilon^2}+\frac{1}{\varepsilon}(\psi(2)+G_1-\ell n\,\hat{m}^2)+\ldots\right] \quad . \tag{5.14}$$

Comparing with the "double scoop" diagram of the previous section

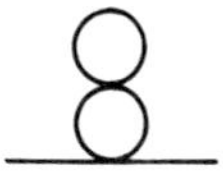 $$-\frac{m^2}{4}\hat{\lambda}^2\left[\frac{1}{\varepsilon^2}+\frac{1}{\varepsilon}(\psi(2)+\psi(1)-2\,\ell n\,\hat{m}^2)+..\right] \quad , \tag{5.15}$$

we observe that its double pole is exactly canceled by the counterterm diagram (5.12) although the simple pole remains. This is obvious from the point of view of diagrams since  + 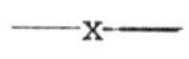  is finite by definition. Adding all the diagrams, we find

 $$= -\frac{\hat{\lambda}^2}{24\varepsilon}p^2+\frac{m^2}{2}\lambda^2\left[\frac{1}{\varepsilon^2}+\frac{1}{2\varepsilon}(F_1+3G_1-1)+\ldots\right]+\mathcal{O}(\lambda^3) \quad , \tag{5.16}$$

where, again, we have not shown the finite part. We are faced again with a divergent expression as $\varepsilon \to 0$. Lo and behold! The $\ell n\,\hat{m}^2$ terms present in the simple poles of individual diagrams have disappeared as well as the Euler constant present in each $\psi(n)$ but absent in the difference

$$\psi(n+1) - \psi(n) = \frac{1}{n} \quad . \tag{5.17}$$

In order to cancel the poles in (5.16) we introduce a new mass counterterm so that the mass counterterm Feynman rule becomes

$$—x— = -\frac{m^2}{2}[\frac{\hat{\lambda}^2}{\epsilon^2} + \frac{1}{\epsilon}(\hat{\lambda} + \frac{\hat{\lambda}^2}{4}(F_1+3G_1-1))+\hat{\lambda}^2F_2+\hat{\lambda}F_1] \qquad (5.18)$$

where $F_2$ is an arbitrary function of $\varepsilon$ and $\hat{m}^2$, but finite as $\varepsilon \to 0$. This term is generated by the additional counterterm

$$\frac{m^2}{4}\phi^2[\frac{\hat{\lambda}^2}{\epsilon^2} + \frac{1}{\epsilon}(\hat{\lambda} + \frac{\hat{\lambda}^2}{4}(F_1+3G_1-1)) + \hat{\lambda}^2F_2+\hat{\lambda}F_1] \qquad . \qquad (5.19)$$

This takes care of only one type of infinity in $\tilde{\Gamma}^{(2)}$. The other one is canceled by adding yet another term to our ever expanding Lagrangian of the form

$$\frac{1}{2}\partial_\mu\phi\partial_\mu\phi[-\frac{\hat{\lambda}^2}{24\epsilon} - \hat{\lambda}^2H_2(\epsilon,\hat{m}^2)] \qquad , \qquad (5.20)$$

$H_2$ being arbitrary and analytic as $\varepsilon \to 0$. Thus with all this patching up, we have been able to eliminate the ultraviolet divergences to order $\lambda^2$. It is clear that we can continue this little game ad nauseam: calculate diagrams to order $\lambda^3$, with the original $\mathcal{L}$ and the counterterms (5.19) and (5.20); then invent new counterterms which to $\mathcal{O}(\lambda^3)$ are chosen to cancel the new divergences, etc. So far in this process, the remarkable thing has been that the counterterms needed to remove the divergences all generate new interactions of the same type as those that were present in the original Lagrangian; we have not been forced to ($\mathcal{O}(\lambda^2)$) to introduce counterterms that correspond to terms of type absent in $\mathcal{L}$. If it can be shown that this noteworthy matching feature continues to all orders in $\lambda$, we will say that the theory is renormalizable. Here we do not attempt a proof for $\lambda\phi^4$ theory, but rather point out where the procedure might fail.

Consider a generic two loop diagram

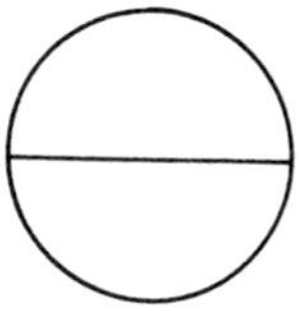

where all external legs have been removed. This diagram is divergent due to the various loop integrations. In accordance with our new rules for the counterterms, it must be added to the three counterterm diagrams, (presented here schematically)

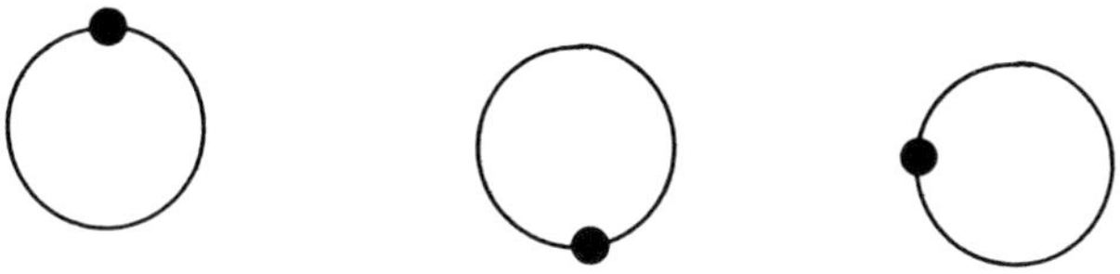

Here each  represents the lower order counterterm vertex needed to cancel the infinity coming from one loop diagrams. These counterterm diagrams will therefore contain a $\frac{1}{\varepsilon}$ coming from  which multiplies a $\ln p^2$ where p is some momentum coming from the loop integration. It would therefore seem that this would generate counterterms of the form

$$\frac{\ln p^2}{\varepsilon} ,$$

which do not correspond to any term in $\mathscr{L}$ because of the $\ln p^2$ residue, which is highly nonlocal in position space. Such terms would clearly throw a monkey-wrench in the works. This is the famous problem of overlapping divergences. A close study of these diagrams shows that the sum of all the diagrams shown above does not contain any poles with logarithmic residue: they cancel against similar poles contained in the two loop diagram. We have in fact seen an example of this miracle when the $\ell n\, \hat{m}^2$ terms canceled from the residue of the simple pole in (5.16) [For more details, the reader is refered to the paper of 't Hooft and M. Veltman, Nuclear Phys. 44B, 189 (1972).] In the proof of renormalizability, it is crucial to be able to prove that these overlapping divergences do, in fact, cancel.

Let us assume that this is so and that to an arbitrarily high order in $\lambda$ only counterterms which match the original $\mathcal{L}$ are needed to render the theory finite. This means that the Lagrangian which gives finite answers has the form

$$\mathcal{L}^{ren} = \mathcal{L} + \mathcal{L}_{c.t.} \quad , \tag{5.21}$$

where $\mathcal{L}$ is our original Lagrangian

$$\mathcal{L} = \frac{1}{2}\, \partial_\mu\phi\partial_\mu\phi + \frac{1}{2}\, m^2\phi^2 + \frac{\lambda}{4!}\, \mu^{2\varepsilon}\phi^4 \tag{5.22}$$

and $\mathcal{L}_{c.t.}$ is the counterterm Lagrangian

$$\mathcal{L}_{c.t.} = \frac{1}{2}\, A\partial_\mu\phi\partial_\mu\phi + \frac{1}{2}\, m^2 B\phi^2 + \frac{\lambda}{4!}\, \mu^{2\varepsilon} C\phi^4 \quad . \tag{5.23}$$

It is (by assumption, but verified to $\mathcal{O}(\lambda^2)$) exactly of the same form as $\mathcal{L}$, but with specially designed A, B and C so that the Green's functions generated by $\mathcal{L}^{ren}$ are finite as $\varepsilon \to 0$. By defining new fields and parameters, we can rewrite $\mathcal{L}^{ren}$ in the form

$$\mathcal{L}^{ren} = \frac{1}{2}\, \partial_\mu\phi_0\partial_\mu\phi_0 + \frac{1}{2}\, m_0^2\phi_0^2 + \frac{\lambda_0}{4!}\, \phi_0^4 \quad , \tag{5.24}$$

where

$$\phi_0 \equiv (1+A)^{1/2}\phi \equiv Z_\phi^{1/2}\phi \quad , \tag{5.25}$$

$$m_0^2 = m^2 \frac{1 + B}{1 + A} = m^2(1+B)Z_\phi^{-1} \quad , \quad (5.26)$$

$$\lambda_0 = \lambda\mu^{2\varepsilon} \frac{1+C}{(1+A)^2} = \lambda\mu^{2\varepsilon}(1+C)Z_\phi^{-2} \quad , \quad (5.27)$$

are called the <u>bare</u> field, mass and coupling constant, respectively. Note that $\mathscr{L}^{ren}$ looks exactly the same as $\mathscr{L}$ except for the parameters and the field. Yet $\mathscr{L}^{ren}$ leads to a finite theory while $\mathscr{L}$ does not. This shows that by cleverly putting all the infinities in $\phi_0$, $m_0$ and $\lambda_0$, we can make the theory finite. The infinities are then absorbed by renormalization. The bare quantities all diverge as $\varepsilon \to 0$ while the (renormalized) quantities $m,\lambda$ all give finite (but arbitrary) values as $\varepsilon \to 0$. The latter are to be identified with the physical parameters of the theory. In the path integral formalism, one integrates over the fields; thus their rescaling by $Z_\phi$ can be absorbed provided one rescales the source accordingly, defining a bare source

$$J_0 = Z_\phi^{-1/2} J \quad , \quad (5.28)$$

or a bare classical field

$$\phi_{c\ell\,0} = Z_\phi \phi_{c\ell} \quad . \quad (5.29)$$

Starting from the new Lagrangian (5.21) we obtain the Green's functions of the previous section with $m$ and $\lambda$ replaced by $m_0$ and $\lambda_0$. However, by expressing the bare parameters in terms of the physical parameters $m$ and $\lambda$ and by suitably renormalizing $J$, we obtain finite Green's functions. For the 1PI Green's functions, this equality reads

$$\tilde{\Gamma}_0^{(n)}(p_1,..,p_n;\lambda_0,m_0,\varepsilon) = Z_\phi^{-n/2}\tilde{\Gamma}^{(n)}(p_1,..,p_n;\lambda,m,\mu,\varepsilon), \quad (5.30)$$

where the $\tilde{\Gamma}^{(n)}$ are finite as $\varepsilon \to 0$. In this equation we can either regard the bare parameters as functions of the renormalized ones or take the bare parameters to be the independent variables; in the latter case the dressed parameters are functions of the bare parameters. In this form we note that the left hand side of (5.30) does not depend on $\mu$ while the right hand side is explicitly as well as implicitly (through $\lambda$ and m) dependent on $\mu$. Therefore, by differentiating (5.30) with respect to $\mu$, we obtain a differential equation that summarizes the magic of renormalization

$$[\mu \frac{\partial}{\partial \mu} + \mu \frac{\partial \lambda}{\partial \mu} \frac{\partial}{\partial \lambda} + \mu \frac{\partial m}{\partial \mu} \frac{\partial}{\partial m} - \frac{n}{2} \mu \frac{\partial \, \ell n \, Z_\phi}{\partial \mu}]\tilde{\Gamma}^{(n)} = 0 \quad . \tag{5.31}$$

The beauty of this equation is that it only involves the renormalized Green's function $\tilde{\Gamma}^{(n)}$ which is finite as $\varepsilon \to 0$. The various derivatives come from the implicit dependence of $\tilde{\Gamma}^{(n)}$ on $\mu$ via $\lambda$ and m. Define the coefficients

$$\beta(\lambda, \frac{m}{\mu}, \varepsilon) \equiv \mu \frac{\partial \lambda}{\partial \mu} \tag{5.32}$$

$$\gamma_d(\lambda, \frac{m}{\mu}, \varepsilon) \equiv \frac{1}{2} \mu \frac{\partial \, \ell n \, Z_\phi}{\partial \mu} \tag{5.33}$$

$$\gamma_m(\lambda, \frac{m}{\mu}, \varepsilon) = \frac{1}{2} \mu \frac{\partial \, \ell n \, m^2}{\partial \mu} \quad . \tag{5.34}$$

They are analytic as $\varepsilon \to 0$ and dimensionless - they depend only on $\lambda$ and $\frac{m}{\mu}$.

On the other hand, $\tilde{\Gamma}^{(n)}$ has an engineering dimension equal to $4 - n + \varepsilon(n-2)$, which can be read off as the sum of its degree of homogeneity in its dimensionful parameter, i.e.,

$$(\mu \frac{\partial}{\partial \mu} + s \frac{\partial}{\partial s} + m \frac{\partial}{\partial m} - [4-n+\varepsilon(n-2)])\tilde{\Gamma}^{(n)}(sp; m, \lambda, \mu, \varepsilon) = 0 \tag{5.35}$$

where we have introduced a scale s for the momenta. This equation, in conjunction with (5.31) can be turned into a scaling equation for $\tilde{\Gamma}^{(n)}$, by eliminating $\mu \frac{\partial}{\partial \mu}$. Taking the limit $\varepsilon \to 0$, we obtain

$$[- s \frac{\partial}{\partial s} + \beta(\lambda, \frac{m}{\mu}) \frac{\partial}{\partial \lambda} + [\gamma_m(\lambda, \frac{m}{\mu}) - 1] m \frac{\partial}{\partial m} - n\gamma_d(\lambda, \frac{m}{\mu}) + 4-n]\tilde{\Gamma}^{(n)}(sp,m,\lambda,\mu) = 0 \;. \tag{5.36}$$

This equation summarizes the behavior of $\tilde{\Gamma}^{(n)}$ as one scales its momenta. (An equation of this type for QED was first obtained by Gell-Mann and Low, Phys. Rev. 95, 1300 (1954).) If we could solve it, it would tell us how the Green's functions at momenta sp are related to the same functions at a reference p. The difficulty in solving (5.36) lies in the fact that the coefficients $\beta$, $\gamma$ and $\gamma_m$ depend on two variables $\lambda$ and $\frac{m}{\mu}$. They can be computed explicitly order by order in perturbation theory, but they are at the moment quite arbitrary because we have not stated what the finite part of the counterterms is. This will be done in the next section where various "renormalization prescriptions" will be investigated. Suffice it to say that these coefficients depend on the way we choose the finite part of the counterterms.

We can express the bare parameters as a Laurent series in the renormalized parameters

$$\lambda_0 = \mu^{2\varepsilon}[a_0(\lambda, \frac{m}{\mu}, \varepsilon) + \sum_{k=1}^{\infty} \frac{a_k(\lambda, \frac{m}{\mu})}{\varepsilon^k}] \tag{5.37}$$

$$m_0^2 = m^2[b_0(\lambda, \frac{m}{\mu}, \varepsilon) + \sum_{k=1}^{\infty} \frac{b_k(\lambda, \frac{m}{\mu})}{\varepsilon^k}] \tag{5.38}$$

$$Z_\phi = c_0(\lambda, \frac{m}{\mu}, \varepsilon) + \sum_{k=1}^{\infty} \frac{c_k(\lambda, \frac{m}{\mu})}{\varepsilon^k} \quad , \tag{5.39}$$

where $a_0$, $b_0$ and $c_0$ are analytic as $\varepsilon \to 0$. Comparing with the counterterms already obtained up to $\mathcal{O}(\lambda^2)$, we find

$$a_0(\lambda, \frac{m}{\mu}, \varepsilon) = \lambda(1 + \frac{3}{2}\hat{\lambda}G_1) + \mathcal{O}(\lambda^3) \tag{5.40}$$

$$b_0(\lambda, \frac{m}{\mu}, \varepsilon) = 1 + \frac{1}{2}(\hat{\lambda}F_1+\hat{\lambda}^2F_2) + \hat{\lambda}^2H_2 + \mathcal{O}(\lambda^3) \tag{5.41}$$

$$c_0(\lambda, \frac{m}{\mu}, \varepsilon) = 1 - \hat{\lambda}^2H_2(\varepsilon, \frac{m}{\mu}) + \mathcal{O}(\lambda^3) \tag{5.42}$$

$$a_1(\lambda, \frac{m}{\mu}) = \frac{3}{2}\frac{\lambda^2}{16\pi^2} + \mathcal{O}(\lambda^3) \tag{5.43}$$

$$b_1(\lambda, \frac{m}{\mu}) = \frac{1}{2}(\hat{\lambda} + \frac{\hat{\lambda}^2}{4}(F_1+3G_1-1)) + \frac{\hat{\lambda}^2}{24} + \mathcal{O}(\lambda^3) \tag{5.44}$$

$$b_2(\lambda, \frac{m}{\mu}) = \frac{1}{2}\hat{\lambda}^2 + \mathcal{O}(\lambda^3) \tag{5.45}$$

$$c_1(\lambda, \frac{m}{\mu}) = -\frac{\hat{\lambda}^2}{24} + \mathcal{O}(\lambda^3) \quad . \tag{5.46}$$

In these formulae, we remark that the a, b and c coefficients depend on $\frac{m}{\mu}$ only through the unknown functions $F_1$, $F_2$, $G_1$, $H_2$. This can be understood heuristically by noting that the counterterms are used to eliminate the divergences that occur at very large momentum (~ mass) scales. There any fixed mass parameter is not expected to play any role. Thus as long as the counterterms have order by order no finite part, we do not expect the residues of their poles to depend on m. This is exactly what the formulae (5.40) - (5.46) reflect. This remark is at the heart of the mass-independent renormalization prescription we discuss in the next section. The dependence of these coefficients on the arbitrary finite parts of the counterterms reflects the (a priori) prescription dependence of the β- and γ-functions. It follows that the solution of the renormalization group equation (5.31) is not to be attempted before a prescription has been chosen. The technical difficulty in finding solutions lies in the dependence of the coefficients on both λ and $\frac{m}{\mu}$. However, as we shall see, there is a prescription where the coefficients become mass independent, greatly simplifying the solution of (5.31).

Otherwise, one has to solve (5.31) in a region where the masses can be neglected, i.e., where momenta are large compared with input mass parameters.

Finally we mention that one can derive another type of renormalization group equation, first obtained by C. Callan [Phys. Rev. D5, 3202 (1972)] and K. Symanzik [Comm. Math. Phys. 23, 49 (1971)]. This kind of equation studies the variation of the $\tilde{\Gamma}$'s with respect to the physical mass. The $\beta$ and $\gamma$ coefficients depend only on $\lambda$; the $\gamma_m$ and $\mu \frac{\partial}{\partial \mu}$ terms do not appear but are replaced by an inhomogeneous term, which can be neglected in the limit of small masses, or equivalently large momenta.

Problems.

A. Compute the value of the extra counterterm diagrams (5.12) and (5.13) including the finite part.

B. Verify that the propagator itself is finite to $\mathcal{O}(\lambda^2)$ (hint - the propagator contains one particle reducible graphs).

**C. In $\lambda\phi^3$ theory, verify that the overlapping divergences from the diagram

are, in fact, canceled by the counterterm diagram that regulates the one loop diagram

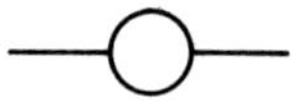

## 6. Renormalization Prescriptions

In the previous section we carried out in detail the renormalization procedure in $\lambda\phi^4$ theory. Besides the arbitrary scale $\mu$, the elimination of divergences brought with it an extra arbitrariness reflected by the functions $F_1$, $F_2$, $G_1$, $H_2$, ... which constitute the finite part of the counterterms. From the structure of the Lagrangian that gives finite answers

$$\mathcal{L}^{\text{ren}} = \mathcal{L} + \mathcal{L}_{\text{c.t.}} \quad , \tag{6.1}$$

it is clear that the finite part of $\mathcal{L}_{\text{c.t.}}$ can be absorbed in a redefinition (or finite renormalization) of the initial parameters appearing in $\mathcal{L}$ since both $\mathcal{L}$ and $\mathcal{L}_{\text{c.t.}}$ have the same structure. It follows that the finite part of the counterterms can be fixed only by defining the parameters that enter in $\mathcal{L}$. There is, however, much arbitrariness in the method used to define m, $\lambda$ and $\phi$, and the choice of method is dictated by convenience, or by the convergence properties of the perturbation theory.

In some cases it is possible to directly relate the renormalized parameters to physically measured quantities. QED is a case in point where the physical electric charge is equated to the vertex function in the Thomson limit.

Specification of the scale at which the renormalized parameters are equated to the relevant Green's functions is largely arbitrary (in Euclidean space) with one important restriction for theories that involve massless particles. These theories give infrared divergent Green's function for zero value of input momenta. It would not be wise to choose the subtraction point at the scale where the Green's function diverges. Such points are to be avoided. Later when the amplitude is continued to Minkowski space, the subtraction scale will appear at a space-like value of input momenta and will not interfere with the singularities that Green's functions must and do have to qualify as transition amplitudes. These appear in the physical region where at least some of the momenta are always timelike.

Let us give several examples of subtractions, also called prescriptions:

A. This first way of fixing parameters is the most common. We define the input parameters by

$$\tilde{\Gamma}^{(2)}(p,m_A) = p^2 + m_A^2 \text{ at } p^2 = 0 \tag{6.2}$$

$$\tilde{\Gamma}^{(4)}(p_1,p_2,p_3,p_4) = -\mu^{2\varepsilon}\lambda_A \text{ at } p_i = 0 \quad . \tag{6.3}$$

In the absence of infrared divergences which occur when $m^2 = 0$, this prescription is well-defined. We have subscripted the input parameters to indicate how they were defined. Note that (6.2) embodies two conditions since it fixes the mass as well as the field normalization. These fix the finite parts of the counterterms. In particular we find that

$$F_1^A = \psi(2) - \ln \hat{m}_A^2; \quad G_1^A = \psi(1) - \ln \hat{m}_A^2; \quad H_2^A = 0, \text{ etc. } . \tag{6.4}$$

Ideally one would prefer to identify the coupling constant with $\tilde{\Gamma}^{(4)}$ at a physical point where the particles are in Minkowski space and on their mass-shell ($p_M^2 = m^2$).

B. One can change the subtraction point at will provided it does not interfere with the continuation to Minkowski space or with infrared singularities. We should add that as long as the subtraction procedure is performed on Euclidean Green functions, it will result in a spacelike subtraction in Minkowski space. Thus our second prescription [H. Georgi and H. D. Politzer, Phys. Rev. D14, 1829 (1976)] is the same as A but carried out at an arbitrary value of p:

$$\tilde{\Gamma}^{(2)}(p,m_B) = p^2 + m_B^2 \text{ at } p^2 = M^2 \quad , \qquad (6.5)$$

$$\tilde{\Gamma}^{(4)}(p_1,p_2,p_3,p_4) = -\mu^{2\varepsilon}\lambda_B \text{ at } p_i p_j = M^2(\delta_{ij}-1/4) \quad , \qquad (6.6)$$

the latter point chosen so that $s = t = u = M^2$. One can, of course, choose any value for s, t and u, and the $p^2$ at which $\tilde{\Gamma}^{(2)}$ is normalized. In this prescription, the unknown functions are now fixed to be

$$F_1^B = \psi(2) - \ln \hat{m}_B^2; \quad H_2^B = 0 \qquad (6.7)$$

$$G_1^B = \psi(1) - \ln \hat{m}_B^2 - \int_0^1 dx \ln [1 + \frac{M^2}{m_B^2} x(1-x)], \text{ etc.} \quad . \qquad (6.8)$$

In this case the scale $\mu$ has been totally absorbed and replaced by the scale $M^2$, which is equally arbitrary. The numerical value of $M^2/m_B^2$ now clearly becomes relevant in choosing M.

The trouble with the type of prescription outlined above is that the renormalization group equation of the last section is not easily solved except in the deep Euclidean region where all masses can presumably be neglected. Yet equating a coupling constant with the value of an amplitude at some scale has some physical appeal even though the identification may take place at an unphysical point. The reason is that it brings in the masses explicitly in the calculation and allows for a ready identification of various physical thresholds.

C. We now present a very beautiful prescription invented by 't Hooft and Weinberg [Nuclear Phys. B61, 455 (1973) and Phys. Rev. D8, 3497 (1973), respectively]. It is very simple to state and allows for a simple solution of the "renormalization group equation" (5.36) of the last section. In it one simply sets all the finite parts of the counterterms equal to zero, order by order in the coupling, i.e.,

$$F_1^C = H_2^C = G_1^C = F_2^C = 0, \text{ etc.} \quad . \tag{6.9}$$

Then, by comparing with eqs. (5.40) - (5.46) of the last section, we notice that all the a, b and c coefficients are independent of m. This prescription is aptly called "mass-independent" renormalization. This mass independence survives to arbitrarily high order; that this is, in fact, true is not hard to understand by means of the following heuristic argument: when the counterterms have no finite part, they just have the "bare bones" structure needed to cancel the infinite behavior at very short distances and no more, but there, at infinite momenta, all masses can presumably be neglected, provided the amplitudes are well-behaved as $p \to \infty$ - hence the mass independence. It is true that in prescriptions A and B the mass dependence appeared only through the finite part of the counterterms. This enormous simplification enables us to compute the $\beta$, $\gamma$ and $\gamma_m$ coefficients appearing in (5.36) in a straightforward way. For instance, we now write

$$\lambda_0 = \mu^{2\varepsilon}[\lambda + \sum_{k=1}^{\infty} \frac{a_k(\lambda)}{\varepsilon^k}] \quad . \tag{6.10}$$

Differentiate with respect to $\mu$ at fixed $\lambda_0$ to obtain

$$0 = 2\varepsilon(\lambda + \sum_{k=1}^{\infty} \frac{a_k(\lambda)}{\varepsilon^k}) + \mu \frac{\partial\lambda}{\partial\mu} (1 + \sum_{k=1}^{\infty} \frac{a'_k(\lambda)}{\varepsilon^k}) \quad , \tag{6.11}$$

with the prime denoting differentiation with respect to $\lambda$. In this formula $\lambda$ and $\mu \frac{\partial\lambda}{\partial\mu}$ are analytic at $\varepsilon = 0$. It follows that

$$\mu \frac{\partial\lambda}{\partial\mu} = - 2\varepsilon\lambda - 2a_1(\lambda) + 2\lambda a'_1(\lambda) \quad , \tag{6.12}$$

or taking the limit $\varepsilon \to 0$

$$\beta(\lambda) = \lim_{\varepsilon\to 0} \mu \frac{\partial\lambda}{\partial\mu} = -2(1 - \lambda \frac{d}{d\lambda})a_1(\lambda) \quad , \tag{6.13}$$

showing that the $\beta$-function appearing in the "renormalization group equation" depends only on $\lambda$ and is determined by the residue of the simple pole in $\varepsilon$. Using (6.12) and the fact that the residues of the various poles in $\varepsilon$ must vanish in (6.11), we find that

$$(1 - \lambda \frac{d}{d\lambda})a_{k+1}(\lambda) = a_k'(\lambda)(1 - \lambda \frac{d}{d\lambda})a_1(\lambda) \quad . \tag{6.14}$$

The meaning of (6.12) is clear: a successful renormalization means that the bare coupling does not depend on what $\mu$ is - a change in $\mu$ is accompanied by a change in $\lambda$ so as to leave (6.10) invariant. Let us now evaluate $\beta$ in perturbation theory. Using (5.43) we find

$$\mu \frac{\partial\lambda}{\partial\mu} = \frac{3\lambda^2}{16\pi^2} + \mathcal{O}(\lambda^3) \quad . \tag{6.15}$$

Neglecting the terms of $\mathcal{O}(\lambda^3)$, we can easily integrate (6.15), obtaining

$$\lambda = \lambda_0 \frac{1}{1 - \frac{3}{16\pi^2}\lambda_0 \ln\frac{\mu}{\mu_0}} \quad , \tag{6.16}$$

where $\lambda_0$ is the value of $\lambda$ at $\mu_0$.

It is clear from (6.15) that $\lambda$ increases with $\mu$. Thus if we start with a small $\lambda_0 (<< 1)$ at a given scale $\mu_0$, the effective coupling constant will increase with $\mu$. In so doing, however, we will have to deal with larger and larger $\lambda$ and will therefore leave the domain of validity of perturbation theory: $\lambda << 1$, or more exactly $\frac{3}{16\pi^2}\lambda_0 \ln\frac{\mu}{\mu_0} << 1$. Thus at shorter distances, we have to add more contributions to the RHS of (6.15).

Thus for $\lambda\phi^4$ theory, perturbation theory becomes more reliable at larger distances, that is in the long range properties of the interaction: the perturbative approach to defining asymptotic states is to be trusted. Should the sign of the RHS of (6.15) ever be found to be negative in a field theory, then perturbation arguments would be misleading for the definition of the asymptotic states, but great for the short distance behavior. (We will see that this situation occurs in quantum chromodynamics (QCD) which describes the interaction among quarks - quarks are useful to describe the short range interaction of two protons but they are not asymptotic states, just the constituents of asymptotic states such as the proton.)

Note that the expression for $\beta(\lambda)$ obtained here is the same as that obtained via the $\zeta$-function evaluation of determinants. This should not be too surprising since the former method was $\mathcal{O}(\hbar)$ and here we have only shown the one loop result which is also $\mathcal{O}(\hbar)$.

Let us now speculate on the possible behaviors of $\lambda$ as a function of $\mu$, outside of perturbation theory. First of all we note that if $\beta(\lambda)$ is given by (6.15) even for large $\lambda$, then it will blow up at a scale

$$\mu = \mu_0 \exp\left[\frac{16\pi^2}{3\lambda_0}\right] \quad , \qquad (6.17)$$

a very large scale if $\lambda_0$ was small to start with. This is called the Landau point after Landau who recognized the same behavior in QED. However, there is no reason to believe that the one-loop contribution to $\beta$ is valid for large $\lambda$. We do not know how to calculate $\beta$ for large $\lambda$, but let us present some possible scenarios, starting from $\beta = 0$ at $\lambda = 0$, the no interaction point:

1. $\beta(\lambda)$ stays positive for large $\lambda$; then $\lambda$ keeps increasing with the scale in a concave or convex curve depending on the sign of $\beta'(\lambda)$. If $\beta(\lambda)$ blows up for some value of $\lambda$, $\lambda$ itself becomes infinite there (Landau point).

2. $\beta(\lambda)$ starts out positive for low $\lambda$ then turns over and becomes negative, crossing the axis at $\lambda_F$:

$$\beta(\lambda_F) = 0 \tag{6.18}$$

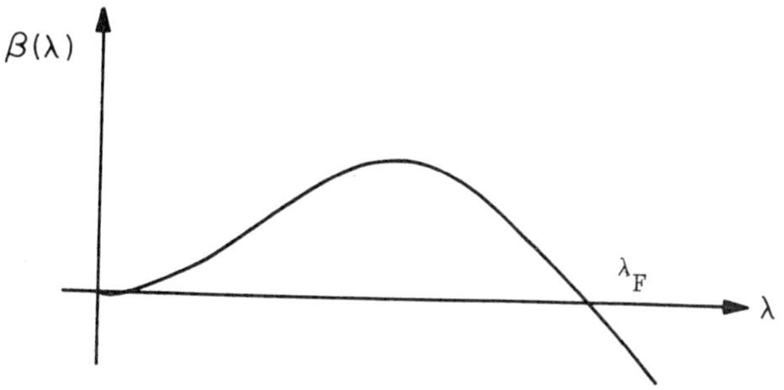


$\lambda_F$ is called a fixed point because if for some reason the coupling was originally at $\lambda_F$, it would remain there. We can analyze the behavior of $\lambda$ near $\lambda_F$ by expanding $\beta$ about $\lambda_F$, yielding the equation

$$\mu \frac{\partial\lambda}{\partial\mu} = (\lambda-\lambda_F)\beta'(\lambda_F) + \ldots \quad . \tag{6.19}$$

We see that the sign of $\beta'(\lambda_F)$ is crucial. If $\beta'(\lambda_F) < 0$, as in the figure, $\mu \frac{\partial\lambda}{\partial\mu}$ is positive for $\lambda$ just below $\lambda_F$, which drives $\lambda$ to a larger value, i.e., towards the fixed point $\lambda_F$, while $\mu \frac{\partial\lambda}{\partial\mu}$ is negative when $\lambda$ is above the fixed point, therefore driving $\lambda$ towards $\lambda_F$. We then see that $\lambda$ will be driven towards $\lambda_F$ as $\mu$ increases: such a fixed point is called ultraviolet stable, because $\lambda$ will approach the value $\lambda_F$ asymptotically as $\mu \to \infty$, from above or from below depending on the starting point $\lambda_0$ which can be either above or below $\lambda_F$. If there were a field theory for which $\beta$ behaved as in the curve then at very short distances $\lambda$ would be more and more like $\lambda_F$. If $\lambda_F$ were small, it would mean that if we start from a small $\lambda < \lambda_F$ we never leave perturbation theory! Alternatively if we started with $\lambda > \lambda_F$ then as the distance decreased, $\lambda$ would be driven to a perturbation region. These two situations are depicted below

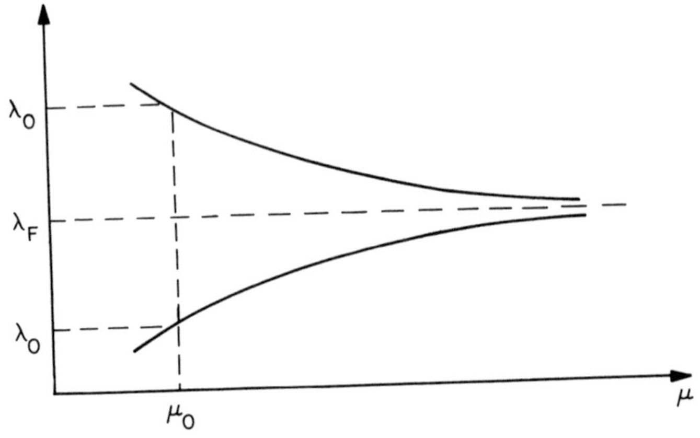


No field theory in four dimensions exhibits this behavior perturbatively ($\lambda_F << 1$).

The point $\lambda = 0$ is a fixed point for which $\beta'(0) > 0$, which means that above it $\mu \frac{\partial\lambda}{\partial\mu}$ is positive driving $\lambda$ away from it as the distance decreases. Such a fixed point is called infrared stable. Finally we note that for small $\lambda$ most field theories behave this way, with $\beta(\lambda)$ starting out positive.

3. $\beta(\lambda)$ starts out negative for low $\lambda$, decreasing in value monotonically. This means that $\lambda$ decreases monotonically with $\ell n\, \mu$. In this case the perturbation approximation becomes better at shorter distances, and $\lambda$ is driven to zero which in this instance becomes an ultraviolet stable fixed point. Such coupling constant behavior for low $\lambda$ is exhibited by gauge theories in four dimensions - a phenomenon known as asymptotic freedom. This will be analyzed in detail when we study gauge theories.

4. $\beta(\lambda)$ starts out negative, turns over and becomes positive, crossing the axis at $\lambda_F$.

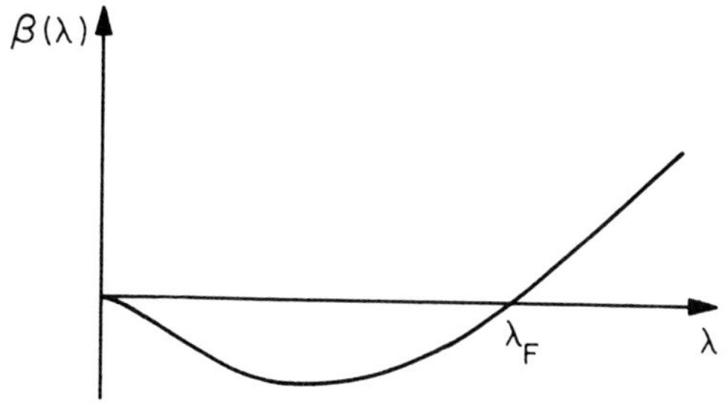


In this case, $\beta'(\lambda_F) > 0$ and $\lambda_F$ is an infrared fixed point. This means that if at $\mu_0$, $\lambda_0 < \lambda_F$, $\lambda$ will be driven towards 0, but if $\lambda_0 > \lambda_F$ it will be driven away from $\lambda_F$ to larger values of $\lambda$:

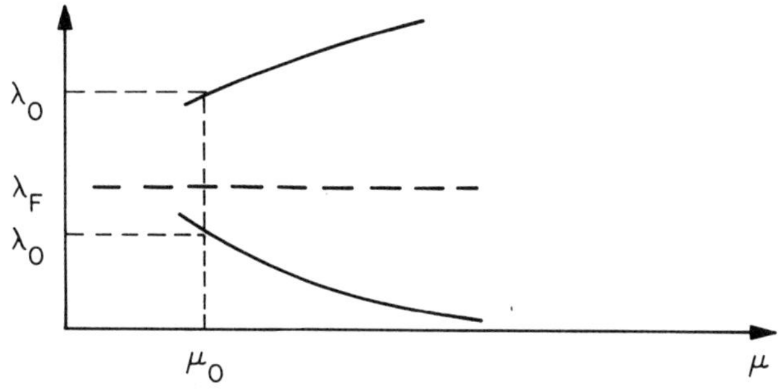


We can derive the variation of the mass with $\mu$, starting from (5.41) with $b_0 = 1$ and $b_k$ independent of $\frac{m}{\mu}$; it is not hard to find that

$$\mu \frac{\partial m^2}{\partial \mu} = 2\lambda m^2 \frac{db_1}{d\lambda} \tag{6.20}$$

so that

$$\gamma_m(\lambda) = \lambda \frac{db_1(\lambda)}{d\lambda} \tag{6.21}$$

leading to the value in $\lambda\phi^4$ theory

$$\gamma_m(\lambda) = \frac{\lambda}{16\pi^2} + \frac{7}{12}\left(\frac{\lambda}{16\pi^2}\right)^2 + \mathcal{O}(\lambda^3) \quad . \quad (6.22)$$

We also get a recursion formula for the residues of the higher poles:

$$\lambda \frac{db_{k+1}}{d\lambda} = b_k \lambda \frac{db_1}{d\lambda} - \frac{db_k}{d\lambda}\left(1 - \lambda \frac{d}{d\lambda}\right) a_1(\lambda) \quad k=1,2,\ldots \quad . \quad (6.23)$$

Finally, we can also derive from the definition of $Z_\phi$ with $c_0 = 1$ and $c_k$ independent of $\frac{m}{\mu}$ the equation

$$\mu \frac{\partial}{\partial\mu} \ell n\, Z_\phi = -2\lambda \frac{dc_1}{d\lambda} \quad , \quad (6.24)$$

leading to

$$\gamma_d(\lambda) = -\lambda \frac{dc_1}{d\lambda} \quad , \quad (6.25)$$

so that in $\lambda\phi^4$ theory

$$\gamma_d(\lambda) = \frac{1}{12}\left(\frac{\lambda}{16\pi^2}\right)^2 + \mathcal{O}(\lambda^3) \quad . \quad (6.26)$$

The $c_k$ coefficients in turn obey

$$\lambda \frac{dc_{k+1}}{d\lambda} = c_k \lambda \frac{dc_1}{d\lambda} - \left(a_1 - \lambda \frac{da_1}{d\lambda}\right) \frac{dc_k}{d\lambda} \, , \quad k=1,2,\ldots \quad . \quad (6.27)$$

These recursion relations are useful in computing the residues of the higher order poles in terms of those of the simple pole. This is where the power of the procedure resides: if the theory is renormalizable, then many coefficients can be calculated indirectly without the aid of Feynman diagrams.

In this 't Hooft-Weinberg mass independent renormalization prescription, the renormalization group equation is easy to integrate because the m dependence has dropped out of the coefficients. It now reads

$$\left[- s \frac{\partial}{\partial s} + \beta(\lambda) \frac{\partial}{\partial\lambda} + (\gamma_m(\lambda)-1) m \frac{\partial}{\partial m} + d_n - n\gamma_d(\lambda)\right] \tilde{\Gamma}^{(n)}(sp;m,\lambda,\mu) = 0 .$$

(6.28)

Introduce scale dependent variables $\bar{\lambda}(s)$ and $\bar{m}(s)$ such that

$$s \frac{\partial \bar{\lambda}(s)}{\partial s} = \beta(\bar{\lambda}(s)) \qquad \bar{\lambda}(s=1) = \lambda \tag{6.29}$$

$$s \frac{\partial \bar{m}(s)}{\partial s} = \bar{m}(s)(\gamma_m(\bar{\lambda}(s))-1) \qquad \bar{m}(s=1) = m \quad . \tag{6.30}$$

Then (6.19) is easy to integrate for it becomes a first order differential equation in s; the result is

$$\tilde{\Gamma}^{(n)}(sp;m,\lambda,\mu) = s^{-d_n} \tilde{\Gamma}^{(n)}(p;\bar{m}(s),\bar{\lambda}(s),\mu) e^{n\int_1^s \frac{ds'}{s'} \gamma_d(\bar{\lambda}(s'))} \quad . \tag{6.31}$$

The interpretation of this equation is that under a change of scale of the external momenta the Green's functions scale in an unexpected way: the coupling constant and mass scale in a nontrivial fashion and the Green's functions develop beside the engineering dimension $d_n$ an anomalous dimension $\gamma_d$ for each outside leg.

Suppose m had been set equal to zero initially. Then the classical theory would have been invariant under dilatations (and conformal transformations - see Chapter I). This is no longer true in quantum theory where the regularization technique introduces a scale, either via a high momentum cutoff or via the $\mu$-parameter of dimensional regularization, which breaks dilatation invariance [see S. Coleman's lecture at the Erice summer school 1971].

Also this shows that the Green's functions at scaled momenta give a behavior ruled by $\bar{\lambda}(s)$ and $\bar{m}(s)$ which therefore govern the physics at large scales.

It is interesting to investigate the behavior of the Green's functions at large s. Suppose the theory has an ultraviolet stable fixed point at $\lambda = \lambda_F$. At large scales $\lambda$ will be driven to $\lambda_F$. Hence $\gamma_m$ and $\gamma_d$ will be driven to $\gamma_m(\lambda_F)$ and $\gamma_d(\lambda_F)$, respectively, so that the solution of (6.21)

$$\bar{m}(s) = m e^{-s\int_1^s d\,\ell n\, s\, \gamma_m(\bar{\lambda}(s))} \tag{6.32}$$

can be integrated to

$$\bar{m}(s) \to m\, e^{+s(-1+\gamma_m(\lambda_F))} \tag{6.33}$$

if we assume that the integral is dominated by the large s range. A similar assumption about the integration over $\gamma_d$ leads to

$$\tilde{\Gamma}^{(n)}(sp;m,\lambda,\mu) \to s^{(-d_n+n\gamma_d(\lambda_F))}\tilde{\Gamma}^{(n)}(p;m\, e^{-s(1-\gamma_m(\lambda_F))},\lambda_F,\mu). \tag{6.34}$$

Thus if $1 - \gamma_m(\lambda_F)$ is positive the mass can be neglected altogether at large scales and also $\gamma_d(\lambda_F)$ appears indeed as an anomalous dimension.

We see from (6.23) that the naive expectation that the masses decouple from the theory at large momenta depends essentially on the integral over the anomalous dimensions which we write as

$$\int_\lambda^{\bar{\lambda}(s)} d\lambda' \frac{\gamma_m(\lambda')}{\beta(\lambda')} \quad . \tag{6.35}$$

This form suggests that unless $\gamma_m(\lambda)$ and $\beta(\lambda)$ have simultaneous zeros, the greater contributions to the integral will come from the fixed points, which "kind of" justifies our earlier assumption.

If the ultraviolet stable fixed point is at $\lambda_F = 0$ (as in gauge theories), then all is well because the large momentum behavior of the theory is given by perturbation theory, that is $\gamma_m(0) = 0$.

In $\lambda\phi^4$ theory we can integrate the various equations by using the lowest perturbative results found for $\gamma_m$ and $\gamma_d$. The results are

$$\bar{m}(s) = m\, e^{-s}\left(\frac{\bar{\lambda}(s)}{\lambda}\right)^{1/3} e^{\frac{7}{12}\frac{\bar{\lambda}(s)-\lambda}{16\pi^2}} \tag{6.36}$$

for the scale dependent mass and

$$\tilde{\Gamma}^{(n)}(sp) \sim s^{-d_n + \frac{n}{36}\frac{\bar{\lambda}(s)-\lambda}{16\pi^2}}\,\tilde{\Gamma}^{(n)}(p) \tag{6.37}$$

for the anomalous dimension. These results are only valid for small mass scales because as we have seen, the perturbative $\beta$ function is positive and, after a certain scale is reached perturbation theory loses its validity.

Problems.

A. Given $\beta(\lambda) = \mu \frac{\partial\lambda}{\partial\mu}$, describe the behavior of the hypothetical field theories for which

1. $\beta(\lambda_F) = \beta'(\lambda_F) = 0$

2. $\beta(\lambda_k) = 0 \qquad \lambda_k = \lambda_F + \frac{a}{k} \qquad k = 1, 2, \ldots \infty.$

B. Given

$$m_0^2 = m^2(1 + \sum_{n=1}^{\infty} \frac{b_n(\lambda)}{\varepsilon^n}) \equiv m^2 Z_m ,$$

show that

$$\gamma_m = -\frac{1}{2}\mu \frac{\partial \ln Z_m}{\partial\mu} = \lambda \frac{db_1(\lambda)}{d\lambda}$$

and that

$$\lambda \frac{db_{n+1}}{d\lambda} = b_n \lambda \frac{db_1}{d\lambda} + \frac{1}{2}\frac{db_n}{d\lambda}\beta(\lambda) \qquad n = 1, 2, \ldots .$$

If $b_n(\lambda) = b_{n,n+1}\lambda^{n+1} + \mathcal{O}(\lambda^2)$, derive a formula relating $b_{n,n+1}$ to $b_{1,2}$.

C. Given

$$\phi_0 = \phi(1 + \sum_{n=1}^{\infty} \frac{c_n(\lambda)}{\varepsilon^n})^{1/2} \equiv \phi Z_\phi^{1/2},$$

show that

$$\gamma_d(\lambda) = \frac{1}{2}\mu \frac{\partial \ln Z_\phi}{\partial\mu} = -\lambda \frac{dc_1}{d\lambda}$$

and that

$$\lambda \frac{dc_{n+1}}{d\lambda} = c_n \lambda \frac{dc_1}{d\lambda} + \frac{1}{2}\frac{dc_n}{d\lambda}\beta(\lambda) .$$

D. Verify the renormalization group equation in $\lambda\phi^4$ theory using the results from perturbation theory for $\beta$, $\gamma_m$ and $\gamma_d$.

## 7. Prescription Dependence of the Renormalization Group Coefficients

In the previous section we computed the value of the $\beta$ and $\gamma$ coefficients using the 't Hooft-Weinberg mass independent renormalization prescriptions to lowest nontrivial order in $\lambda$. We have drawn conclusions of physical import from their behavior - yet the alert reader might wonder to what extent these coefficients depend on the prescription for renormalization: for a general prescription of type A or B these coefficients will, in general, depend on masses through the finite parts of the counterterms.

Starting from eqs. (5.37) - (5.39) we find by differentiating with respect to $\mu$ at fixed bare parameters that

$$0 = 2\varepsilon[a_0 + \frac{a_1}{\varepsilon}] + \mu \frac{\partial\lambda}{\partial\gamma} [a_0' + \frac{a_1'}{\varepsilon}] + (\frac{\partial m}{\partial\mu} - \frac{m}{\mu})[\dot{a}_0 + \frac{\dot{a}_1}{\varepsilon}]$$

$$+ \text{higher poles} \qquad , \qquad (7.1)$$

from which we deduce that

$$\mu \frac{\partial\lambda}{\partial\mu} = \varepsilon(-2\lambda + \frac{3\lambda^2}{16\pi^2} G_1) + \frac{3\lambda^2}{16\pi^2} + \frac{3}{2}\frac{m}{\mu}\frac{\lambda^2}{16\pi^2}\dot{G}_1 + \mathcal{O}(\lambda^3). \qquad (7.2)$$

In these formulae, the prime and dot denote differentiation with respect to $\lambda$ and $\frac{m}{\mu}$, respectively. Comparing with (6.15), we see the explicit prescription dependence coming through $G_1$. In particular we note that the prescription dependence enters in the lowest order because of the mass. In deriving eq. (7.2) it becomes evident that the order $\lambda^3$ in (7.2) depends directly on $G_1$ without multiplicative mass factors. In a similar way we can show that

$$\gamma_m = \mu \frac{\partial \,\ell n\, m}{\partial\mu} = (\frac{1}{2}\frac{\lambda}{16\pi^2} + \frac{1}{4}\frac{m}{\mu}\frac{\lambda}{16\pi^2}\dot{F}_1) + \varepsilon\frac{1}{2}\frac{\lambda}{16\pi^2} F_1 + \mathcal{O}(\lambda^3),$$

$$(7.3)$$

as well as

$$\gamma_d = \mu \frac{\partial}{\partial\mu} \ln Z_\phi^{1/2} = \frac{1}{48} \left(\frac{\lambda}{16\pi^2}\right)^2 - \frac{m}{2\mu} \left(\frac{\lambda}{16\pi^2}\right)^2 \dot{H}_2 + \varepsilon \frac{1}{2} \left(\frac{\lambda}{16\pi^2}\right)^2 H_2$$

$$+ \mathcal{O}(\lambda^3) \quad . \quad (7.4)$$

We let $\varepsilon \to 0$ and finally obtain

$$\beta(\lambda, \frac{m}{\mu}) = \frac{3\lambda^2}{16\pi^2} + \frac{3}{2} \frac{m}{\mu} \frac{\lambda^2}{16\pi^2} \dot{G}_1 + \mathcal{O}(\lambda^3) \quad , \quad (7.5)$$

$$\gamma_m(\lambda, \frac{m}{\mu}) = \frac{1}{2} \frac{\lambda}{16\pi^2} + \frac{1}{4} \frac{m}{\mu} \frac{\lambda}{16\pi^2} \dot{F}_1 + \mathcal{O}(\lambda^2) \quad , \quad (7.6)$$

$$\gamma_d(\lambda, \frac{m}{\mu}) = \frac{1}{48} \left(\frac{\lambda}{16\pi^2}\right)^2 - \frac{m}{2\mu} \left(\frac{\lambda}{16\pi^2}\right)^2 \dot{H}_2 + \mathcal{O}(\lambda^3) \quad . \quad (7.7)$$

These formulae indicate that when masses are present, the prescription dependence enters in the lowest order. Then it is only when the mass can be neglected that we can say that the lowest order renormalization equation coefficients are independent of the prescription. Typically when prescriptions of type A or B are used, one solves the renormalization group equation in a regime where the mass can be neglected. If $\mu$ is the renormalization point, we can take $\frac{m}{\mu}$ to be exceedingly small by choosing a very large $\mu$.

Suppose we have two renormalization prescriptions. They must be related to one another by a finite renormalization since they differ only in the definition of the renormalized parameters. Hence the parameters in one prescription will be related to those in the other as follows

$$\lambda' = T(\lambda, \frac{m}{\mu}) = \lambda + \mathcal{O}(\lambda^2) \quad , \quad (7.8)$$

$$Z_m'(\lambda', (\frac{m}{\mu})') = Z_m(\lambda, \frac{m}{\mu}) U(\lambda, \frac{m}{\mu}); \; U = 1 + \mathcal{O}(\lambda) \quad , \quad (7.9)$$

$$Z_\phi'(\lambda', (\frac{m}{\mu})') = Z_\phi(\lambda, \frac{m}{\mu}) V(\lambda, \frac{m}{\mu}); \; V = 1 + \mathcal{O}(\lambda^3) \quad . \quad (7.10)$$

In particular, it follows that, where A is some function of $\lambda$ and $\frac{m}{\mu}$,

$$\beta'(\lambda', (\frac{m}{\mu})') = \beta(\lambda, \frac{m}{\mu})A' + (\frac{m}{\mu})(\gamma_m - 1)\dot{A} \quad . \qquad (7.11)$$

In the deep Euclidean region this reduces to an equation of the form

$$\beta'(\lambda') = \beta(\lambda)A'(\lambda) \quad , \qquad (7.12)$$

which shows that a fixed point at $\lambda = \lambda_F$ results in a fixed point at $\lambda'_F = A(\lambda_F) \neq \lambda_F$. Note that this result is not perturbative. Hence, as can be expected, the presence (or absence) of a fixed point is prescription independent. It can also be shown that the sign of the first derivative of $\beta$ at a fixed point is prescription independent (see problem).

The other renormalization group parameters $\gamma_m$ and $\gamma$ also have some features which are prescription independent, notably their numerical value at a fixed point in the deep Euclidean region.

Problems.

A. Verify eqs. (7.2), (7.3) and (7.4).

B. Verify to lowest order the renormalization group equation for $\tilde{\Gamma}^{(2)}$ and $\tilde{\Gamma}^{(4)}$, keeping the finite parts $F_1$, $F_2$, $H_2$ and $G_1$.

C. Show that the sign of $\frac{d\beta}{d\lambda}$ at a fixed point $\lambda_F$ is prescription independent, neglecting masses.

*D. Relate $\gamma'(\lambda')$ and $\gamma_m'(\lambda')$ to $\gamma(\lambda)$, $\gamma_m(\lambda)$ and $\beta(\lambda)$ where the prime and unprimed system refers to two mass independent renormalization schemes [F, G and H functions are first taken independent of $\frac{m}{\mu}$, but can have numerical values]. Show that the value of $\gamma$ and $\gamma_m$ at $\lambda_F$ is prescription independent.

## 8. Continuation to Minkowski Space; Analyticity

At this stage we have obtained the finite Green's functions at the price of introducing an arbitrary scale. However, we know from the renormalization group that if we alter this scale, nothing happens to the Green's functions because the change is compensated by a concomitant change in the renormalized parameters and field. There now remains to express the Green's functions in Minkowski space in order to make contact with reality.

This is achieved by means of analytic continuation. Consider an Euclidean Green's function which depends on momenta $\bar{p}_1, \ldots, \bar{p}_N$. We first change all the time components of the $\bar{p}$ by making them imaginary

$$\bar{p} = (\bar{p}_0, \bar{p}_i) \rightarrow p = (p_0 = i\bar{p}_0,\ p_i = \bar{p}_i) \qquad .$$

As an example let us examine what happens to the propagator. According to the above

$$\frac{1}{\bar{p}^2+m^2} \rightarrow \frac{1}{-p^2+m^2} \qquad ,$$

since we are using the Minkowski metric $g_{00} = -g_{ii} = +1$. This replacement is not entirely satisfactory because the Minkowski space expression develops a pole at $p^2 = m^2$, i.e., when

$$p_0 = \pm \sqrt{\vec{p}\cdot\vec{p}+m^2} \qquad .$$

The continuation process can be regarded as proceeding from the imaginary to the real axis in the $p_0$ plane in a clockwise fashion; indeed the ability to perform this continuation rests on the avoidance of any pole. It follows that the poles in $p_0$ must be taken to be just below (above) the positive (negative) real axis, that is the continuation is from

$$\frac{1}{p^2+m^2} \quad \text{to} \quad \frac{-1}{p^2-m^2+i\varepsilon} \qquad ,$$

where $\varepsilon > 0$, and the limit $\varepsilon \to 0^+$ is to be taken at the end of all calculations. In more complicated cases the poles will be chosen so that their locations do not interfere with the clockwise rotation of the imaginary time axis into the real time axis. In this $-i\varepsilon$ prescription we recognize the familiar device used to make the path integral convergent which we discussed earlier. Another way to look at this is to compare the Feynman rules in Minkowski and Euclidean spaces, say for $\phi^4$ theory:



$$\begin{cases} \dfrac{1}{\bar{p}^2+m^2} & \text{(Euclidean)} \\[2ex] \dfrac{i}{p^2-m^2+i\varepsilon} & \text{(Minkowski)} \end{cases} \tag{8.1}$$

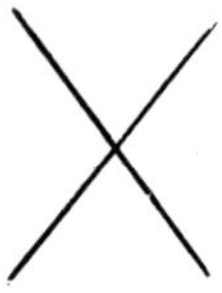

$$\begin{cases} -\lambda & \text{(Euclidean)} \\ -i\lambda & \text{(Minkowski)} \end{cases} \tag{8.2}$$

$$\text{loop integration:} \begin{cases} \dfrac{d^4\bar{k}}{(2\pi)^4} & \text{(Euclidean)} \\[2ex] \dfrac{d^4k}{(2\pi)^4} & \text{(Minkowski)} \end{cases} \quad . \tag{8.3}$$

Consider a Feynman diagram with L loops, V vertices and I internal lines. The difference between their computation in Minkowski and Euclidean space will be: a factor of (-i) for each vertex and propagator, and a factor of i for each loop because $d^4k = i d^4\bar{k}$. In addition $m^2$ in the Euclidean Green's function is replaced by $m^2 - i\varepsilon$ in the corresponding Minkowski space expression. Thus we arrive at

$$G_M^{(n)}(p_1,\ldots,p_n;m^2) = (i)^{L+V+I}(-1)^I \, G_E^{(n)}(\bar{p}_1 = p_1,\ldots,\bar{p}_n = p_n;m^2-i\varepsilon) \quad , \tag{8.4}$$

where we replace in the Euclidean function the momenta by their Minkowski space continuation, i.e., $\bar{p} = (\bar{p}_0, \bar{p}_i)$ is replaced by $p = (i\bar{p}_0, \bar{p}_i)$. Using the topological relation $L = I - V + 1$, $(V \neq 0)$, we obtain

$$G_M^{(n)}(p, m^2) = (i) G_E^{(n)}(p = \bar{p}; m^2 - i\varepsilon) \qquad , \qquad (8.5)$$

valid for any Feynman diagram with $V \neq 0$. The exception to this rule is the original propagator expression $(V = 0)$ for which $i$ is replaced by $-i$, as can be seen from (8.4) by putting $L = V = 0$, $I = 1$.

As an example of the procedure, consider the four-point function in Minkowski space. We have

$$\tilde{\Gamma}_M^{(4)}(p_1,p_2,p_3,p_4) = i[-\lambda - \frac{\lambda^2}{2.16\pi^2} \int_0^1 dx \, \ell n \, [\frac{m^2 - i\varepsilon - sx(1-x)}{m^2 - i\varepsilon + M^2 x(1-x)}]$$
$$+ \text{ t- and u-channels}] \qquad . \qquad (8.6)$$

In the above we have multiplied the Euclidean expression by $i$, replaced $\bar{p}^2$ by $-p^2$, $m^2$ by $m^2 - i\varepsilon$. The subtraction was performed on the Euclidean Green's function at $\bar{p}_i \bar{p}_j = M^2(\delta_{ij} - 1/4)$ so that $\tilde{\Gamma}_E^{(4)} = -\lambda$ at this symmetric point (prescription B). Note that the subtraction point appears as a parameter and it does not get changed in the continuation process. The $-i\varepsilon$ in the denominator is not necessary since it can never vanish. However, the numerator does suffer a change of sign for some value of $s = (p_1+p_2)^2$. When this occurs the logarithm develops a cut in the complex s-plane. Consider the argument of the log

$$F(s,x) \equiv m^2 - i\varepsilon - sx(1-x) \qquad . \qquad (8.7)$$

Now $x(1-x)$ is positive definite and varies between 0 and 1/4. Thus the least value of $s$ for which $F$ vanishes is

$$s_0 = 4m^2 \qquad (8.8)$$

where $x(1-x)$ assumes its largest value. At this point $\tilde{\Gamma}^{(4)}$ develops a branch point. Traditionally one attaches to it a cut in the complex s-plane extending from $4m^2$ to $+\infty$ along the positive real s-axis. In a similar way the u- and t-channel contributions give cuts starting at $u_0$ and $t_0 = 4m^2$. In view of the relation

$$s + t + u = 4m^2 \quad , \tag{8.9}$$

these cuts are not all independent. These branch points can be physically understood if we interpret $\tilde{\Gamma}^{(4)}$ as the scattering amplitude for two particles with momenta $p_1$ and $p_2$ to scatter into two particles of momenta $p_3$ and $p_4$:

$$p_1 + p_2 \rightarrow p_3 + p_4 \quad .$$

Assuming that incoming and outgoing particles are on their "mass-shells"

$$p_a^2 = m^2 \qquad a = 1, 2, 3, 4 \quad ,$$

it is easy to see that $s_0 = 4m^2$ corresponds to the two particles having their minimum energies $E_a = m$. It is only beyond $s_0$ that the two particles have enough energy to scatter nontrivially into two others. Consequently $s_0$ is called a physical two-particle threshold. Below it $\tilde{\Gamma}^{(4)}$ is a real function of its arguments, but it develops an imaginary part for $s > s_0$.

To recapitulate: by continuing $\tilde{\Gamma}^{(4)}$ in Minkowski space, we see a nontrivial analytical structure emerging; this is of course the structure demanded by unitarity and causality which will enable us to regard the Minkowski space Green's functions as transition amplitudes.

Another example of a nontrivial analytic structure emerging as a result of continuation in Minkowski space is given by the "setting sun" diagram. There it is easy to see that the best way to find the branch points is to look at the argument of the logarithm in the parametric integral. In this case the argument of interest is

$$A = -y(1-y)p^2 + m^2(1-y + \frac{y}{x(1-x)}) \qquad . \qquad (8.10)$$

It will vanish when

$$p^2 = m^2(\frac{1}{y} + \frac{1-y}{x(1-x)}) \qquad , \qquad (8.11)$$

and the location of the branch point will be given by the <u>least</u> such value of $p^2$. In order to find its value we have to extremize the parametric expression that multiplies $m^2$. In general for the two-point function, branch points will appear at the minimum value of $p^2$ for which

$$p^2 = m^2 f(x_1, x_2, \ldots, x_N) \qquad (8.12)$$

where $x_1 \ldots x_N$ are the Feynman parameters needed for an N-loop diagram. The branch point is then located at

$$p^2 = m^2 f(x_1^0, \ldots, x_N^0) \qquad , \qquad (8.13)$$

where the points $x_i^0$ are determined by

$$\frac{\partial f}{\partial x_i} = 0 \qquad \text{at } x_i = x_i^0 \qquad . \qquad (8.14)$$

[it must be checked that $x_i = x_i^0$ is, in fact, a minimum]. Such equations are called the Landau equations after Landau who introduced a systematic procedure to hunt for the branch points of Feynman diagrams. Applying this procedure to our case, we find from (8.14) that the minimum occurs at

$$x = \frac{1}{2} \qquad y = \frac{1}{3} \qquad ,$$

so that the branch point is located at

$$p^2 = 9m^2 \qquad . \qquad (8.15)$$

If you recall the form of the "setting sun",  we see that it corresponds to the minimum energy needed to excite three particles so it is called the three-particle threshold.

Hence the propagator has in Minkowski space the following singularity structure: - a pole at $p^2 = m^2$ (appropriately displaced by the $i\varepsilon$ prescription, - a branch point at $p^2 = 9m^2$ with a cut taken traditionally along the real positive $p^2$ axis extending to $p^2 = +\infty$. When higher orders of $\lambda$ are included, it is expected that branch points at higher values of $p^2$ will be encountered. This singularity structure is (of course) consistent with the interpretation of $G^{(2)}$ as a propagator.

Problems.

*A. Using diagrams and physical arguments, find the location of branch points in $\tilde{\Gamma}^{(4)}$ including $\mathcal{O}(\lambda^4)$.

*B. Repeat problem A for the propagator.

*C. Show that $\tilde{\Gamma}^{(4)}$ satisfies a dispersion relation that expresses its real part in terms of its imaginary part. [Use only its perturbative value up to $\mathcal{O}(\lambda^2)$.]

## 9. Cross-Sections and Unitarity

We are now almost at the end of the road. We are about to identify the Minkowski space Green's functions with transition amplitudes. However, not all functions can be transition amplitudes for they must satisfy certain requirements, notably those of unitarity and causality. As you might expect the Green's functions of the previous section are acceptable candidates.

In order to state precisely the requirements, let us review the S-matrix formalism and apply it to $\lambda\phi^4$ theory.

Suppose it were possible to define states far away from the region of interaction; in particular in the very distant past or future. Such a concept clearly makes sense in the case of short range forces, as for instance in weak and strong interactions. When long range forces are involved, the concept is trickier and special care must be exercised in the definition of such states. Schematically let the states be described by kets $|\alpha;\pm T\rangle$ where T is a very large time and $\alpha$ represents a complete set of observables. These states obey completeness and orthogonality

$$\sum_{\alpha} |\alpha,\pm T\rangle\langle\pm T,\alpha| = 1 \tag{9.1}$$

$$\langle\alpha,\pm T|\beta,\pm T\rangle = \delta_{\alpha\beta} \quad . \tag{9.2}$$

If the system were a harmonic oscillator, $\alpha$ would denote the occupation number, etc. It is crucial to note that these relations hold only for a given time and therefore involve no dynamics, only kinematics. If, when T is large, the interaction can be turned off (short range forces) the states should be easy to recognize because they diagonalize the unperturbed Hamiltonian.

In $\lambda\phi^4$ theory, when $m^2 \neq 0$, there is no difficulty in recognizing these states. They are made out of the one-particle Wigner states labeled by m and $\vec{p}$ with the energy identified with $+\sqrt{\vec{p}^2+m^2}$. If we adopt a more relativistic-looking notation and denote those states by $|p\rangle$, where p stands for the momentum four vector, they are required to satisfy

$$\int \frac{d^4p}{(2\pi)^3} |p\rangle\, \theta(p_0)\delta(p^2-m^2)\langle p| = 1 \quad , \qquad (9.3)$$

$$\langle p|p'\rangle = 2(2\pi)^3\sqrt{\vec{p}^2+m^2}\;\delta(\vec{p}-\vec{p}\,') \quad . \qquad (9.4)$$

Then any multiparticle state will be a superposition of noninteracting one-particle states:

$$|\alpha,\pm\infty\rangle \sim |p_1,p_2,\ldots,p_n\rangle = |p_1\rangle \otimes |p_2\rangle \cdots \otimes |p_n\rangle \quad . \qquad (9.5)$$

In $\lambda\phi^4$ theory there is some justification for believing these states describe the asymptotic states because the large scale behavior of the coupling is such that the free Feynman propagator accurately describes signal propagation, and we know that $\Delta_F$ does propagate one particle states of the type described above. [Here we are a bit cavalier since strictly speaking $\Delta_F$ propagates both positive and negative energy states.] We note in passing that if the large distance behavior of the coupling constant were such that it grew with distance, the identification of asymptotic states would have had to be made only for constructs that can escape this formidable force. This is supposedly the case for QCD where quarks are subject to such a force. Hence they cannot serve as asymptotic states. However, this force only attacks objects with color and allows for the definition of asymptotic states that have no color (hadrons).

A question of physical interest is the computation of the transition amplitude

$$T_{\alpha\beta} = \langle\alpha,+\infty|\beta,-\infty\rangle \quad . \qquad (9.6)$$

With Heisenberg we define an S-matrix with the property

$$|\beta,+\infty\rangle = \hat{S}|\beta,-\infty\rangle \quad . \tag{9.7}$$

Its job is to contain all the dynamical information about the evolution of the physical states in time. Completeness of the states at $+\infty$ and $-\infty$

$$1 = \sum_\beta |\beta,+\infty\rangle\langle+\infty,\beta| = \sum_\beta \hat{S}|\beta,-\infty\rangle\langle-\infty,\beta|\hat{S}^+$$

$$= \hat{S}\hat{S}^+ \quad , \tag{9.8}$$

implies that $\hat{S}$ is unitary. [You can show that $\hat{S}^+\hat{S} = 1$ as well.] In physical terms the unitarity of S means that the system cannot disappear into nothing [black holes?]. Most of the time nothing will happen when states are scattered - they are much more likely to miss each other than to interact. For this reason we set

$$\hat{S} = 1 + i\hat{R} \quad , \tag{9.9}$$

with R containing the interesting information. Hence it follows that

$$T_{\alpha\beta} = \langle\alpha,-\infty|\hat{S}^+|\beta,-\infty\rangle \tag{9.10}$$

$$= \delta_{\alpha\beta} - i\,\langle\alpha,-\infty|\hat{R}^+|\beta,-\infty\rangle \quad . \tag{9.11}$$

Since the interaction is Lorentz invariant, it is convenient to take this into account and write

$$T_{\alpha\beta} = \delta_{\alpha\beta} - i(2\pi)^4\delta^{(4)}(p_\alpha-p_\beta)\langle\alpha,-\infty|\hat{T}^+|\beta,-\infty\rangle \quad , \tag{9.12}$$

where $p_\alpha(p_\beta)$ is the sum of the momenta in the final (initial) state. The transition probability over all of space-time is then given by

$$\omega_{\alpha\beta} = [(2\pi)^4\delta^{(4)}(p_\alpha-p_\beta)]^2\langle\alpha,-\infty|\hat{T}^+|\beta,-\infty\rangle\langle\beta,-\infty|\hat{T}|\alpha,-\infty\rangle. \tag{9.13}$$

The square of the $\delta$ function is quickly understood since $(2\pi)^4\delta^{(4)}(0)$ is the volume of space-time [as can be seen by putting the system in a box].

It follows that the transition probability per element of space-time is

$$\Omega_{\alpha\beta} = (2\pi)^4 \delta^{(4)}(p_\alpha - p_\beta) |\langle\alpha|\hat{T}|\beta\rangle|^2 \quad . \tag{9.14}$$

This form is valid for states that satisfy (9.2). In our case the momentum states are not normalized to 1 but according to (9.4). Hence we find, after dividing by the normalization, that

$$\Omega(p_\alpha | p_\beta) \equiv \Omega p_\alpha p_\beta = \frac{(2\pi)^4 \delta^{(4)}(p_\alpha - p_\beta)}{(2E_\alpha)(2E_\beta)} |\langle\alpha|\hat{T}|\beta\rangle|^2 \quad , \tag{9.15}$$

where $E_\alpha (E_\beta)$ stands for the product of the energies in the $\alpha$-($\beta$-) state, each energy being given by

$$E_i = \sqrt{\vec{p}_i^{\,2} + m^2} \quad . \tag{9.16}$$

In scattering experiments one is usually interested in the scattering cross-section of two particles (target and projectile) into many. It is given by

$$d\sigma(a+b \to 1+2+..+N) = \frac{1}{v_{ab}} \Omega(p_a, p_b | p_1, p_2, .., p_N) \frac{d^3p_1 d^3p_2 .. d^3p_N}{(2\pi)^{3N}} \quad , \tag{9.17}$$

where $v_{ab}$ is the relative velocity of particles a and b. For equal mass particles it is conveniently expressed as

$$v_{ab} = \frac{\sqrt{(p_a \cdot p_b)^2 - m^4}}{E_a E_b} \quad . \tag{9.18}$$

Putting it all together we obtain

$$d\sigma(a+b \to 1+2+..+N) = \frac{(2\pi)^4 \delta^{(4)}(p_a + p_b - p_1 - \cdots - p_N)}{4\sqrt{(p_a \cdot p_b)^2 - m^4}} |\langle p_a p_b|\hat{T}|p_1 \cdots p_N\rangle|^2 \prod_{i=1}^{N} \frac{d^3p_i}{2(2\pi)^3 E_i} \quad . \tag{9.19}$$

Note that the measure $\frac{d^3p}{2E}$ is relativistic since

$$\frac{d^3p}{2E} = d^4p\theta(p_0)\delta(p^2-m^2) \quad . \tag{9.20}$$

A special case of interest is elastic scattering where N = 2. Define the center of mass frame where

$$\vec{p}_a + \vec{p}_b = \vec{p}_1 + \vec{p}_2 = 0 \quad . \tag{9.21}$$

Then simple kinematics yields

$$d\sigma(a+b\to 1+2) = \frac{|T|^2}{64\pi^2 s}\, d\Omega \quad , \tag{9.22}$$

where $d\Omega = d\phi d(\cos\theta)$, $\theta$ being the angle between the ingoing and outgoing directions:

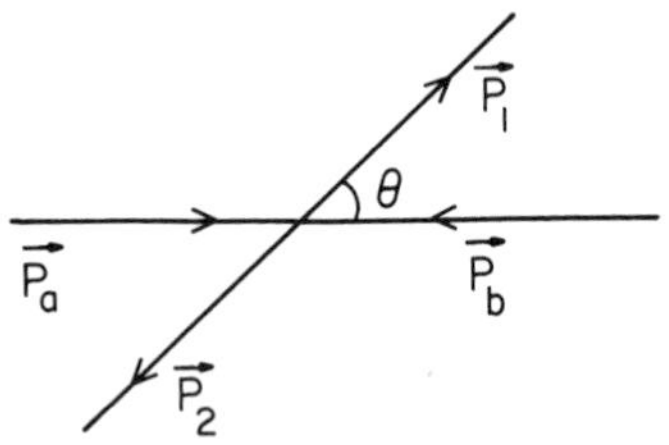


and s is Mandelstam's variable

$$s = (p_a+p_b)^2 \quad .$$

As you might have expected the renormalized Green's functions will be identified with the matrix elements of $\hat{T}$. Hence it is important to translate the requirements of unitarity into conditions on T and verify that they are met by our identification.

From the unitarity condition on $\hat{S}$, it follows that

$$\hat{R} - \hat{R}^+ = i\hat{R}^+\hat{R} = i\hat{R}\hat{R}^+ \quad . \tag{9.23}$$

This operator equation summarizes the restrictions on R due to unitarity.

For instance, let us take its matrix element between two particle states, labeled $|1,2\rangle$ and $\langle 3,4|$:

$$\langle 3,4|\hat{R}|1,2\rangle - \langle 3,4|\hat{R}^{+}|1,2\rangle = i\langle 3,4|\hat{R}\hat{R}^{+}|1,2\rangle \quad . \tag{9.24}$$

It is not hard to see that when the external particles are spinless

$$\langle 3,4|\hat{R}^{+}|1,2\rangle = \langle 1,2|\hat{R}^{+}|3,4\rangle \quad . \tag{9.25}$$

Using

$$\langle 1,2|\hat{R}^{+}|3,4\rangle = (\langle 3,4|\hat{R}|1,2\rangle)^{*} \quad , \tag{9.26}$$

we arrive at

$$2\mathrm{Im}\langle 3,4|\hat{R}|1,2\rangle = \langle 3,4|\hat{R}\hat{R}^{+}|1,2\rangle \quad . \tag{9.27}$$

Now the RHS of this equation can be rewritten by introducing a set of intermediate states. Since we want to limit ourselves to interactions that involve an even number of states (invariance under $\phi \to -\phi$), the lowest energy intermediate state is the two particle state $|a,b\rangle = |a\rangle|b\rangle$. Hence, using (9.3), we arrive at

$$2\mathrm{Im}\langle 3,4|\hat{R}|1,2\rangle = \int \frac{d^4a\, d^4b}{(2\pi)^6}\, \theta(a_0)\theta(b_0)\delta(a^2-m^2)\delta(b^2-m^2)\langle 3,4|\hat{R}|a,b\rangle\langle a,b|\hat{R}^{+}|1,2\rangle + \ldots \tag{9.28}$$

where ... denotes the sum over 4-, 6-, ... particle intermediate states. In terms of the T-matrix of (9.12), this equation becomes

$$2\mathrm{Im}\langle 3,4|\hat{T}|1,2\rangle = \int \frac{d^4a\, d^4b}{(2\pi)^2}\, \theta(a_0)\theta(b_0)\delta(a^2-m^2)\delta(b^2-m^2)\delta(a+b-1-2)$$
$$\langle 3,4|\hat{T}^{+}|a,b\rangle\langle a,b|T|1,2\rangle + \ldots \quad . \tag{9.29}$$

Since both a and b are on their mass-shells, the RHS will be nonzero only when the $|1,2\rangle$ initial state has enough energy to produce the $|a,b\rangle$ intermediate state, that is when $s = (p_1+p_2)^2 \geq 4m^2$. Thus we see that as a consequence of unitarity and completeness the T matrix elements are real for $s < 4m^2$ and acquire an imaginary part after the two-particle threshold is crossed, and for all other higher thresholds there is a further contribution to the imaginary part of the matrix element of T.

Let us now compare this with the four-point function obtained from perturbation theory.

$$\tilde{\Gamma}^{(4)} = -i[\lambda + \frac{\lambda^2}{2\cdot 16\pi^2}\int dx\, \ell n\, [\frac{m^2-i\varepsilon-sx(1-x)}{m^2+M^2x(1-x)}] + (s\to t) + (s\to u) + \mathcal{O}(\lambda^3)]. \quad (9.30)$$

We have seen that the parametric integral does develop an imaginary part and has a branch point at $s = 4m^2$, all of it consistent with the unitarity equation (9.29). So we are led to identify Green's functions with (-i) times the T matrix. In this case

$$\tilde{\Gamma}^{(4)}(1,2,3,4) = -i\langle 3,4|\hat{T}|1,2\rangle \quad . \quad (9.31)$$

We can easily check that this is true: we have already calculated the imaginary part of $\tilde{\Gamma}^{(4)}$; it is of order $\lambda^2$ and appears only for $s \geq 4m^2$. On the other hand, we can calculate it from the unitarity equation (9.29), by putting in its RHS the lowest perturbation vertex, leading to

$$2\mathrm{Im}(i\tilde{\Gamma}^{(4)}) = \frac{\lambda^2}{(2\pi)^2}\int d^4a\; d^4b\,\theta(a_0)\theta(b_0)\delta(a^2-m^2)\delta(b^2-m^2)\delta(1+2-a-b)+\mathcal{O}(\lambda^4). \quad (9.32)$$

We leave its verification to the reader. This yields only the lowest order contribution to the imaginary part - the 4-, 6-, ... particle thresholds will add to the imaginary part but in higher orders of $\lambda$.

With this identification between Green's functions and scattering amplitudes there emerges a new way of computing the imaginary part of diagrams using the unitarity equation. It is useful in perturbation theory because of the quadratic nature of the RHS of (9.29); for if one has computed $\langle a,b|T|1,2\rangle$ to order $\lambda^k$, it will give the imaginary part to order

$\lambda^{k+1}$. This remark is important because of the optical theorem which relates the imaginary part of the forward scattering amplitude to the total cross section. This theorem is simply obtained by putting $|3,4\rangle = |1,2\rangle$ in (9.29) and comparing the RHS with the integrated form of (9.22).

We can arrive at a diagrammatic representation of the unitarity constraint, if we recall the meaning of the Feynman propagator. We have

$$\Delta_F(x) = \frac{i}{(2\pi)^4} \int d^4k \, \frac{e^{ikx}}{k^2-m^2+i\varepsilon}$$

$$= \theta(x_0) \int \frac{d^4k}{(2\pi)^3} \, \theta(-k_0)\delta(k^2-m^2)$$

$$+ \theta(-x_0) \int \frac{d^4k}{(2\pi)^3} \, \theta(+k_0)\delta(k^2-m^2) \quad ,$$

which expresses the fact that $\Delta_F$ includes propagation of both positive and negative energy states depending on the sign of $x_0$. Now if we invent a new set of rules where the full propagator $\Delta_F$ is replaced by $\theta(k_0)\delta(k^2-m^2)$, we can arrive at a pictographic way to compute imaginary parts and therefore total cross-sections. This new rule would apply only in Minkowski space, of course. While we have the old rule

$$\text{———} \Rightarrow \frac{i}{p^2-m^2+i\varepsilon} \quad ,$$

invent a new one for the cut propagator

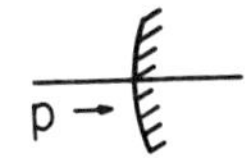


$$\rightarrow (2\pi)\theta(k_0)\delta(p^2-m^2) \quad .$$

Note that the cut propagator is not symmetric, as the shading shows. The reason is that since we have to compute $TT^+$ on the RHS of (9.29), portions of the diagram to the left of the cut must correspond to the conjugate of diagrams to the right of it although they may be different. The interested reader is referred to Diagrammar by 't Hooft and Veltman, in Particle Interactions at Very High Energies, part B; D. Speiser et al., editors, Plenum Press, N.Y. 1974, for more details.

Thus our equation (9.32) would read diagrammatically

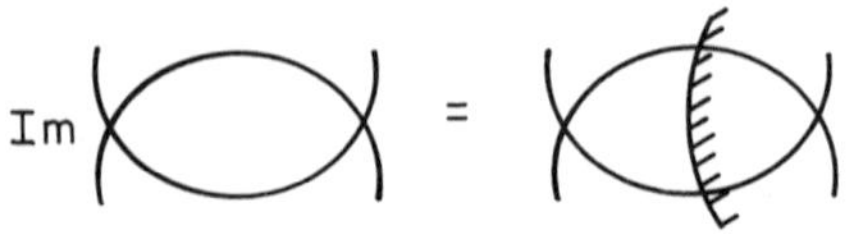


The end result is that one can derive general cutting equations that express the imaginary part of diagrams in terms of the sum of all the possible cuts. [This is not as bad as it sounds since many cut diagrams vanish by energy conservation for cutting a Feynman propagator restricts the energy flow to one direction.]

This concludes our study of perturbative $\lambda\phi^4$ theory.

Problems.

A. Show that $\frac{d\sigma}{d\Omega} = \frac{1}{s} \frac{|T|^2}{64\pi^2}$ starting from eq. (9.19).

B. Show that for elastic scattering of spinless particles

$$\langle 3,4|\hat{S}|1,2\rangle = \langle 1,2|\hat{S}|3,4\rangle .$$

C. Compute $\mathrm{Im}\, i\tilde{\Gamma}^{(4)}$ by means of the unitarity equation and compare with the result previously obtained from perturbation theory.

D. Show that in general for $\varepsilon > 0$

$$\frac{1}{x+i\varepsilon} = -i\pi\delta(x) + P(\frac{1}{x}),$$

where $P(\frac{1}{x})$ is the Cauchy principal part of $\frac{1}{x}$ defined by

$$P(\frac{1}{x}) = -i \int_{-\infty}^{+\infty} dy\ e^{ixy}[\theta(y)-\theta(-y)].$$

*E. Compute the imaginary part of the setting sun by using the unitarity equation

Im[i 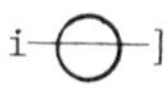 ] = 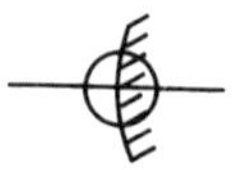

**F. Given $\mathcal{L}_E = \frac{1}{2}\partial_\mu\phi\partial_\mu\phi + \frac{1}{2}m^2\phi^2 + \frac{h}{3!}\phi^3 + \frac{\lambda}{4!}\phi^4$,

a) derive the Feynman rules

b) find the change of m, h and $\lambda$ with scale to $\mathcal{O}(\hbar)$

c) solve the equations of b) and interpret the result physically.

You may use any renormalization prescription, but the mass independent prescription is strongly advised.

## V. PATH INTEGRAL FORMULATION WITH FERMIONS

### 1. Integration over Grassmann Numbers

In Chapter I we gave several examples of Action functionals involving Fermi fields, that is, fields transforming as half-integer spin representations of the Lorentz group. It was then pointed out that the Fermi fields should be taken to be anticommuting classical fields, and that this was a classical identification which did not imply any quantization. If we reason by analogy with the quantization of, say, the scalar field, we are led to considering a "path" integral over anticommuting fields. This can at best be a formal concept devoid of any direct physical meaning, but as is usual with such things, the final answer will be of interest, although the method of derivation might raise a few eyebrows!

To start with, consider the case of one Grassmann (anticommuting) "variable" $\theta$. It satisfies [$\{,\}$ denotes the anticommutator]

$$\{\theta,\theta\} = 0 \qquad \text{or} \qquad \theta^2 = 0 \quad . \tag{1.1}$$

One defines the differential operator $\frac{d}{d\theta}$ by means of

$$\{\frac{d}{d\theta},\theta\} = 1 \quad . \tag{1.2}$$

Any function of $\theta$, $f(\theta)$, will have a simple expansion

$$f(\theta) = a + \beta\theta \quad , \tag{1.3}$$

which terminates because of (1.1). For convenience take $\beta$ to be of Grassmann type and a to be a real commuting number. [From now on Grassmann variables will be denoted by Greek letters.] Then it follows that

$$\frac{df}{d\theta} = -\beta \quad , \tag{1.4}$$

so that

$$\frac{d^2f}{d\theta^2} = 0 \tag{1.5}$$

where in the last two equations we have taken $\frac{d}{d\theta}\, a = \{\frac{d}{d\theta}, \beta\} = 0$. From (1.5) we see that

$$\{\frac{d}{d\theta}, \frac{d}{d\theta}\} = 0 \quad , \tag{1.6}$$

which means that there is no inverse differentiation. This is awkward because one often likes to think of integration and differentiations as inverse operations. So we are warned that integration has to be introduced in formal terms. It is defined to be an operation denoted by $\int d\theta\ldots$ with the properties

$$\int d\theta = 0 \qquad \int d\theta\theta = 1 \quad ; \tag{1.7}$$

it acts exactly like differentiation. This choice permits the integration to satisfy the criterion of invariance under a translation of the integration variable by a constant.

This world of one Grassmann variable is rather dull, so let us consider N Grassmann variables $\theta_i$ $i = 1,..N$ which obey

$$\{\theta_i, \theta_j\} = 0 \qquad i,j = 1\ldots N \quad . \tag{1.8}$$

Introduce their respective derivative operators $\frac{\partial}{\partial\theta_i}$ by means of

$$\{\frac{\partial}{\partial\theta_i}, \theta_j\} = \delta_{ij} \tag{1.9}$$

and

$$\{\frac{\partial}{\partial\theta_i}, \frac{\partial}{\partial\theta_j}\} = 0 \quad . \tag{1.10}$$

Any normal (i.e., non-Grassmann) function of the $\theta_i$'s can be written as

$$f(\theta_i) = a + \beta_i\theta_i + c_{ij}\theta_i\theta_j + \ldots + c\theta_1\theta_2\ldots\theta_N \quad , \qquad (1.11)$$

where the last coefficient is Grassmann or normal depending on N and the nature of f. Integration is defined in the same way as for one variable

$$\int d\theta_i = 0 \qquad \int d\theta_i\theta_i = 1 \qquad \text{(i not summed)} \quad . \qquad (1.12)$$

When the measure of integration and integrand involve more than one variable we conventionally perform the integrations according to a nested procedure. Thus for instance

$$\int d\theta_1 d\theta_2\theta_1\theta_2 = -\int d\theta_1(d\theta_2\theta_2)\theta_1 = -1 \quad . \qquad (1.13)$$

For instance, consider the integral

$$I_N(M) = \int d\theta_1\ldots d\theta_N \; e^{-\theta^T M\theta} \quad , \qquad (1.14)$$

where M is an antisymmetric N x N matrix with normal elements, $m_{ij}$, and the exponential is defined according to its power series. When N = 2 we have

$$I_2(M) = \int d\theta_1 d\theta_2[1-2m_{12}\theta_1\theta_2] \qquad (1.15)$$

$$= 2m_{12} = 2\sqrt{\det M} \quad . \qquad (1.16)$$

When N is odd one can show that I vanishes; this is consistent with identifying I with the square root of the determinant, since the determinant of odd-dimensional antisymmetric matrices vanishes. To guess at the general formula, let us examine the case N = 4. It is easy to see that the relevant terms in the expansion of the exponential are

$$e^{-\theta^T M\theta} = \ldots + \frac{1}{2!}(\theta^T M\theta)^2 + \ldots$$

$$= 4\theta_1\theta_2\theta_3\theta_4[m_{12}m_{34}-m_{13}m_{24}+m_{14}m_{23}] + \ldots \quad , \qquad (1.17)$$

leading to

$$I_4(M) = 4[m_{12}m_{34} - m_{13}m_{24} + m_{14}m_{23}] \qquad (1.18)$$

$$= 4\sqrt{\det M} \quad , \qquad (1.19)$$

so that the general formula turns out to be

$$I_N(M) = (2)^{N/2}\sqrt{\det M} \quad . \qquad (1.20)$$

This is the first formula of interest. Compare it with the equivalent formula for boson (normal) fields where the square root of the determinant appears in the denominator.

Secondly, let us consider

$$I_N(M;\vec{\chi}) \equiv \int d\theta_1 .. d\theta_N \; e^{-\theta^T M\theta + \chi_i\theta_i} \quad , \qquad (1.21)$$

where the $\chi_i$ are Grassmann numbers

$$\{\chi_i,\chi_j\} = 0, \qquad \{\chi_i,\theta_j\} = 0 \quad . \qquad (1.22)$$

To simplify matters, let us evaluate (1.21) directly when N = 2. There

$$e^{-\theta^T M\theta + \chi^T\theta} = 1 - 2m_{12}\theta_1\theta_2 - \chi_1\chi_2\theta_1\theta_2 \quad , \qquad (1.23)$$

so that

$$I_2(M;\vec{\chi}) = 2(m_{12} + \frac{1}{2}\chi_1\chi_2) \quad . \qquad (1.24)$$

This result could have been obtained more easily by formally completing the squares in the exponent and shifting the variable of integration as if we were dealing with normal integration, i.e., by letting

$$\theta' = \theta + \frac{1}{2}M^{-1}\chi \quad , \qquad (1.25)$$

and rewriting

$$I_N(M,\vec{\chi}) = \int d\theta_1 .. d\theta_N \; e^{-\theta'^T M\theta' + \frac{1}{4}\chi^T M^{-1}\chi} \tag{1.26}$$

$$= e^{\frac{1}{4}\chi^T M^{-1}\chi} \; I_N(M) \quad . \tag{1.27}$$

Specializing to N = 2, we arrive at (1.24). The object of this little exercise is two-fold: to derive (1.27) and to show that shifting of variables is allowed for Grassmann integration because of the definition (1.7).

The above formulae can be generalized to integration over complex Grassmann variables. As an example, let

$$\eta = \frac{1}{\sqrt{2}}(\theta_1 + i\theta_2); \qquad \eta^* = \frac{1}{\sqrt{2}}(\theta_1 - i\theta_2) \quad , \tag{1.28}$$

so that

$$d\theta_1 d\theta_2 = d\eta^* d\eta \tag{1.29}$$

and

$$\theta^T M\theta = -2i\eta^* m_{12}\eta \quad . \tag{1.30}$$

If we introduce a 1 x 1 matrix $2m_{12}$, application of (1.16) yields

$$\int d\eta^* d\eta \; e^{i\eta^* M\eta} = \det M; \qquad M = 2m_{12} \quad . \tag{1.31}$$

This formula can be generalized to N complex Grassmann numbers. In an analogous fashion, we can show that

$$\int d\eta^* d\eta \; e^{i\eta^* M\eta + i\zeta^*\eta + i\zeta\eta^*} = \det M \; e^{-i\zeta^*(M)^{-1}\zeta} \quad , \tag{1.32}$$

where $\zeta$ and $\zeta^*$ are complex Grassmann numbers. Formulae (1.27) and (1.32) are very important for evaluating path integrals over fermions coupled to external Grassmann sources.

Problems.

A. Develop a general proof for (1.20).

B. Prove (1.27) for N = 4 by explicit computation.

C. Prove (1.31) when M' is a 2 x 2 matrix.

D. Prove (1.32).

E. Show that

$$\int d\alpha d\beta \; e^{\alpha M \beta} = \det M,$$

where $\alpha$ and $\beta$ are independent Grassmann variables.

## 2. Path Integral of Free Fermi Fields

In Minkowski space there are three ways to describe free spin 1/2 particles. A) By means of the Weyl Lagrangian

$$\mathcal{L}_W = \psi_L^\dagger \sigma \cdot \partial \psi_L \quad , \tag{2.1}$$

which describes via the two component complex spinor $\psi_L$ a left-handed massless particle, together with its right-handed antiparticle (e.g., the massless left-handed neutrinos and right-handed antineutrinos), both related by the discrete CP transformation

$$\text{CP:} \qquad \psi_L \to \sigma_2 \psi_L^* \quad . \tag{2.2}$$

B) By means of the Majorana Lagrangian

$$\mathcal{L}_M = \psi_L^\dagger \sigma \cdot \partial \psi_L - \frac{im}{2} (\psi_L^\dagger \sigma_2 \psi_L + \psi_L^\dagger \sigma_2 \psi_L^*) \quad , \tag{2.3}$$

which describes a massive Weyl spinor. It is then interpreted as a spin 1/2 self-conjugate particle with spin up and spin down degrees of freedom. It can also be expressed in terms of the four component Majorana field

$$\Psi_M = \begin{pmatrix} \psi_L \\ \\ -\sigma_2 \psi_L^* \end{pmatrix} \quad , \tag{2.4}$$

in terms of which the Majorana Lagrangian becomes

$$\mathcal{L}_M = \frac{1}{2} \bar{\Psi}_M \gamma \cdot \partial \Psi_M + i \frac{m}{2} \bar{\Psi}_M \Psi_M \quad . \tag{2.5}$$

C) The Dirac Lagrangian

$$\mathcal{L}_D = \psi_L^\dagger \sigma \cdot \partial \psi_L + \psi_R^\dagger \bar{\sigma} \cdot \partial \psi_R + im(\psi_R^\dagger \psi_L + \psi_L^\dagger \psi_R) \tag{2.6}$$

describes a particle with two degrees of freedom and its distinct antiparticle (e.g., the electron and the positron). It has twice as many degrees of freedom as the Weyl or Majorana Lagrangian, and conserves P, in addition to CP. It can be conveniently expressed in terms of the four component Dirac spinor

$$\Psi_D = \begin{pmatrix} \psi_L \\ \\ \psi_R \end{pmatrix} \tag{2.7}$$

as

$$\mathcal{L}_D = \bar{\Psi}_D(\gamma\cdot\partial + im)\Psi_D \quad . \tag{2.8}$$

For each of these Lagrangians we can build a generating functional when we include sources to external coupling. The Weyl fields $\psi_L$ can be coupled to sources in the forms

$$\chi_L^T\sigma_2\psi_L + \text{h.c.} \qquad \text{and} \qquad \chi_R^\dagger\psi_L + \text{h.c.} \quad . \tag{2.9}$$

These two couplings are equivalent under the replacement $\chi_R = \sigma_2\chi_L^*$. Thus it suffices to consider only one type of source coupling.

In the Weyl case we consider the functional

$$W[\chi_L,\chi_L^\dagger] = N\int\mathcal{D}\psi_L\mathcal{D}\psi_L^\dagger\, e^{i\int d^4x[\mathcal{L}_W + i\chi_L^T\sigma_2\psi_L + i\psi_L^\dagger\sigma_2\chi_L^*]} \quad . \tag{2.10}$$

As in all free theories it can be readily evaluated. Introduce the Fourier transforms

$$\psi_L(x) = \int\frac{d^4p}{(2\pi)^2}\, e^{ip\cdot x}\,\tilde{\psi}_L(p), \qquad \text{etc.} \quad , \tag{2.11}$$

as in Chapter III. The exponent now reads

$$iS_W = -\int d^4p[\tilde{\psi}_L^\dagger(p)\sigma\cdot p\tilde{\psi}_L(p) + \tilde{\chi}_L^T(-p)\sigma_2\tilde{\psi}_L(p) + \tilde{\psi}_L^\dagger(p)\sigma_2\tilde{\chi}_L^*(-p)] \quad . \tag{2.12}$$

We rewrite it in the form

$$- \int d^4p\{[\tilde{\psi}_L^\dagger(p)+\tilde{\phi}_L^\dagger(p)]\sigma\cdot p[\tilde{\psi}_L(p)+\tilde{\phi}_L(p)] + \tilde{\phi}_L^\dagger(p)\sigma\cdot p\tilde{\phi}_L(p)\} \quad , \qquad (2.13)$$

where $\tilde{\phi}_L(p)$ is the solution of the equations of motion

$$\tilde{\phi}_L(p) = \frac{\bar{\sigma}\cdot p}{p^2}\sigma_2\tilde{\chi}_L^*(-p) \quad . \qquad (2.14)$$

In this form we see that integration over $\psi_L$ can be readily performed after shifting the integration variable by $\tilde{\phi}_L$, resulting in a change in the arbitrary normalization:

$$W[\chi_L,\chi_L^\dagger] = N' e^{- \int d^4p\, \tilde{\chi}_L^\dagger(p) \frac{\sigma\cdot p}{p^2} \tilde{\chi}_L(p)} \quad , \qquad (2.15)$$

where we have used [see (4.37) of Chapter I]

$$\bar{\sigma}\cdot p\ \sigma\cdot p = p^2 \quad , \qquad (2.16)$$

and

$$\sigma_2\bar{\sigma}^T\cdot p\sigma_2 = \sigma\cdot p \quad . \qquad (2.17)$$

If we set as in the boson case

$$W = e^{iZ} \quad , \qquad (2.18)$$

where Z is the generator of connected Green's functions, we find that

$$Z[\chi_L,\chi_L^\dagger] = -i \int d^4x\ \chi_L^\dagger(x)(i\bar{\sigma}\cdot\partial)^{-1}\chi_L(x) \quad . \qquad (2.19)$$

Thus we extract the two point connected Green's function

$$G^{(2)}(x_1,x_2) = -i(i\bar{\sigma}^\mu \frac{\partial}{\partial x_2^\mu})^{-1} \delta^{(4)}(x_1-x_2) \quad , \qquad (2.20)$$

or in momentum space

$$\tilde{G}^{(2)}(p) = - \frac{i}{\bar{\sigma}\cdot p} = -i\frac{\sigma\cdot p}{p^2} \quad , \qquad (2.21)$$

which is, of course, the inverse propagator. As it stands, this expression is meaningless until we have specified the pole prescription at $p^2 = 0$. We can interpret it in analogy with the boson case but it should be noted that in the fermion case we have no nice convergence argument to introduce the $-i\varepsilon$ prescription since we are dealing with formal Grassmann integration. Thus it would seem that we have to set up the problem in Euclidean space in order to justify the same pole prescription as we used for bosons.

The other two cases are treated along similar lines. In the Majorana case we start with

$$W_M[X_M] = N \int \mathscr{D}\Psi_M \, e^{i \int d^4x [\mathscr{L}_M + i\bar{\Psi}_M X_M]} \quad , \qquad (2.22)$$

and by completing the squares, arrive at the expression

$$W_M[X_M] = N' \, e^{-\frac{1}{2} \int d^4p \, \bar{X}_M (\not{p}+m)^{-1} X_M} \quad , \qquad (2.23)$$

where

$$\not{p} = \gamma^\mu p_\mu \qquad (2.24)$$

leading to the propagator

$$\tilde{G}_M^{(2)}(p) = \frac{-i}{\not{p}+m} = -i \, \frac{\not{p}-m}{p^2-m^2} \quad , \qquad (2.25)$$

using $\not{p}\not{p} = p^2$. Again the pole prescription has to be added in explicitly. The Dirac case is treated the same way, starting with

$$W_D[\zeta,\bar{\zeta}] = N \int \mathscr{D}\Psi_D \mathscr{D}\bar{\Psi}_D \, e^{i \int d^4x [\mathscr{L}_D + i\bar{\Psi}_D\zeta + i\bar{\zeta}\Psi_D]} \quad , \qquad (2.26)$$

where now $\zeta$ and $\bar{\zeta}$ are four component Dirac Grassmann sources. A similar reasoning leads to

$$W_D = N' \, e^{-\int d^4p \, \bar{\zeta}_D (\not{p}+m)^{-1} \zeta_D} \quad , \qquad (2.27)$$

from which we extract the propagator

$$G_D^{(2)}(p) = \frac{-i}{\not{p}+m} = -i \, \frac{\not{p}-m}{p^2-m^2} \quad , \qquad (2.28)$$

where the $-i\varepsilon$ prescription has to be added on.

As in the boson case, one might set up the generating functional directly in Euclidean space, and then perform the continuation into Minkowski space on the Green's functions.

In Euclidean space the Lorentz group becomes compact, which means (see Chapter I) that it is composed of two truly inequivalent SU(2) factors. However, the derivative operator still transforms as the (1/2, 1/2) representation so that if we now want to make a Lorentz scalar linear in the derivative we need two different fields $\psi_L \sim (1/2, 0)$ and $\psi_R \sim (0, 1/2)$ to build vector quantities transforming as (1/2, 1/2). We can build two such real vectors with components

$$(i\psi_L^\dagger \psi_R + i\psi_R^\dagger \psi_L , -\psi_L^\dagger \vec{\sigma} \psi_R + \psi_R^\dagger \vec{\sigma} \psi_L) \qquad (2.29)$$

and

$$(\psi_L^\dagger \psi_R - \psi_R^\dagger \psi_L , \; i\psi_L^\dagger \vec{\sigma} \psi_R + i\psi_R^\dagger \vec{\sigma} \psi_L) \quad , \qquad (2.30)$$

remembering that because $\psi_L$ and $\psi_R$ are Grassmann numbers

$$(\psi_L^\dagger \psi_R)^* = \psi_L^T \psi_R^* = -\psi_R^\dagger \psi_L \quad . \qquad (2.31)$$

If we introduce the four component Euclidean Dirac spinor

$$\Psi_E \equiv \begin{pmatrix} \psi_L \\ \\ \psi_R \end{pmatrix} \quad , \quad (2.32)$$

we can rewrite the vectors in the form

$$\Psi_E^\dagger \bar{\gamma}_\mu \Psi_E \quad \text{and} \quad \Psi_E^\dagger \bar{\gamma}_5 \bar{\gamma}_\mu \Psi_E \quad , \quad (2.33)$$

respectively, where $\bar{\gamma}_\mu$ are the Euclidean $\gamma$-matrices

$$\bar{\gamma}_0 = \begin{pmatrix} 0 & i \\ \\ i & 0 \end{pmatrix} \quad , \quad \bar{\gamma}_i = \begin{pmatrix} 0 & -\vec{\sigma} \\ \\ \vec{\sigma} & 0 \end{pmatrix} \quad (2.34)$$

which satisfy

$$\{\bar{\gamma}_\mu, \bar{\gamma}_\nu\} = -2\delta_{\mu\nu} \quad , \quad (2.35)$$

and

$$\bar{\gamma}_5 = \begin{pmatrix} 1 & 0 \\ \\ 0 & -1 \end{pmatrix} \quad . \quad (2.36)$$

The possible mass terms are (in Euclidean space there is only one type of mass)

$$\psi_L^\dagger \psi_L \quad \text{and} \quad \psi_R^\dagger \psi_R \quad , \quad (2.37)$$

so that the Euclidean Lagrangian is given by

$$\mathcal{L}_E = \psi_E^\dagger \bar{\gamma}_\mu \bar{\partial}_\mu \psi_E + im\psi_E^\dagger \psi_E \quad ; \quad (2.38)$$

it is carefully chosen to be real

$$\mathcal{L}_E^* = \mathcal{L}_E \quad . \quad (2.39)$$

The corresponding generating functional is

$$W_E[\zeta_E,\zeta_E^+] = N \int \mathcal{D}\psi_E^+ \mathcal{D}\psi_E \; e^{-\int d^4\bar{x}[\mathcal{L}_E + i\zeta_E^+\psi_E + i\psi_E^+\zeta_E]} \tag{2.40}$$

$$= N' \; e^{i\int d^4 p \tilde{\zeta}_E^+(p)[\bar{\not p}+m]^{-1}\tilde{\zeta}_E(p)} \quad , \tag{2.41}$$

leading to the propagator

$$G_E(p) = \frac{-i}{\bar{\not p}+m} = i\,\frac{\not p - m}{\bar{p}^2+m^2} \quad , \tag{2.42}$$

where we have used $\bar{\not p}\bar{\not p} = -\bar{p}^2$. We remark that it has the expected $\bar{p}^2 + m^2$ denominator. It is satisfying to see that it has the same structure as the Dirac propagator in Minkowski space.

There does not seem to be any such correspondence for Weyl fields: one cannot construct a first order equation for a field transforming as (1/2, 0) in Euclidean space starting from an invariant Lagrangian that contains this field alone (as we just saw it can easily be done when two Weyl fields are considered). Should the FPI be well defined only in Euclidean space, as axiomaticians would have it, then there seems to exist a very real problem when dealing with Weyl fields as in the theory of weak interactions or in its unification with QCD. It must be emphasized that there is nothing apparently wrong with the field theory of Weyl fields in Minkowski space (excepting possible anomalies) as far as perturbation theory is concerned - it could well be that a more complete treatment might yield surprises, requiring the doubling of the Weyl fermions, which would restore at some (large) scale a vectorlike structure for the weak interactions.

Alternatively, since fermi fields appear only quadratically in renormalizable Lagrangians, their integration leads only to determinants. Thus one might argue that any Euclidean functional that leads to the correct (as determined by its continuation in Minkowski space) determinant is all that is needed. This approach requires the doubling of the number of independent Grassmann fields (see S. Coleman's Erice lecture "The Uses of Instantons" and problem E).

Problems.

A. Evaluate the generating functional for Majorana and Dirac fields in Minkowski space.

B. Show that in Euclidean space the derivative operator $\partial_\mu$ transforms according to the (1/2, 1/2) representation.

C. Given in Euclidean space $\psi_L \sim (1/2, 0)$, build explicitly the quadratic form which behaves as (1, 0).

D. Show that the Euclidean space spinor Lagrangian has the curious property that its mass term is invariant under the so-called chiral transformation

$$\Psi_E \rightarrow e^{i\alpha\gamma_5}\Psi_E \quad ,$$

while the kinetic term is not!

E. Formally define

$$\mathcal{L} = X(\not\partial + im)\psi,$$

where X and $\psi$ are independent 4 component Grassmann fields. Then show how to judiciously integrate in order to obtain the usual Dirac determinant. Discuss chiral invariance in this example.

## 3. Feynman Rules for Spinor Fields

The Feynman rules for free Fermi fields have already been discussed in the previous section. Here we derive the rules for interacting spinors. Spinors can interact in a variety of ways subject to the conservation of spin which requires that all interaction vertices include an even pair of spinor fields. We have given in Chapter I examples of interacting theories with fermions.

The number of possible fermion interactions is drastically reduced when we impose the constraint of renormalizability which demands as a necessary condition that the number of primitively divergent diagrams be finite. Let us therefore compute the superficial degree of divergence D of an arbitrary Feynman diagram with fermions.

Consider a diagram with L loops, $I_b$ boson internal lines, $I_f$ fermion internal lines, V vertices each with $N_b$ boson and $N_f$ fermion lines, $E_b$ external boson lines and $E_f$ external fermion lines. As we just remarked, $N_f$ and $E_f$ must be even. The number of loops L is given by

$$L = I - V + 1 = I_b + I_f - V + 1 \quad . \tag{3.1}$$

The superficial degree of divergence in d dimensions is

$$D_d = dL - I_f - 2I_b \quad , \tag{3.2}$$

since each internal spinor line contributes only one inverse power of momentum. Furthermore, the total number of fermion lines is given by

$$N_f V = E_f + 2I_f \tag{3.3}$$

with a similar relation for boson lines

$$N_b V = E_b + 2I_b \quad . \tag{3.4}$$

These enable us to express $D_d$ in the form

$$D_d = d - \frac{1}{2}(d-1)E_f - \frac{1}{2}(d-2)E_b - V[d - \frac{1}{2}(d-1)N_f - \frac{1}{2}(d-2)N_b] \ . \qquad (3.5)$$

When $N_f = E_f = 0$ this reduces to the previously obtained expression with only bosons present. In two dimensions, it reduces to

$$D_2 = 2 - \frac{1}{2}E_f - V(2 - \frac{1}{2}N_f), \qquad [2 \text{ dimensions}] \qquad (3.6)$$

which shows that

$$N_f \leq 4 \quad , \qquad [2 \text{ dimensions}] \qquad (3.7)$$

otherwise the divergence will grow with the number of vertices. Hence there is a restriction on the type of allowed fermion interaction, even in two dimension: It must not be of a degree higher than $(\psi)^4$. We can understand this fact in another way: unlike boson fields which are dimensionless in two dimensions, spinor fields have dimension -1/2 so that $\psi^4$ is the highest interaction which does not necessitate the introduction of a dimensionful coupling constant.

In four dimensions, we have

$$D_4 = 4 - \frac{3}{2}E_f - E_b - V[4 - \frac{3}{2}N_f - N_b] \qquad . \qquad (3.8)$$

If we do not want the number of primitively divergent graphs to grow with the number of vertices, we must require

$$4 - \frac{3}{2}N_f - N_b \geq 0 \qquad , \qquad (3.9)$$

where $N_f$ is even. The possible solutions are

$$N_b = 0 \qquad N_f = 2 \qquad ,$$

which is like a mass insertion and not an interaction vertex,

$$N_f = 0 \qquad N_b = 2,\ 3,\ 4 \quad ,$$

which give $\phi^2$, $\phi^3$, $\phi^4$ interactions we have previously analyzed. The only new solution involving both fermions and bosons is

$$N_f = 2 \qquad N_b = 1 \quad , \tag{3.10}$$

which gives

$$D_4 = 4 - \frac{3}{2} E_f - E_b \quad . \tag{3.11}$$

This new solution which describes the only fermion interaction allowed by renormalizability is incredibly restrictive: renormalizable fermion interactions must involve at most two spinor fields and one boson field. Thus fermions in four-dimensions appear only quadratically in $\mathcal{L}$! This fact can also be understood in another way: in four dimensions fermions have dimension -3/2, bosons -1. Hence the only nontrivial coupling of dimension 4 is the one with two fermions and one boson:

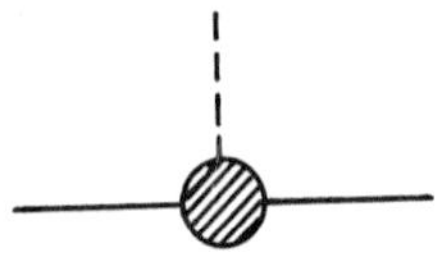

This remarkable fact greatly simplifies our analysis of interacting theories with spinors. Given two spin 1/2 fields we can form either a spin 0 or a spin 1 combination. The couplings to a spin 0 field are the Yukawa couplings - they appear in many guises: in Minkowski space we have couplings for Dirac fields

$$i\bar{\Psi}_D\Psi_D\phi, \qquad \bar{\Psi}_D\gamma_5\Psi_D\phi' \quad , \tag{3.12}$$

where $\phi$ and $\phi'$ are scalar and pseudoscalar fields, respectively. For Weyl fields we have

$$i\psi_L^T\sigma_2\psi_L\phi_1, \qquad i\psi_L^\dagger\sigma_2\psi_L^*\phi_2 \qquad , \tag{3.13}$$

where $\phi_1$ and $\phi_2$ have no defined parity. In Euclidean space the possible couplings are

$$i\Psi_E^\dagger\Psi_E\phi, \qquad \Psi_E^\dagger\gamma_5\Psi_E\phi' \qquad \text{[Euclidean space]} \qquad . \tag{3.14}$$

The couplings to spin 1 give for a Dirac particle in Minkowski space

$$i\bar{\Psi}_D\gamma_\mu\Psi_D A^\mu, \qquad \bar{\Psi}_D\gamma_\mu\gamma_5\Psi_D A_p^\mu \qquad , \tag{3.15}$$

where $A^\mu$ is a vector field, $A_p^\mu$ an axial vector field; for Weyl fields we have

$$i\psi_L^\dagger\sigma_\mu\psi_L B^\mu, \qquad i\psi_R^\dagger\bar{\sigma}_\mu\psi_R B'^\mu \qquad , \tag{3.16}$$

where $B_\mu$ and $B'_\mu$ have no well defined parity properties. In Euclidean space the vector couplings are

$$i\Psi_E^\dagger\bar{\gamma}_\mu\Psi_E A_\mu, \qquad \Psi_E^\dagger\bar{\gamma}_\mu\bar{\gamma}_5\Psi_E A_{p\mu} \qquad . \tag{3.17}$$

Interaction of spinor and vector fields will be extensively discussed in later chapters. The Feynman rule for the Yukawa vertices is simply the dimensionless Yukawa couplings themselves

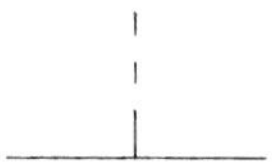

$$\longleftrightarrow \qquad if: \qquad if\bar{\Psi}\Psi\phi \tag{3.18}$$

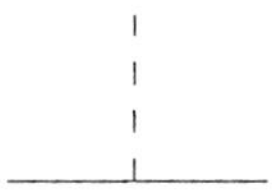

$$\longleftrightarrow \qquad f'\gamma_5: \qquad f'\bar{\Psi}\gamma_5\Psi\phi_p \qquad . \tag{3.19}$$

In the above, the dotted line denotes a boson field and the solid line a spinor field with the spinor indices suppressed.

The Grassmann nature of the spinor field is reflected in one crucial change in the Feynman rules: wherever a closed fermion line (loop) appears in a diagram, one should charge the diagram a minus sign, as the following example will illustrate.

Consider the expression (say, in Euclidean space)

$$W = \frac{\det[\partial_\mu \partial_\mu + \lambda A(x)]}{\det[\partial_\mu \partial_\mu + \lambda A(x)]} = 1 \tag{3.20}$$

where A(x) is a scalar field and the determinants are to be understood as functional determinants. We can express them in terms of path integrals, the one in the denominator giving a path integral over boson fields and the one in the numerator a path integral over Grassmann fields. The result is

$$W = \int \mathcal{D}\phi \mathcal{D}\phi^* \mathcal{D}\Psi \mathcal{D}\Psi^* \, e^{i\langle\psi^*(\partial^2+\lambda A(x))\psi\rangle + \langle\phi^*(\partial^2+\lambda A(x))\phi\rangle} \quad . \tag{3.21}$$

In this form it looks like a theory of a Grassmann field $\psi$ interacting with a complex scalar field through the external field A(x). The Feynman rules are

———— : $\frac{i}{p^2}$ for the Grassmann line

-------- : $\frac{1}{p^2}$ for the boson field line

 : $\lambda$ ;  : $\lambda$

where A appears as the wavy line. In this theory, the A propagator will be corrected by the "vacuum polarization" diagrams

which, according to the previous Feynman rules are exactly the same. But, these diagrams cannot possibly alter the A line because we know the theory to be trivial, starting from W = 1! Hence it must be that rather than adding, the two diagrams cancel: the closed Grassmann loop must acquire a minus sign relative to the closed boson loop.

Hence, wherever Grassmann (spinor) fields are encountered, the Feynman rule says that one must multiply a diagram with n distinct closed fermion loops by $(-1)^n$. A related way of understanding this fact is to remark that the cut fermion loop must by generalized Fermi statistics be antisymmetric under the interchange of its legs since it is related by unitarity to a physical amplitude.

Finally we note that the Yukawa coupling can induce renormalizable self-interactions among the scalar fields (see problem).

Problems.

A. Find the dimensions for which there are renormalizable theories involving fermions.

B. Given the Lagrangian

$$\mathcal{L} = \bar{\Psi}_D(\gamma\cdot\partial+im)\Psi_D + if\bar{\Psi}_D\phi\Psi_D + \frac{1}{2}\,\partial_\mu\phi\partial^\mu\phi + \frac{1}{2}\,m^2\phi^2$$

1) Derive the Feynman rules,

2) Find the scale dependence of the Yukawa coupling constant at the one loop level,

3) Discuss the renormalizability of this Lagrangian; in particular analyze all the one loop diagrams and discuss the ensuing counterterm structure. Is this Lagrangian renormalizable as written? If not, amend it.

## 4. Evaluation and Scaling of Fermion Determinants

Let us start with the Euclidean space generating functional for the theory that describes a scalar field in interaction with a four-component spinor field

$$W_E[\zeta^\dagger,\zeta,J] = e^{-Z_E} = \int \mathcal{D}\phi\mathcal{D}\Psi^\dagger\mathcal{D}\Psi\; e^{-S_E[\phi,\psi^\dagger,\psi,J,\zeta^\dagger,\zeta]} \quad , \qquad (4.1)$$

with

$$S_E = \int d^4\bar{x}[\frac{1}{2}\,\bar{\partial}_\mu\phi\bar{\partial}_\mu\phi + \frac{1}{2}\,m^2\phi^2 + \frac{\lambda}{4!}\,\phi^4 - J\phi + \Psi^\dagger(\not{\partial}+im'+if\phi)\Psi - i\zeta^\dagger\Psi - i\Psi^\dagger\zeta] \quad . \qquad (4.2)$$

In this expression the spinor fields appear quadratically and can therefore be functionally integrated. This step leaves us with the expression ($S_E[\phi,J]$ is given by eq. (4.5) of Chapter III)

$$W_E[\zeta^\dagger,\zeta,J] = \int \mathcal{D}\phi\; e^{-S_E[\phi,J]}\det(\not{\partial}+im'+if\phi)\; e^{<\zeta^\dagger(\not{\partial}+im'+if\phi)^{-1}\zeta>} , \qquad (4.3)$$

where we have used (1.31) after completing the squares. We can rewrite it in the form

$$W_E = e^{<\zeta^\dagger(\not{\partial}+im'+if\frac{\delta}{\delta J})^{-1}\zeta>}\int \mathcal{D}\phi\; e^{-S_E[\phi,J]}\det(\not{\partial}+im'+if\phi) \quad . \qquad (4.4)$$

This is a pretty formula which is useful in the perturbative evaluation of $W_E$. In this section we would rather concentrate on the saddle point evaluation of $W_E$. To start with, let us expand $S_E$ around a classical field configuration $\phi_0$, $\Psi_0$, $\Psi_0^\dagger$. The functional expansion is of the form

$$S_E = S_E\Big|_0 + \langle\eta^\dagger \frac{\delta S}{\delta\Psi^\dagger}\Big|_0\rangle + \langle\frac{\delta S}{\delta\Psi}\Big|_0 \eta\rangle + \langle\frac{\delta S}{\delta\phi}\Big|_0 \rho\rangle$$

$$+ \frac{1}{2}\langle\rho_1 \frac{\delta^2 S}{\delta\phi_1\delta\phi_2}\Big|_0 \rho_2\rangle_{1,2} + \langle\eta_1^\dagger \frac{\delta^2 S}{\delta\psi_1^\dagger\delta\psi_2}\eta_2\rangle_{1,2}$$

$$+ \langle\rho_1 \frac{\delta^2 S}{\delta\phi_1\delta\psi_2}\Big|_0 \eta_2\rangle_{1,2} + \langle\eta_1^\dagger \frac{\delta^2 S}{\delta\psi_1^\dagger\delta\phi_2}\Big|_0 \rho_2\rangle_{1,2} + \cdots \qquad (4.5)$$

where $\eta = \Psi - \Psi_0$, and $\rho = \phi - \phi_0$.

As in Section 4 of Chapter III, we expand around field configurations that leave the Action stationary, i.e., solve the classical equations of motion:

$$\frac{\delta S}{\delta\Psi^\dagger}\Big|_0 = (\not{\partial}+im'+if\phi_0)\Psi_0 - i\zeta = 0 \qquad (4.6)$$

$$\frac{\delta S}{\delta\Psi}\Big|_0 = \Psi_0^\dagger(-\overleftarrow{\not{\partial}}+im'+if\phi_0) - i\zeta^\dagger = 0 \qquad (4.7)$$

$$\frac{\delta S}{\delta\phi}\Big|_0 = (-\partial^2+m^2 + \frac{\lambda}{3!}\phi_0^2)\phi_0 + if\Psi_0^\dagger\Psi_0 - J = 0 \qquad . \qquad (4.8)$$

By eliminating these linear terms we have an approximate expression for $S_E$ that is quadratic in $\eta$, $\eta^\dagger$ and $\rho$, the differences of the fields away from their stationary values. Explicitly

$$S_E \simeq S_E\Big|_0 + \langle\eta^\dagger(\not{\partial}+im+if\phi_0)\eta\rangle + \frac{1}{2}\langle\rho(-\partial^2+m^2 + \frac{\lambda}{2}\phi_0^2)\rho\rangle$$

$$+ if\langle\eta^\dagger\rho\Psi_0\rangle + if\langle\Psi_0^\dagger\rho\eta\rangle \qquad . \qquad (4.9)$$

Since we want to functionally integrate over $\eta$, $\eta^\dagger$ and $\rho$, we complete the squares, obtaining

$$S_E \simeq S_E\Big|_0 + \langle\eta'^{\dagger}(\not\partial+im'+if\phi_0)\eta'\rangle + \frac{1}{2}\langle\rho[-\partial^2+m^2+\frac{\lambda}{2}\phi_0^2+2f^2\Psi_0^{\dagger}(\not\partial+im'+if\phi_0)^{-1}\Psi_0]\rho\rangle ,$$

(4.10)

where

$$\eta' = \eta + if(\not\partial+im'+if\phi_0)^{-1}\rho\Psi_0 \quad . \tag{4.11}$$

Moreover in this approximation

$$\mathcal{D}\Psi = \mathcal{D}\eta = \mathcal{D}\eta' \tag{4.12}$$

$$\mathcal{D}\phi = \mathcal{D}\rho \quad , \tag{4.13}$$

since the Jacobian of the transformation is 1. This allows us to functionally integrate (4.10) using the formulae of Appendix A and of Section 1 of this chapter. The result is

$$W_E \simeq e^{-S_E|_0} \det(\not\partial+im'+if\phi_0)[\det(-\partial^2+m^2+\frac{\lambda}{2}\phi_0^2+2f^2\Psi_0^{\dagger}(\not\partial+im'if\phi_0)^{-1}\Psi_0)]^{-1/2} ,$$

(4.14)

where in the second determinant the inverse operator acts on and through $\Psi_0$.

As in the saddle point approximation for scalar field theory, $S_E\Big|_0$ generates all the tree diagrams when viewed as a functional of the sources J, $\zeta$ and $\zeta^{\dagger}$, while the determinants give the one loop contributions which are of first order in $\hbar$.

Let us perform a functional Legendre transformation between the sources J, $\zeta$ and $\zeta^{\dagger}$ and the new classical sources

$$\phi_{c\ell}(x) = -\frac{\delta Z_E}{\delta J(x)} \simeq -\frac{\delta S_E}{\delta J(x)} + \mathcal{O}(\hbar) \tag{4.15}$$

$$\Psi_{c\ell}(x) = -\frac{\delta Z_E}{\delta\zeta^{+}(x)} \simeq -\frac{\delta S_E}{\delta\zeta^{+}(x)} + \mathcal{O}(\hbar) \quad , \tag{4.16}$$

and introduce the effective Action

$$\Gamma_E[\phi_{c\ell},\Psi_{c\ell},\Psi^\dagger_{c\ell}] \equiv Z_E[J,\zeta,\zeta^\dagger] - \langle J\phi_{c\ell}\rangle - i\langle\zeta^\dagger\Psi_{c\ell}\rangle - i\langle\Psi^\dagger_{c\ell}\zeta\rangle \quad , \tag{4.17}$$

which generates the one particle irreducible Green's functions. In the classical approximation, it is nothing but the classical Action with the classical sources of (4.15) and (4.16) playing the role of the fields:

$$\Gamma_E = \langle\Psi^\dagger_{c\ell}(\not\partial+im+if\phi_{c\ell})\Psi_{c\ell}\rangle + \frac{1}{2}\langle\phi_{c\ell}(-\partial^2+m^2+\frac{\lambda}{12}\phi^2_{c\ell})\phi_{c\ell}\rangle + \mathcal{O}(\hbar) \quad . \tag{4.18}$$

The first quantum mechanical corrections to (4.18) are given by the determinants of (4.14). We will evaluate some of their properties by using the $\zeta$-function technique of Chapter III. These determinants are more complicated because of the spinor indices and of the inverse operator appearing in the second determinant of (4.14).

Let us specialize to constant field configurations, and neglect all masses. Then

$$\frac{1}{\not\partial+if\phi_0} = (\not\partial-if\phi_0)\frac{1}{[-\partial^2+f^2\phi_0^2]} \quad , \tag{4.19}$$

so that the argument of the scalar field determinant becomes

$$[(-\partial^2+\frac{\lambda}{2}\phi_0^2)(-\partial^2+f^2\phi_0^2)+2f^2\Psi^\dagger_0(\not\partial-if\phi_0)\Psi_0]\frac{1}{(-\partial^2+f^2\phi_0^2)} \quad . \tag{4.20}$$

In addition, for constant $\phi_0$, the fermion determinant becomes (see problem)

$$\det(\not\partial+if\phi_0) = [\det(-\partial^2+f^2\phi_0^2)]^2 \quad . \tag{4.21}$$

Using the property of determinants on (4.20), we can rewrite the generating functional as

$$W_E \simeq e^{-S_E\big|_0} [\det(-\partial^2+f^2\phi_0^2)]^{5/2}[\det((-\partial^2 + \frac{\lambda}{2}\phi_0^2)(-\partial^2+f^2\phi_0^2)+2f^2\Psi_0^\dagger(\not\partial-if\phi_0)\Psi_0)]^{-1/2}, \quad (4.22)$$

which is valid only for constant $\phi_0$ and $\Psi_0$. To simplify matters further, let us assume that $\Psi_0$ is chiral, i.e.,

$$\Psi_0^\dagger\gamma_\mu\Psi_0 = 0 \quad (\text{or } \psi_R = 0); \qquad \Psi^\dagger\Psi \neq 0 \quad . \quad (4.23)$$

Then the argument of the second determinant can be rewritten in the form

$$(-\partial^2+A)(-\partial^2+B) \quad , \quad (4.24)$$

where A and B are constants involving $\phi_0$ and $\Psi_0^\dagger\Psi_0$ which satisfy

$$A + B = (f^2 + \frac{\lambda}{2})\phi_0^2 \quad (4.25)$$

$$AB = \frac{1}{2}\lambda f^2\phi_0^4 - 2if^3\Psi_0^\dagger\phi_0\Psi_0 \quad . \quad (4.26)$$

We have thus reduced the problem to evaluating determinants of the form $\det(-\partial^2+C)$ where C is a constant. Summarizing the results of Chapter III, we know that

$$\det(-\partial^2+C) = e^{-\zeta'_{[-\partial^2+C]}(0)} \quad , \quad (4.27)$$

with

$$\zeta_{[-\partial^2+C]}(s) = \frac{\mu^4}{16\pi^2}\left(\frac{C}{\mu^2}\right)^{2-s}\frac{\Gamma(s-2)}{\Gamma(s)}\int d^4x \quad (4.28)$$

or

$$\zeta'_{[-\partial^2+C]}(0) = -\frac{1}{32\pi^2}C^2\left(-\frac{3}{2} + \ln\frac{C}{\mu^2}\right) \quad . \quad (4.29)$$

Thus armed, it is easy to read off the one loop contribution to the effective potential [see problem]. Here we merely quote the result when $\Psi_0 = 0$:

$$V(\phi_0) = \frac{\lambda^2}{256\pi^2}\phi_0^4[-\frac{3}{2} + \ell n\frac{\lambda\phi_0^2}{2\mu^2}] - \frac{f^4}{8\pi^2}\phi_0^4[-\frac{3}{2} + \ell n\frac{f^2\phi_0^2}{\mu^2}] \quad , \qquad (4.30)$$

where the first term is the same as in the pure scalar case - it comes from the boson loops; the second term comes from the contribution of the closed fermion loops to the potential, and the relative minus sign comes from the closed fermion loop.

The scaling properties of these determinants are equally straightforward to evaluate. Recall from Chapter III that under a scale change

$$\det[e^{-2a}(-\partial^2+C)] = e^{-2a\zeta^{(0)}_{[-\partial^2+C]}} \det(-\partial^2+C) \quad ; \qquad (4.31)$$

with $\zeta$ given by (4.28), so that

$$\det{}^n[e^{-2a}(-\partial^2+C)] = e^{-2na\int d^4x \frac{C^2}{8\cdot 16\pi^2}} \det{}^n(-\partial^2+C) \quad , \qquad (4.32)$$

where we have treated the constant C as if it changed under a scale transformation with the same dimension as $-\partial^2$. For a more rigorous treatment, see Chapter III, Section 6. Thus, the one loop scaling correction is

$$\Gamma_E \to \Gamma_E + \frac{\hbar a}{128\pi^2}\int d^4\bar{x}(-2(\frac{5}{2})(f^2\phi^2)^2+(-2)(-\frac{1}{2})[A^2+B^2]) \quad . \qquad (4.33)$$

Rewriting

$$\begin{aligned} A^2 + B^2 &= (A+B)^2 - 2AB \\ &= (f^2 + \frac{\lambda}{2})^2\phi_0^4 - \lambda f^2\phi_0^4 + 4if^3\Psi_0^\dagger\phi_0\Psi_0 \end{aligned} \quad , \qquad (4.34)$$

we see that the effect of a scale change on $\Gamma_E$ is to generate terms of the same type as in the classical Lagrangian, which produces a change in the dimensionless coupling constants

$$\frac{\lambda}{4!} \to \frac{\lambda'}{4!} = \frac{\lambda}{4!} - \frac{\hbar a}{128\pi^2}\left(\frac{\lambda^2}{4} - 5f^4\right) \quad , \qquad (4.35)$$

$$f \to f' = f - \frac{\hbar a}{32\pi^2} f^2 \quad . \qquad (4.36)$$

This provides an example of the scale dependence of a theory with several coupling constants. The new feature is that the scale changes form a coupled system. This coupling phenomenon is easy to understand from a diagrammatic point of view since the closed fermion loop obviously contributes to the $\phi^4$ coupling, while to $\mathcal{O}(\hbar)$ the fermion coupling is affected only by the presence of the original fermion vertex:

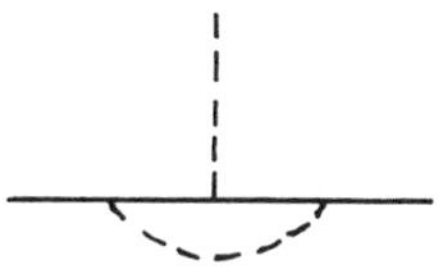

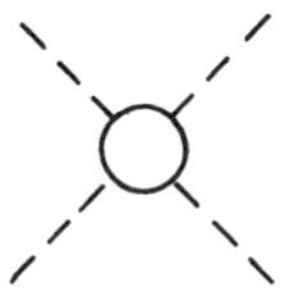

in these the dashed (solid) line represents a scalar (spinor) field.

It must be noted that a scale change will also generate additions to the fermion and scalar kinetic terms, but in our approximation of constant fields these did not show up in the changes (4.35) and (4.36). For instance, the fermion determinant

$$\det(\partial\!\!\!/ + if\phi)$$

cannot be written in the form (4.21) unless $\phi$ is independent of x. In general, its scaling will involve a kinetic term; it corresponds to the fact that the fields themselves acquire at the one-loop level anomalous dimensions coming from the diagram

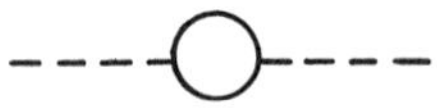

In the pure $\phi^4$ theory, the scalar field acquired an anomalous dimension only at the two-loop level, and therefore the effect did not show up in the scaling of the determinant. Thus eqs. (4.35) and (4.36) have to be corrected for wave function renormalization.

Problems.

A. Show that $\Gamma_E[\phi_{c\ell}, \Psi_{c\ell}, \Psi^\dagger_{c\ell}]$ is the classical Action when terms of $\mathcal{O}(\hbar)$ are neglected.

B. Show that in four-dimensions

$$\det(\not\partial+im) = [\det(-\partial^2+m^2)]^2 \quad .$$

C. Find the one-loop contribution to the potential including the fermion contribution.

*D. Using diagrams, derive the expression for the scale dependence of $\lambda$ and f (including wave function renormalization) at the one loop level, and compare with (4.35) and (4.36).

# VI. GAUGE SYMMETRIES: THE YANG-MILLS CONSTRUCTION

## 1. Global and Local Symmetries

In Chapter I we gave examples of Lagrangians involving fields of spin 0 and 1/2 but refrained from presenting any theory involving higher spin fields in interaction. The reason for this omission stems from the fact that fields of spin 1, 3/2 and 2 can be introduced in a very beautiful way just by requiring that whatever symmetries present in the spin 1/2 - 0 system be generalized to vary arbitrarily from point to point in spacetime. Spin 1 fields correspond to generalizing internal (i.e., non-Lorentz) symmetries; a spin 2 field occurs when spacetime symmetries (global Poincaré invariance) are made local in spacetime; spin 3/2 and 2 fields appear when globally supersymmetric theories are generalized to be locally supersymmetric.

We will not present the local generalization of spacetime symmetries, but will focus on building theories which are locally invariant under internal symmetries, following Yang and Mills, Phys. Rev. 96, 191 (1954). Maxwell's electrodynamics provides the earliest example of a theory with a local symmetry. It was E. Noether who first realized the generality of the concept of "gauging" (i.e., "making local") symmetries [in Nachr. Kgl. Ges. Wiss., Göttingen 235 (1918)]. The gauging procedure in its modern form was formulated by H. Weyl in the 1920's.

Consider the simplest possible Lagrangian involving a spinor field

$$\mathcal{L}_0 = \frac{1}{2}\, \psi_L^\dagger \sigma \cdot \overleftrightarrow{\partial} \psi_L = \psi_L^\dagger \sigma^\mu \partial_\mu \psi_L + \text{surface term} \qquad (1.1)$$

which we know to be invariant under the phase transformation

$$\psi_L(x) \to e^{i\alpha} \psi_L(x) \quad , \qquad (1.2)$$

where $\alpha$ is a constant. The basic idea behind "gauging" this phase symmetry is to make $\mathcal{L}$ invariant under phase transformations just like (1.2) but with $\alpha$ depending arbitrarily on $x_\mu$, i.e.,

$$\psi_L(x) \rightarrow e^{i\alpha(x)}\psi_L(x) \quad . \quad (1.3)$$

What stands in the way of $\mathcal{L}$ being invariant under the local phase transformation (1.3) is the presence of the derivative operator $\partial_\mu$, since under (1.3) we have

$$\partial_\mu\psi_L(x) \rightarrow \partial_\mu e^{i\alpha(x)}\psi_L(x) = e^{i\alpha(x)}[\partial_\mu+i\partial_\mu\alpha(x)]\psi_L(x) \quad , \quad (1.4)$$

so that

$$\mathcal{L}_0 \rightarrow \mathcal{L}_0 + i\psi_L^\dagger\sigma^\mu\psi_L\partial_\mu\alpha(x) \quad . \quad (1.5)$$

The trick behind generalizing $\mathcal{L}$ is the invention of a new operator which generalizes $\partial_\mu$, call it $\mathcal{D}_\mu$, with the property that under a local phase transformation

$$\mathcal{D}_\mu\psi_L \rightarrow e^{i\alpha(x)}\mathcal{D}_\mu\psi_L \quad , \quad (1.6)$$

or in operator language

$$\mathcal{D}_\mu \rightarrow e^{i\alpha(x)}\mathcal{D}_\mu e^{-i\alpha(x)} \quad . \quad (1.7)$$

This new derivative operator is called a <u>covariant derivative</u>. Then it trivially follows that the new Lagrangian

$$\mathcal{L} \equiv \psi_L^\dagger\sigma^\mu\mathcal{D}_\mu\psi_L \quad (1.8)$$

is invariant under (1.3). This is all fine but we have to build this covariant derivative. We look for an expression of the form

$$\mathcal{D}_\mu = \partial_\mu + iA_\mu(x) \quad , \quad (1.9)$$

where $A_\mu(x)$ is a function of x. Then the covariance requirement

$$\mathcal{D}_\mu \to \mathcal{D}'_\mu = \partial_\mu + iA'_\mu(x) = e^{i\alpha(x)}(\partial_\mu + iA_\mu(x))e^{-i\alpha(x)} \tag{1.10}$$

becomes a transformation property of $A_\mu$

$$A_\mu(x) \to A'_\mu(x) = A_\mu(x) - \partial_\mu \alpha(x) \quad . \tag{1.11}$$

Now then the new Lagrangian

$$\mathcal{L} = \psi_L^\dagger \sigma^\mu (\partial_\mu + iA_\mu(x))\psi_L = \mathcal{L}_0 + i\psi_L^\dagger \sigma^\mu \psi_L A_\mu(x) \tag{1.12}$$

is invariant under the simultaneous local transformations

$$\psi_L(x) \to e^{i\alpha(x)}\psi_L$$

$$A_\mu(x) \to A_\mu(x) - \partial_\mu \alpha(x) \quad . \tag{1.13}$$

The global symmetry of $\mathcal{L}_0$ is generalized to a local symmetry or gauged at the price of introducing a new vector field $A_\mu(x)$ which interacts with the conserved current. We can see that the new field $A_\mu(x)$ has the same dimensions as $\partial_\mu$; it can therefore be identified with a canonical field in four dimensions [in other number of dimensions, one has to understand $A_\mu$ as being multiplied by a dimensionful coupling before it can be thus identified]. Furthermore $A_\mu(x)$ is real since $i\partial_\mu$ is hermitean.

It is therefore easy to write a kinetic term for the field $A_\mu(x)$ in such a way that preserves the gauge invariance (1.13), by noting that the combination

$$F_{\mu\nu} = \partial_\mu A_\nu - \partial_\nu A_\mu \tag{1.14}$$

is invariant. It has dimension -2 and therefore we can build out of it a new Lagrangian

$$\mathcal{L} = -\frac{1}{4g^2} F_{\mu\nu}F^{\mu\nu} \quad , \qquad (1.15)$$

where we have introduced a dimensionless constant g which can be absorbed by letting $A_\mu = gA'_\mu$, so that in terms of $A'_\mu$ it appears in the coupling between $A'_\mu$ and the current $\psi_L^+\sigma^\mu\psi_L$ in (1.12). The factor of -1/4 corresponds to the conventional definition of g. Of course, as you might have guessed (1.15) is the Maxwell Lagrangian. We now have a fully interacting theory of spin 1 and spin 1/2 fields, described by

$$\mathcal{L} = -\frac{1}{4g^2} F_{\mu\nu}F^{\mu\nu} + \psi_L^\dagger\sigma^\mu(\partial_\mu + iA_\mu(x))\psi_L \quad . \qquad (1.16)$$

Although pretty, this theory is not renormalizable (as we shall see later) because of a tricky complication appropriately called the (Adler-Bell-Jackiw) anomaly, which has to do with the left handed nature of $\psi_L$. It causes no problem if we couple $A_\mu$ gauge invariantly to a Dirac four component field, leading to

$$\mathcal{L}_{QED} = -\frac{1}{4g^2} F_{\mu\nu}F^{\mu\nu} + \bar{\psi}\gamma^\mu(\partial_\mu + iA_\mu(x))\psi \qquad (1.17)$$

which describes QED when $A_\mu(x)$ is identified with the photon, $\psi$ with the electron and g with the electric charge. $\mathcal{L}_{QED}$ is invariant under the local symmetry

$$\psi(x) \rightarrow e^{i\alpha(x)}\psi(x)$$

$$A_\mu(x) \rightarrow A_\mu(x) - \partial_\mu\alpha(x)$$

and in the absence of a mass term for $\psi$ under the global chiral transformation

$$\psi(x) \rightarrow e^{i\beta\gamma_5}\psi(x) \quad . \qquad (1.18)$$

[This chiral symmetry is not exact (the anomaly again) in quantum field theory even in the absence of the electron mass but it does not cause any problem since no gauge field is coupled to it.] Gauge invariance forbids any mass term for $A_\mu$.

Before generalizing this construction to more complicated symmetries, let us review the different types of global symmetries.

The Lagrangian for N real scalar fields $\phi_1, \ldots, \phi_N$

$$\mathscr{L} = \frac{1}{2} \sum_{i=1}^{N} \partial_\mu \phi_i \partial^\mu \phi_i = \frac{1}{2} \partial_\mu \Phi^T \partial^\mu \Phi \tag{1.19}$$

is invariant under global rotations in N dimensions, O(N) under which the N-dimensional column vector $\Phi$ changes as

$$\Phi \rightarrow \Phi' = R\Phi \quad , \tag{1.20}$$

where R is a rotation matrix (proper and improper). Since $\Phi^T\Phi$ (the length of $\Phi$) is O(N) invariant, the matrix R obeys

$$R^T R = RR^T = 1 \quad . \tag{1.21}$$

Proper rotation matrices can be written in the form

$$R = e^{\frac{i}{2} \omega^{ij} \Sigma_{ij}} \quad , \tag{1.22}$$

where $\omega^{ij} = -\omega^{ji}$ are the real $\frac{N(N-1)}{2}$ parameters of the rotation group, and the $\Sigma_{ij}$ are the $\frac{N(N-1)}{2}$ generators of the rotation group. By considering the infinitesimal change in $\Phi$ and by requiring that the group properties be satisfied

$$\delta\Phi = \frac{i}{2} \omega^{ij} \Sigma_{ij} \Phi \quad , \tag{1.23}$$

one can prove that the $\Sigma_{ij}$ satisfy a Lie algebra

$$[\Sigma_{ij}, \Sigma_{k\ell}] = i\delta_{ik}\Sigma_{j\ell} + i\delta_{j\ell}\Sigma_{ik} - i\delta_{i\ell}\Sigma_{jk} - i\delta_{jk}\Sigma_{i\ell} \quad . \tag{1.24}$$

In the above we have derived the Lie algebra for SO(N) by using the N x N $\Sigma_{ij}$ matrices which act on the N-dimensional vector $\Phi$. It is easy to see from (1.21) and (1.22) that they are real and antisymmetric. However, one can build many kinds of matrices which satisfy (1.24). This is because there are many kinds of ways of representing SO(N). We have chosen to do it in the N dimensional representation but we could as well have described it in the adjoint representation which has the same number of dimensions as there are parameters in the group. In the case of SO(N) it can be represented by an antisymmetric second rank tensor $A_{ij} = -A_{ji}$. A convenient way of treating the adjoint is produced by treating $A_{ij}$ as the matrix elements of an antisymmetric matrix A. Then the rotations take the form

$$A \to A' = RAR^T; \qquad A^T = -A \qquad , \qquad (1.25)$$

where R is the N x N matrix given by (1.22). Then it is easy to build an invariant Lagrangian with A as scalar fields

$$\mathcal{L} = \frac{1}{4}\,\mathrm{Tr}(\partial_\mu A^T \partial^\mu A) \qquad . \qquad (1.26)$$

The symmetric "quadrupole" representation $S_{ij} = +S_{ji}$ can be handled in a similar way when the trace of S is recognized to be SO(N) invariant. Starting from the $\underset{\sim}{N}$ representation of SO(N), one can construct more complicated representations described by higher rank tensors. An arbitrarily high rank tensor is in general a combination of tensors that transform irreducibly (among themselves) under the group. For instance consider a third rank tensor $T_{ijk}$ and take i,j,k = 1,...10 for convenience. It is decomposed into irreducible representations of SO(10) as follows:

- the totally antisymmetric part $T_{[ijk]}$ with $\frac{10\cdot9\cdot8}{1\cdot2\cdot3} = 120$ components
- the totally symmetric part $T_{(ijk)}$ has $\frac{10\cdot11\cdot12}{1\cdot2\cdot3} = 220$ components,

but it contains, by contracting two indices, a vector $T_{iij}$ with 10 dimensions. Hence $T_{(ijk)}$ is decomposed as an irreducible 210 dimensional representation plus a 10 dimensional representation of $SO_{10}$.

- tensors with mixed symmetry among the indices: antisymmetric under the interchange of two indices with 320 components (= 45 x 10 - 10 - 120), and symmetric in two indices only, with 320 + 10 components.

Thus in summary we have obtained the SO(10) decomposition of the third rank reducible tensor with 1,000 components into its irreducible parts

$$1000 = 120 + 220 + 10 + 10 + 320 + 320 \quad .$$

This type of construction is straightforward (if tedious); the only subtlety occurs when N is even in which case one can use the Levi-Civitá tensor $\varepsilon_{ij\ldots k}$ with N entries to split the N/2-rank totally antisymmetric tensor in half. Furthermore when N/2 is even, this results in splitting the N/2-rank antisymmetrized tensor into two real and inequivalent representations. When N/2 is odd the procedure results in two representations which are the conjugate of one another. For instance in SO(10), the 5th rank totally antisymmetrized tensor has 252 components, which split into the 126-dimensional representation and its conjugate, the ε-symbol acting as a conjugation.

The procedure of taking tensor products of vectors does not generate all representations because SO(N) has in addition spinor representations (e.g., SO(3) has half-integer spin representations). When N = 2n+1, n = 1,2,..., SO(N) has only one real fundamental spinor representation of dimension $2^n$, e.g., SO(3) has a two dimensional real representation, SO(5) a real four-dimensional spinor representation, etc., out of which all representations can be built. When N = 2n, n = 2,4,6, SO(N) has two real and inequivalent fundamental spinor representations each with $2^{n-1}$ dimensions. Finally for N = 2n, n = 3,5,..., SO(N) has two fundamental complex spinor representations conjugate to one another. For instance SO(6) has a 4 and a $\overline{4}$ conjugate to one another, etc. All representations can be constructed from these spinor representations, which means that they are in this sense more fundamental than the vector representation.

Consider now the kinetic term for N two component spinor fields

$$\mathcal{L}_F = \frac{1}{2}\,\psi_L^{\dagger a}\sigma^{\mu}\overleftrightarrow{\partial}_{\mu}\psi_{La} \quad , \qquad (1.27)$$

where a runs from 1 to N and the sum over a is implied. For a = 1 we have seen that (1.27) is invariant under a phase transformation. When a > 1, $\mathcal{L}$ is invariant under a much larger symmetry: consider the change (suppressing the a index)

$$\psi_L \to U\psi_L \quad , \qquad (1.28)$$

where U is an N x N matrix; and then

$$\psi_L^{\dagger} \to \psi_L^{\dagger}U^{\dagger} \quad . \qquad (1.29)$$

Clearly if U is x-independent and unitary

$$UU^{\dagger} = U^{\dagger}U = 1 \quad . \qquad (1.30)$$

$\mathcal{L}_F$ is invariant under the transformation (1.28). The unitarity condition implies that U can be expressed in terms of an hermitean N x N matrix

$$U = e^{iH}; \qquad H = H^{\dagger} \quad . \qquad (1.31)$$

This hermitean matrix depends on $N^2$ real parameters. Note that by taking H proportional to the identity matrix we recover the earlier phase invariance. The extra new transformations are then generated by the traceless part of H, expressed in terms of $N^2 - 1$ parameters by

$$H = \sum_{A=1}^{N^2-1} \omega^A T^A, \qquad T^{A^\dagger} = T^A \tag{1.32}$$

where the $\omega^A$ are real parameters and the $T^A$ are hermitean traceless N x N matrices. They generate SU(N), the unitary group in N dimensions, and satisfy the appropriate Lie algebra

$$[T^A, T^B] = if^{ABC}T^C \quad , \tag{1.33}$$

where $f^{ABC}$ are real totally antisymmetric coefficients called the structure constants of the algebra [this relation is similar to (1.24), but has different f's]. Some celebrated examples are

<u>N = 2</u>: $T^A = \frac{1}{2}\sigma^A$, $\sigma^A$: Pauli spin matrices A=1,2,3

<u>N = 3</u>: $T^A = \frac{1}{2}\lambda^A$, $\lambda^A$: Gell-Mann matrices A=1,...,8.

The Gell-Mann matrices satisfy the normalization condition

$$\mathrm{Tr}(\lambda^A\lambda^B) = 2\delta^{AB} \tag{1.34}$$

and are given by

$$\lambda_1 = \begin{pmatrix} 0 & 1 & 0 \\ 1 & 0 & 0 \\ 0 & 0 & 0 \end{pmatrix} \quad \lambda_2 = \begin{pmatrix} 0 & -i & 0 \\ i & 0 & 0 \\ 0 & 0 & 0 \end{pmatrix} \quad \lambda_3 = \begin{pmatrix} 1 & 0 & 0 \\ 0 & -1 & 0 \\ 0 & 0 & 0 \end{pmatrix}$$

$$\lambda_4 = \begin{pmatrix} 0 & 0 & 1 \\ 0 & 0 & 0 \\ 1 & 0 & 0 \end{pmatrix} \quad \lambda_5 = \begin{pmatrix} 0 & 0 & -i \\ 0 & 0 & 0 \\ i & 0 & 0 \end{pmatrix} \quad \lambda_6 = \begin{pmatrix} 0 & 0 & 0 \\ 0 & 0 & 1 \\ 0 & 1 & 0 \end{pmatrix}$$

$$\lambda_7 = \begin{pmatrix} 0 & 0 & 0 \\ 0 & 0 & -i \\ 0 & i & 0 \end{pmatrix} \quad \lambda_8 = \frac{1}{\sqrt{3}}\begin{pmatrix} 1 & 0 & 0 \\ 0 & 1 & 0 \\ 0 & 0 & -2 \end{pmatrix} \quad . \tag{1.35}$$

Under $SU_N$, then, we have the following fundamental representations

$$\psi \sim N \text{ means that} \qquad \delta\psi = i\omega^A T^A \psi \tag{1.36}$$

$$\psi' \sim \bar{N} \text{ means that} \qquad \delta\psi' = -i\omega^A T^{A*} \psi' \quad , \tag{1.37}$$

where the last transformation property is obtained by requiring that $\psi'^{\dagger}\psi$ be invariant. The tensor structure of SU(N) is simpler than that of SO(N): associate a lower (upper) index a to a quantity transforming as the N ($\bar{N}$) representation of SU(N), i.e., $\psi_a \sim N$, $\psi^a \sim \bar{N}$. Starting from these as building blocks we can generate all the representations of SU(N) by taking tensor products. One representation of interest is the adjoint representation $M^a_b$ where M is traceless hermitean and contains $N^2 - 1$ components; as its indices indicate it is constructed out of the product of $\underset{\sim}{N}$ and $\underset{\sim}{\bar{N}}$: $N \times \bar{N} = (N^2-1) + 1$. It is convenient to express it as a matrix M which then transforms as

$$M \to UMU^{\dagger} \tag{1.38}$$

or

$$\delta M = i\omega^A [T^A, M] \quad . \tag{1.39}$$

Alternatively we could have represented the adjoint representation by an $N^2 - 1$ dimensional real vector in which case the representation matrices $T^A$ would have been $(N^2-1) \times (N^2-1)$ dimensional.

Other types of representations can be built as tensors with arbitrary numbers of upper and lower indices. Upper and lower indices can be contracted to make singlets but not lower or upper indices among themselves. Thus for instance $T_{ab}$ can be broken down to its irreducible components just by the symmetry scheme of the ab indices: symmetric and antisymmetric. For example take SU(5):

$$T_{ab} = T_{(ab)} + T_{[ab]}$$

$$\underset{\sim}{5} \times \underset{\sim}{5} = \underset{\sim}{15} + \underset{\sim}{10} \quad .$$

Here (...) means total symmetry, [...] total antisymmetry.

We have seen that by considering the kinetic terms of fermion and scalar fields, we can generate Lagrangians invariant under unitary and orthogonal transformations. It is also possible to generate symplectic group invariance by means of a kinetic term. We note that for a Grassmann Majorana field the possible candidate for a kinetic term

$$\frac{1}{4} \bar{\psi}_M \gamma_5 \gamma^\mu \overleftrightarrow{\partial}_\mu \psi_M \tag{1.40}$$

is identically zero when there is only one Majorana field. However, consider the case of an even number of Majorana fields, $\psi_{Mi}$ $i = 1 \ldots 2N$, coupled by means of an antisymmetric numerical matrix $E_{ij} = -E_{ji}$ with entries

$$E_{ij} = \begin{cases} +1 & i > j \\ 0 & i = j \\ -1 & i < j \end{cases} \quad . \tag{1.41}$$

In this case we can form invariant expressions of the form (1.40) and obtain a nonzero result provided we antisymmetrize in the running i indices. In this way we arrive at the nonvanishing kinetic term

$$\mathcal{L} = \frac{1}{4} \bar{\psi}_{Mi} E_{ij} \gamma_5 \gamma^\mu \overleftrightarrow{\partial}_\mu \psi_{Mj} \quad . \tag{1.42}$$

It is invariant under transformations that leave an antisymmetric quadratic expression invariant. These transformations form the symplectic group Sp(2N), with the $\psi_{Mi}$ transforming as the real 2N dimensional defining representation. As the presence of the antisymmetric tensor $E_{ij}$ indicates, the Sp(2N) singlet resides in the antisymmetric product of two 2N representations. [As a consequence, one cannot form a kinetic term for scalar fields transforming as the 2N of Sp(2N).] In fact, for symplectic groups irreducible representations appear in the symmetric tensor product, and reducible representations in the antisymmetrized product [exactly the opposite of rotation groups]. In particular the adjoint representation is given by $(\underset{\sim}{2N} \times \underset{\sim}{2N})_{sym.}$ and therefore contains N(2N+1) elements. A case of interest arises when N = 1. Then the Lagrangian (1.42) is seen to be invariant under SU(2) because the E-matrix can be identified with the Levi-Civitá symbol $\varepsilon_{ij}$. This is no accident: the Lie algebras of Sp(2), SU(2) and SO(3) are the same. In fact, we note that Sp(2N) has the same number of dimensions as SO(2N+1). By matching representations, we see that we have another identification

$$SO(5) \sim Sp(4)$$

since the $\underset{\sim}{4}$ of Sp(4) has the same dimensions as the spinor of SO(5). However, SO(7) is not the same as Sp(6) since SO(7) has no six-dimensional representation. [Moreover even when the Lie algebras match, their global properties may be different.]

In conclusion we note that by building different types of kinetic terms, we have been able to generate Lagrangians invariant under O(N), U(N) and Sp(2N). Other Lie groups called exceptional groups do not appear in our list because they are not defined in terms of quadratic invariants only, such as kinetic and mass terms; their specification appears at the level of higher order invariants. These can appear in the interaction part of the Lagrangian. So exceptional symmetries are interaction symmetries. Here we do not give example of Lagrangians invariant under exceptional groups but merely list them: $G_2$ with 14 generators, rank 2 and only real representations generated by the fundamental seven-dimensional representation; $F_4$ with 52 generators, rank 4, only real representations generated by the fundamental 26-dimensional representation; $E_6$ with 78 generators, rank 6, and real and complex representations generated by the fundamental 27 or $\overline{27}$ representation; $E_7$ with 133 generators, rank 7, only real representations generated by the 56; and finally $E_8$ with 248 generators has the unique feature of having the adjoint 248-dimensional representation as its fundamental.

Problems.

A. Given a Lie algebra

$$[T^A, T^B] = if^{ABC} T^C \qquad A,B,C = 1, \ldots K$$

where the $f^{ABC}$ are totally antisymmetric real coefficients. The $f^{ABC}$ can be regarded as K K x K matrices $(f^A)^{BC}$. Using the Jacobi identity show that these K x K matrices obey the same Lie algebra as the $T^A$ when a proper factor of -i is provided.

B. Given a complex third rank tensor $T_{abc}$ a,b,c = 1,...5, decompose it into its SU(5) irreducible components.

C. For SU(N), given N-1 different fields $T^1_a, T^2_a, \ldots T^{N-1}_a$, show that it is always possible to build out of their product a field transforming as the $\bar{N}$ representation (i.e., with an upper index).

D. For SU(3) express the product of two Gell-Mann matrices in terms of the Gell-Mann matrices.

E. Show that the Lie algebras of SU(2) and SO(3), SU(4) and SO(6) are isomorphic.

F. Given $\mathscr{L} = \frac{1}{2} \partial_\mu \Phi^T \partial^\mu \Phi$ where $\Phi$ is a column vector with N real scalar fields. Find the Noether currents and charges. What conditions must be imposed on the fields if the Noether charges are to satisfy the Lie algebra of SO(N)?

G. Show explicitly that the Lagrangian (1.42) is SU(2) invariant when N = 2.

## 2. Construction of Locally Symmetric Lagrangians

In the previous section we have shown how to build local phase invariance into a Lagrangian. Now we show how to do the same for the more complicated non-Abelian Lie symmetries we just discussed.

In the following we will use the Lagrangian for N complex two component spinor fields to illustrate the construction. However, the reader must be cautioned that the "gauging" of unitary symmetries with left-handed fields results in a nonrenormalizable theory because of the Adler-Bell-Jackiw anomaly. Since we concern ourselves at this stage only with classical considerations, we temporarily ignore this subtlety. The uneasy reader can carry out the same construction with N Dirac four component spinors if he wishes.

As we just saw, the Lagrangian

$$\mathscr{L} = \psi_L^{\dagger a} \sigma^\mu \partial_\mu \psi_{La} \tag{2.1}$$

where a is summed from 1 to N, is invariant under global U(N) transformations

$$\psi_L(x) \rightarrow U\psi_L(x) ; \qquad U^\dagger U = UU^\dagger = 1 \tag{2.2}$$

with

$$U = e^{i\alpha} e^{i\omega^A T^A} \quad , \tag{2.3}$$

in which the $T^A$ are the $N^2 - 1$ traceless hermitean matrices that generate SU(N). We now want to generalize (2.1) to incorporate invariance under local transformations of the form (2.2), i.e.,

$$\psi_L(x) \rightarrow U(x)\psi_L(x) \tag{2.4}$$

where now

$$U = e^{i\alpha(x)} e^{i\omega^A(x) T^A} \quad . \tag{2.5}$$

Note that it is pretty much a matter of choice how much of the global symmetry one wants to gauge. For instance, we could have limited ourselves only to gauging any subgroup of SU(N). Here we gauge the whole thing! When U depends on x, the derivative term $\partial_\mu\psi_L$ no longer transforms as it should; indeed

$$\partial_\mu\psi_L(x) \to \partial_\mu U(x)\psi_L(x) = [\partial_\mu U(x)]\psi_L(x) + U(x)\partial_\mu\psi_L(x) \tag{2.6}$$

$$\neq U\partial_\mu\psi_L(x) \quad . \tag{2.7}$$

So we look for a generalization of the derivative which does not spoil the invariance of $\mathscr{L}$. We define accordingly the covariant derivative $\mathscr{D}_\mu$ by demanding that

$$\mathscr{D}_\mu\psi_L(x) \to U(x)\mathscr{D}_\mu\psi_L(x) \tag{2.8}$$

or in operator form

$$\mathscr{D}_\mu \to \mathscr{D}'_\mu = U(x)\mathscr{D}_\mu U^+(x) \quad . \tag{2.9}$$

We emphasize that $\mathscr{D}_\mu$ is, in this case, an N x N matrix so if we wanted to show all indices we would write (2.8) as

$$[\mathscr{D}_\mu\psi_L(x)]_a \to [U(x)]_a^b(\mathscr{D}_\mu)_b^c\psi_{Lc}(x) \quad . \tag{2.10}$$

Then if we can find such a $\mathscr{D}_\mu$, the new Lagrangian

$$\mathscr{L}' = \frac{1}{2}\psi_L^\dagger\sigma^\mu\mathscr{D}_\mu\psi_L \tag{2.11}$$

is locally invariant under U(N). Since $\mathscr{D}_\mu$ is to generalize $\partial_\mu$, let us try the ansatz

$$\mathscr{D}_\mu = \partial_\mu + iA_\mu(x) \tag{2.12}$$

where we have suppressed the U(N) indices; $A_\mu(x)$ is an N x N hermitean matrix with vector elements since $i\partial_\mu$ is hermitean and a vector:

$$A_\mu(x) = B_\mu(x)1 + A_\mu^B(x)T^B \qquad , \qquad (2.13)$$

the $T^B$ being the $N^2 - 1$ hermitean generators of SU(N). The transformation requirement (2.9) implies that

$$\partial_\mu + iA'_\mu(x) = U(x)[\partial_\mu + iA_\mu(x)]U^\dagger(x)$$

$$= \partial_\mu + U(x)[\partial_\mu U^\dagger(x)] + iU(x)A_\mu(x)U^\dagger(x) \qquad (2.14)$$

or

$$A'_\mu(x) = -iU(x)[\partial_\mu U^\dagger(x)] + U(x)A_\mu(x)U^\dagger(x) \qquad . \qquad (2.15)$$

It is easy to see that the fields $B_\mu(x)$ and $A_\mu^B(x)$ transform separately. Indeed taking the trace of (2.15), we find

$$B'_\mu(x) = -\frac{i}{N}\,Tr(U(x)[\partial_\mu U^\dagger(x)]) + B_\mu(x) \qquad . \qquad (2.16)$$

It can be shown (see problem) that

$$Tr(U(x)[\partial_\mu U^\dagger(x)]) = -iN\partial_\mu\alpha(x) \qquad , \qquad (2.17)$$

leading to

$$B'_\mu(x) = -\partial_\mu\alpha(x) + B_\mu(x) \qquad (2.18)$$

which is the transformation previously obtained. Now by multiplying (2.15) by $T^C$ and taking the trace, we obtain the change in the $N^2 - 1$ fields $A_\mu^C$ using the trace properties of the T-matrices

$$TrT^A = 0 \qquad Tr(T^AT^B) = \frac{1}{2}\,\delta^{AB} \qquad , \qquad (2.19)$$

where we have used a conventional normalization. It is perhaps easier to consider infinitesimal "gauge transformations" (2.15). Setting

$$U(x) = 1 + i\omega^AT^A + \ldots \qquad , \qquad (2.20)$$

we arrive at

$$\delta A_\mu(x) = A'_\mu(x) - A_\mu(x) = -T^B\partial_\mu\omega^B(x) + i\omega^B(x)[T^B,A_\mu(x)] + \mathcal{O}(\omega^2) \quad . \tag{2.21}$$

Multiply by $T^C$ and take the trace, using (2.19) and (2.13) we find

$$\delta A^C_\mu(x) = -\partial_\mu\omega^C(x) + 2i\omega^B(x)\mathrm{Tr}([T^B,A_\mu(x)]T^C) + \mathcal{O}(\omega^2) \ . \tag{2.22}$$

The $T^A$'s obey the Lie algebra of SU(N)

$$[T^A,T^B] = if^{ABC}T^C \quad , \tag{2.23}$$

from which we finally obtain

$$\delta A^C_\mu(x) = -\partial_\mu\omega^C(x) - \omega^B(x)A^D_\mu(x)f^{BDC} + \mathcal{O}(\omega^2) \quad . \tag{2.24}$$

The remarkable thing about the gauge transformation (2.24) is that it is expressed in a way that does not depend on the representation of the fermion fields we started with.

The variation (2.21) can be rewritten very elegantly in terms of the covariant derivative: under an SU(N) transformation,

$$\omega^A T^A \equiv \omega \rightarrow U\omega U^\dagger \quad . \tag{2.25}$$

Hence the covariant derivative acting on $\omega$ is given by (see problem)

$$\mathcal{D}_\mu\omega = \partial_\mu\omega + i[A_\mu,\omega] \quad . \tag{2.26}$$

Comparison with (2.21) yields

$$\delta A_\mu(x) = -\mathcal{D}_\mu\omega \quad , \tag{2.27}$$

which shows that even if $A_\mu(x)$ does not transform covariantly under SU(N) because of the $U\partial_\mu U^\dagger$ term, its infinitesimal change does since it can be expressed in terms of a covariant derivative.

So far we have enlarged our Lagrangian in order to have local U(N) symmetry. The price has been the introduction of $N^2$ vector fields to build the covariant derivative. In order to give these new fields an existence of their own, we should include their kinetic terms hopefully in a way that does not break the original local symmetry. For the $B_\mu(x)$ field corresponding to the overall phase transformations, we just repeat the steps of the previous paragraph. So let us rather concentrate on the $N^2 - 1$ fields which come with local SU(N) invariance. The trick in constructing a kinetic term invariant under (2.15) is in building things out of the covariant derivative $\mathcal{D}_\mu$.

Consider the hermitean quantity

$$F_{\mu\nu} \equiv -i[\mathcal{D}_\mu,\mathcal{D}_\nu] \quad . \qquad (2.28)$$

It is assured to transform covariantly since $\mathcal{D}_\mu$ does, that is

$$F_{\mu\nu}(x) \to U(x)F_{\mu\nu}(x)U^\dagger(x) \quad . \qquad (2.29)$$

Using the expression for $\mathcal{D}_\mu$ in the fundamental representation (2.12) and omitting the $B_\mu$ field, we obtain

$$\begin{aligned} F_{\mu\nu} &= -i[\partial_\mu + iA_\mu, \partial_\nu + iA_\nu] \\ &= \partial_\mu A_\nu - \partial_\nu A_\mu + i[A_\mu, A_\nu] \end{aligned} \quad . \qquad (2.30)$$

Since $F_{\mu\nu}(x)$ is an hermitean N x N traceless matrix, we can expand it in the $T^B$'s

$$F_{\mu\nu}(x) = F^B_{\mu\nu}(x)T^B \quad , \qquad (2.31)$$

with

$$F^B_{\mu\nu}(x) = \partial_\mu A^B_\nu(x) - \partial_\nu A^B_\mu(x) - f^{BCD}A^C_\mu(x)A^D_\nu(x) \quad , \qquad (2.32)$$

where we have used (2.13) without $B_\mu$ and (2.23). These $F_{\mu\nu}$'s are, of course, the Yang-Mills generalization of the field strengths of electromagnetism. They are not all independent for they obey the Bianchi identities

$$\mathcal{D}_\mu F_{\rho\sigma} + \mathcal{D}_\rho F_{\sigma\mu} + \mathcal{D}_\sigma F_{\mu\rho} = 0 \quad , \tag{2.33}$$

where the $\mathcal{D}_\mu$'s acting on the $F_{\mu\nu}$'s are to be understood in the sense of (2.25) since the $F_{\mu\nu}$'s transform as members of the adjoint of SU(N). These identities are a direct consequence of the Jacobi identity for the covariant derivative

$$[\mathcal{D}_\mu,[\mathcal{D}_\rho \mathcal{D}_\sigma]] + [\mathcal{D}_\rho,[\mathcal{D}_\sigma,\mathcal{D}_\mu]] + [\mathcal{D}_\sigma,[\mathcal{D}_\mu,\mathcal{D}_\rho]] = 0 \quad . \tag{2.34}$$

These are just kinematic constraints which are trivially satisfied by the field strengths.

It is now easy to build an invariant kinetic term. It is given by

$$\mathcal{L}_{YM} = - \frac{1}{2g^2} \mathrm{Tr}(F_{\mu\nu}F^{\mu\nu}) \quad , \tag{2.35}$$

with the normalization (2.19) for the T-matrices; it generalizes the Maxwell Lagrangian, and can be seen to have the proper dimensions - g is a dimensionless coupling.

We remark that $\mathcal{L}_{YM}$ does not depend on the representation of the fermions, and therefore stands on its own as a highly nontrivial theory. Furthermore, by taking the $f^{ABC}$ structure functions to be those of the other Lie groups, we can obtain the corresponding Yang-Mills theories for these other Lie groups.

The discerning reader may have wondered why we did not consider the other invariant

$$I = \text{Tr}\ \varepsilon^{\mu\nu\rho\sigma} F_{\mu\nu} F_{\rho\sigma} \tag{2.36}$$

as a candidate for the kinetic term. After all it is Lorentz and gauge invariant and has the proper dimension. The answer is that it can be expressed as a pure divergence. To see this we write

$$I = 4\varepsilon^{\mu\nu\rho\sigma}\text{Tr}([\partial_\mu A_\nu + iA_\mu A_\nu][\partial_\rho A_\sigma + iA_\rho A_\sigma]) \tag{2.37}$$

$$= 4\varepsilon^{\mu\nu\rho\sigma}\text{Tr}[\partial_\mu A_\nu \partial_\rho A_\sigma + 2iA_\mu A_\nu \partial_\rho A_\sigma] \quad , \tag{2.38}$$

where we have eliminated the AAAA term using the cyclic property of the trace. Now

$$\varepsilon^{\mu\nu\rho\sigma}\text{Tr}(A_\mu A_\nu \partial_\rho A_\sigma) = \frac{1}{3}\,\partial_\rho \varepsilon^{\mu\nu\rho\sigma}\text{Tr}(A_\mu A_\nu A_\sigma) \quad , \tag{2.39}$$

so that

$$I = 4\partial_\rho\{\varepsilon^{\mu\nu\rho\sigma}\text{Tr}[A_\sigma \partial_\mu A_\nu + \frac{2i}{3} A_\sigma A_\mu A_\nu]\} \quad , \tag{2.40}$$

using $\varepsilon^{\mu\nu\rho\sigma}\partial_\rho \partial_\mu A_\nu = 0$. Thus we arrive at

$$\varepsilon^{\mu\nu\rho\sigma}\text{Tr}(F_{\mu\nu} F_{\rho\sigma}) = 4\partial_\rho W^\rho \tag{2.41}$$

with

$$W^\rho = \varepsilon^{\rho\sigma\mu\nu}\text{Tr}[A_\sigma \partial_\mu A_\nu + \frac{2i}{3} A_\sigma A_\mu A_\nu] \quad . \tag{2.42}$$

It means that by taking I as the kinetic Lagrangian, we could not generate any equation of motion for $A_\mu$ since it would only affect the Action at its end points. We can, however, add it to $\mathcal{L}_{YM}$, resulting in a canonical transformation on $A_\mu$.

Problems.

A. Given a Weyl field transforming as the $\underset{\sim}{6}$ of SU(3), build the SU(3) covariant derivative acting on it in terms of the Gell-Mann matrices.

B. Show that

$$\mathrm{Tr}[U^{\dagger}(x)\partial_{\mu}U(x)] = iN\partial_{\mu}\alpha(x) \quad .$$

C. If $\omega$ transforms as the adjoint representation of SU(N), show that its covariant derivative is given by $\mathcal{D}_{\mu}\omega = \partial_{\mu}\omega + i[A_{\mu},\omega]$ and transforms in the same way as $\omega$ where $A_{\mu}$ is the matrix of gauge fields.

D. Show from the gauge transformation properties of $A_{\mu}$ that the field strength $F_{\mu\nu} = \partial_{\mu}A_{\nu} - \partial_{\nu}A_{\mu} + i\,[A_{\mu},A_{\nu}]$ does indeed transform as the adjoint of SU(N).

E. Starting from $\mathcal{L} = \frac{1}{2}\,\partial_{\mu}\Phi^{T}\partial^{\mu}\Phi$, where $\Phi(x)$ is an N column vector of real scalar fields, generalize it to be locally invariant under SO(N), duplicating the procedure in the text. How many vector fields must be introduced? Show that their infinitesimal change under an SO(N) transformation can also be expressed in terms of the covariant derivative acting on the gauge parameters.

## 3. The Pure Yang-Mills Theory

In this section we study the classical properties of the Yang-Mills Action given by

$$S^{Y.M.} = -\frac{1}{2g^2}\int d^4x\, Tr(F_{\mu\nu}F^{\mu\nu}) \quad , \qquad (3.1)$$

where

$$F_{\mu\nu} = \partial_\mu A_\nu - \partial_\nu A_\mu + i[A_\mu, A_\nu] \quad , \qquad (3.2)$$

and

$$A_\mu(x) = A_\mu^B(x)T^B \quad . \qquad (3.3)$$

The $T^B$ matrices generate one of the Lie algebras

$$[T^B, T^C] = if^{BCD}T^D \quad , \qquad (3.4)$$

with the indices B,C,D running from 1 to K, the dimension of the Lie algebra, itself defined by the totally antisymmetric structure constants $f^{BCD}$. As a consequence of (3.4), the $T^B$ matrices are traceless; they are normalized to satisfy

$$Tr(T^BT^C) = \frac{1}{2}\delta^{BC} \quad . \qquad (3.5)$$

The possible Lie algebras have been classified in E. Cartan's thesis; they are the classical algebras SU(N), dimension $N^2 - 1$, $N \geq 2$; SO(N), dimension $\frac{N(N-1)}{2}$, $N > 2$; Sp(2N), dimension N(2N+1), $N > 1$; and the exceptional Lie algebras $G_2(14)$, $F_4(52)$, $E_6(78)$, $E_7(133)$ and $E_8(248)$ with their dimensions indicated in parenthesis.

One can also write the Yang-Mills Action independently of the $T^B$ matrices,

$$S^{Y.M.} = -\frac{1}{4g^2}\int d^4x(F_{\mu\nu}^B F^{\mu\nu B}) \quad , \qquad (3.6)$$

where now

$$F^B_{\mu\nu} = \partial_\mu A^B_\nu - \partial_\nu A^B_\mu - f^{BCD}A^C_\mu A^D_\nu \quad . \qquad (3.7)$$

It follows that

$$g^2 S^{Y.M.} = \int d^4x[-\frac{1}{2}\partial_\mu A^B_\nu \partial^\mu A^{\nu B} + \frac{1}{2}\partial_\mu A^B_\nu \partial^\nu A^{\mu B}$$

$$+ g f^{BCD} A^C_\mu A^D_\nu \partial^\mu A^{\nu B} - \frac{g^2}{4} f^{BCD} f^{BEF} A^C_\mu A^D_\nu A^{\mu E} A^{\nu F}] \quad . \qquad (3.8)$$

The first two terms are recognized to be of the same type as in Maxwell's Lagrangian (except for the summation). However, the next two show that the vector fields have highly nontrivial cubic and quartic interactions among themselves.

The derivation of the equations of motion proceeds most easily in the matrix form. We start by varying the Action

$$\delta S = -\frac{1}{g^2}\int d^4x \, Tr(F_{\mu\nu}\delta F^{\mu\nu}) \quad , \qquad (3.9)$$

where

$$\delta F^{\mu\nu} = \partial^\mu \delta A^\nu + i\delta A^\mu A^\nu + iA^\mu \delta A^\nu - (\mu \leftrightarrow \nu) \quad . \qquad (3.10)$$

Hence, using the antisymmetry of $F_{\mu\nu}$,

$$\delta S = -\frac{2}{g^2}\int d^4x \, Tr[F_{\mu\nu}(\partial^\mu \delta A^\nu + i\delta A^\mu A^\nu + iA^\mu \delta A^\nu)] \quad . \qquad (3.11)$$

Next we integrate the first term by parts, throwing away the surface term due to the vanishing of the variation at the boundaries. Using the cyclic properties of the trace, we arrive at

$$\delta S = \frac{2}{g^2}\int d^4x \, Tr[(\partial^\mu F_{\mu\nu} + i[A^\mu, F_{\mu\nu}])\delta A^\nu] \qquad (3.12)$$

from which we read off the equation of motion in matrix form

$$\partial^\mu F_{\mu\nu} + i[A^\mu, F_{\mu\nu}] = 0 \quad . \tag{3.13}$$

Since $F_{\mu\nu}$ transforms according to the adjoint representation, this equation can be expressed directly in terms of the covariant derivative

$$\mathcal{D}^\mu F_{\mu\nu} = 0 \quad , \tag{3.14}$$

which shows that it is itself covariant. In addition the $F_{\mu\nu}$ fields satisfy the kinematic (Bianchi) constraints (as do the $F_{\mu\nu}$ of electromagnetism)

$$\mathcal{D}^\mu \tilde{F}_{\mu\nu} = 0 \quad , \tag{3.15}$$

where

$$\tilde{F}_{\mu\nu} = \frac{1}{2} \varepsilon_{\mu\nu\rho\sigma} F^{\rho\sigma} \tag{3.16}$$

is the dual of $F_{\mu\nu}$. We emphasize that (3.15) is not an equation of motion since it is trivially solved by expressing $F_{\mu\nu}$ in terms of the potentials.

From the equation of motion (3.13) it is clear that one can define a current $j_\nu$ which is conserved; indeed the expression

$$j_\nu = -\partial^\mu F_{\mu\nu} = i[A^\mu, F_{\mu\nu}] \tag{3.17}$$

does satisfy

$$\partial^\nu j_\nu = 0 \quad , \tag{3.18}$$

leading to the conserved charges [here in matrix form $Q^A T^A$]

$$Q \equiv \int d^3x j_0 \tag{3.19}$$

$$= - \int d^3x \partial^i F_{i0} \tag{3.20}$$

$$= - \oint d^2\sigma^i F_{i0} \quad , \tag{3.21}$$

where the last integral is over the space surface at infinity. Now the current $j_\nu$ has atrocious transformation properties under a gauge transformation, but the charges Q, as can be seen from (3.21), transform nicely under a very large class of gauge transformations. From (3.21) we have

$$Q \to Q' = -\oint d^2\sigma^i U F_{i0} U^+ \quad , \qquad (3.22)$$

where the U's are on the bounding surface at infinity. Thus by requiring that we limit ourselves to U's which are constant in space at spatial infinity, we can take them out of the surface integral and obtain a covariant transformation for the conserved charges. We should add that this current is the Noether current obtained by canonical methods.

We can couple the Yang-Mills system by adding to $S^{Y.M.}$ a term of the form

$$\frac{2}{g} \int d^4x \, \mathrm{Tr}(A^\mu J_\mu) \quad , \qquad (3.23)$$

where $J_\mu(x)$ is an external source written here in matrix form:

$$J_\mu(x) = J_\mu^B(x) T^B \quad . \qquad (3.24)$$

Then the equations of motion read

$$\mathscr{D}^\mu F_{\mu\nu} = J_\nu \quad . \qquad (3.25)$$

From this equation, we can require that $J_\nu$ transform covariantly in order to preserve the covariance of the equation of motion:

$$J^\mu \to U J^\mu U^\dagger \quad . \qquad (3.26)$$

Furthermore it is not hard to see that $J^\mu$ must be covariantly conserved as a consequence of the equation of motion (see problem)

$$\mathscr{D}^\mu J_\mu = \partial^\mu J_\mu + i[A^\mu, J_\mu] = 0 \quad . \qquad (3.27)$$

We remark that the Noether current is <u>not</u> $J_\mu$ but rather

$$j_\mu = -\partial^\rho F_{\rho\mu} + J_\mu \quad . \qquad (3.28)$$

Now if we go back to the extra term (3.23) we see that it is not invariant under a gauge transformation. Assuming that $J^\mu$ transforms covariantly, we find that

$$\delta \int d^4x \, \mathrm{Tr}(A_\mu J^\mu) = -\int d^4x \, \mathrm{Tr}(J^\mu \partial_\mu \omega)$$

$$= \int d^4x \, \mathrm{Tr}(\omega \partial^\mu J_\mu) \quad , \qquad (3.29)$$

which means that we can restore invariance if the external source $J^\mu$ is conserved. In Maxwell's theory this is no problem since $J^\mu$ does not transform under a change of gauge. But in Yang-Mills the statement $\partial_\mu J^\mu = 0$ is not covariant. This means that coupling to sources in this way breaks gauge invariance. This should not be too surprising. After all, we have seen that by reversing our earlier construction, the way to couple $A_\mu$ in a gauge invariant way is to add a kinetic term for the fields which make up the source $J^\mu$. Having an external nondynamical source will not do.

Nevertheless one is free to examine the solutions of the classical equations (3.25) together with the constraint (3.27), but remember that coupling Yang-Mills to nondynamical external sources is a shady business.

Let us now return to the sourceless equations of motion (3.14). There are in Minkowski space many solutions of this equation. Just as in electrodynamics, there are plane wave solutions to this equation (see problem). They have infinite energy (but finite energy density). However, unlike in Maxwell's theory, they cannot be superimposed to produce finite energy solutions because of the nonlinear nature of this theory, unless they move in the same direction.

There exist many other very interesting finite energy solutions to this equation but they involve some sort of singularities and therefore imply the existence of singular sources (see problem).

In Euclidean space, the Yang-Mills equation of motion has an exceedingly rich structure. Euclidean space can be regarded as Minkowski space with an imaginary time, and in Quantum Mechanics, processes with imaginary time evolution formally correspond to tunneling which happens instantly in real time. Hence 't Hooft called the nonsingular solutions of the sourceless Yang-Mills equation in Euclidean space <u>instantons</u>. On the other hand, we have seen earlier that the Feynman Path Integral may be better defined in Euclidean space. Hence the study of Euclidean space solutions is doubly interesting.

Let us concentrate on Euclidean space solutions which have finite action following Belavin, Polyakov, Schwartz and Tyupkin, Phys. Lett. 59B, 85 (1975). In Euclidean space we have

$$\mathrm{Tr}[(F_{\mu\nu}-\tilde{F}_{\mu\nu})(F_{\mu\nu}-\tilde{F}_{\mu\nu})] \geq 0 \quad , \quad (3.30)$$

since it is the sum of squares. It follows that

$$\mathrm{Tr}(F_{\mu\nu}F_{\mu\nu}+\tilde{F}_{\mu\nu}\tilde{F}_{\mu\nu}) \geq 2\ \mathrm{Tr}(F_{\mu\nu}\tilde{F}_{\mu\nu}) \quad (3.31)$$

or, using

$$\mathrm{Tr}(\tilde{F}_{\mu\nu}\tilde{F}_{\mu\nu}) = \mathrm{Tr}(F_{\mu\nu}F_{\mu\nu}) \quad , \quad (3.32)$$

that

$$\mathrm{Tr}\ F_{\mu\nu}F_{\mu\nu} \geq \mathrm{Tr}\ F_{\mu\nu}\tilde{F}_{\mu\nu} \quad , \quad (3.33)$$

which establishes upon integration a lower bound for the value of the Yang-Mills Euclidean Action. Clearly equality is achieved when

$$F_{\mu\nu} = \tilde{F}_{\mu\nu} \qquad , \qquad (3.34)$$

corresponding to self dual solutions. Antiself-dual solutions also correspond to a lower bound. It is easy to see that self-dual or antiself-dual solutions have zero Euclidean energy momentum tensor (see problem). The integral of the right-hand side of the inequality (3.33) can be rewritten as the integral of a divergence (see eq. (2.41))

$$\int d^4x \, Tr(F_{\mu\nu}\tilde{F}_{\mu\nu}) = 4 \int d^4x \, \partial_\mu W_\mu \qquad , \qquad (3.35)$$

where

$$W_\mu = \varepsilon_{\mu\nu\rho\sigma} Tr[A_\nu \partial_\rho A_\sigma + \frac{2i}{3} A_\nu A_\rho A_\sigma] \qquad , \qquad (3.36)$$

so that

$$S_E^{Y.M.} = \frac{1}{2g^2} \int d^4x \, TrF_{\mu\nu}F_{\mu\nu} \geq \frac{2}{g^2} \oint_S d^3\sigma_\mu W_\mu \qquad , \qquad (3.37)$$

where the last term is integrated over the bounding surface at Euclidean infinity. Hence the minimum value of the action will depend on the properties of the gauge fields at infinity.

Now in order for $S^{Y.M.}$ to be finite, it must be that $F^B_{\mu\nu}$ decreases sufficiently fast at Euclidean infinity

$$F^B_{\mu\nu}(x) \underset{|x|\to\infty}{\to} 0 \qquad , \qquad (3.38)$$

which means in general that $A_\mu$ tends to a configuration

$$A_\mu = -iU\partial_\mu U^\dagger , \qquad \text{for } x^2 \to \infty \qquad (3.39)$$

which is obtained from $A_\mu = 0$ by a gauge transformation; it therefore gives $F_{\mu\nu} = 0$.

Now recall that $S^{Y.M.}$ is bounded from below by a quantity which depends entirely on the behavior of the potentials at Euclidean infinity. In fact, substituting (3.39) into (3.36) we see that on S

$$W_\mu = \frac{1}{3}\,\varepsilon_{\mu\nu\rho\sigma}\,\mathrm{Tr}[U\partial_\nu U^\dagger U\partial_\rho U^\dagger U\partial_\sigma U^\dagger] \qquad , \qquad (3.40)$$

where we have used the antisymmetry of ρ and σ and $UU^\dagger = 1$. Thus

$$S_E^{Y.M.} \geq \frac{2}{3g^2}\oint_S d^3\sigma_\mu\,\varepsilon_{\mu\nu\rho\sigma}\,\mathrm{Tr}[U\partial_\nu U^\dagger U\partial_\rho U^\dagger U\partial_\sigma U^\dagger] \qquad , \qquad (3.41)$$

which depends entirely on the group element U(x)! We have the remarkable result that the (minimum) value of the Euclidean Action depends on the properties of U(x) only and not on the details of the field configurations at finite x.

Let us specialize to the case of SU(2). There the group elements U(x) depend on three parameters, call them $\phi_1$, $\phi_2$, $\phi_3$, which are themselves x-dependent. On the other hand, the surface of integration S is the surface of a sphere with very large (~ infinite) radius. Thus we can think of U as a mapping between the three group parameters and the three coordinates which label the surface of our sphere, that is of a three-sphere onto a three-sphere. Such mappings are characterized by their homotopy class. It roughly corresponds to the number of times one sphere is mapped onto the other. For instance, homotopy class 1 means that the surface of the sphere $S_3^\infty$ at Euclidean infinity is mapped only once on the surface of the sphere $S_3$ of the group manifold labeled by the angles $\phi_i$. In general, homotopy class n means that n points of $S_3^\infty$ are mapped into one point of $S_3$, etc.

If we set

$$\partial_\mu U^\dagger = \sum_{a=1}^{3} \frac{\partial\phi_a}{\partial x^\nu} \frac{\partial}{\partial\phi_a} U^\dagger = \partial_\mu \phi^a \partial_a U^\dagger \tag{3.42}$$

we arrive at

$$S_E^{Y.M.} \geq \frac{2}{3g^2} \oint_S d^3\sigma_\mu \varepsilon_{\mu\nu\rho\sigma} \partial_\nu \phi^a \partial_\rho \phi^b \partial_\sigma \phi^c \, \mathrm{Tr}(U\partial_a U^\dagger U\partial_b U^\dagger U\partial_c U^\dagger) \quad , \tag{3.43}$$

or, using the antisymmetry of the $\varepsilon$ symbol

$$S_E^{Y.M.} \geq \frac{4}{g^2} \oint_S d^3\sigma_\mu \varepsilon_{\mu\nu\rho\sigma} \partial_\nu \phi^1 \partial_\rho \phi^2 \partial_\sigma \phi^3 \, \mathrm{Tr}(U\partial_1 U^\dagger U\partial_2 U^\dagger U\partial_3 U^\dagger) \quad . \tag{3.44}$$

In this form we see clearly the Jacobian of the transformation between variables that label the surface S and the angle $\phi_a$. But, as we have just discussed, this map is characterized by its homotopy class n, when $S_3^\infty$ is mapped n times onto the group manifold of $SU_2$. By parametrizing U in terms of, say, Euler angles, it is straightforward to arrive at

$$S_E^{Y.M.} \geq \frac{8\pi^2}{g^2} n \quad , \tag{3.45}$$

where n is an integer, given by

$$n = \frac{1}{16\pi^2} \int d^4x \, \mathrm{Tr}(F_{\mu\nu} \tilde{F}_{\mu\nu}) \quad ; \tag{3.46}$$

it is called the Pontryargin index.

Thus Euclidean solutions with finite action are labeled by their homotopy class which gives the lower bound for the (Euclidean) Action. The lower bound is attained when the field configurations are either dual or antiself dual, i.e., when

$$F_{\mu\nu} = \pm \tilde{F}_{\mu\nu} \quad . \tag{3.47}$$

As an example, consider the original instanton solution; there the Euclidean SU(2) potential is given by

$$A_\mu(x) = \frac{-ix^2}{x^2+\lambda^2} U\partial_\mu U^\dagger \quad , \quad (3.48)$$

where

$$U = \frac{1}{\sqrt{x^2}} (x_0 - i\vec{x}\cdot\vec{\sigma}) \quad , \quad (3.49)$$

where the $\sigma$ matrices act in the SU(2) space, and

$$x^2 = x_0^2 + \vec{x}\cdot\vec{x} \quad . \quad (3.50)$$

It satisfies the requirement (3.39) for finite action [$\lambda^2$ is a constant]. It can be shown that it is self-dual and that the form of U implies Pontryargin index + 1.

Finally, let us mention that in Yang-Mills theories, functions which transform under gauge transformations cannot in general be taken to be constant because they can be given an x-dependence by a gauge transformation. The closest one can define is a covariant constant which satisfies

$$\mathcal{D}_\mu\phi = (\partial_\mu + iA_\mu)\phi = 0 \quad , \quad (3.51)$$

where we have suppressed all group indices. In solving for $\phi$, we are going to unearth a very interesting object: the path ordered integral. To see it we note that

$$\phi(x+dx) = \phi(x) + dx^\mu\partial_\mu\phi + \ldots \quad , \quad (3.52)$$

where $dx_\mu$ is an arbitrarily small displacement. Using (3.51), we obtain

$$\begin{aligned}\phi(x+dx) &= \phi(x) - idx^\mu A_\mu\phi(x) + \ldots \\ &= e^{-idx^\mu A_\mu}\phi(x) + \mathcal{O}(dx)^2 \quad . \quad (3.53)\end{aligned}$$

Since under a gauge transformation

$$\phi(x) \to U(x)\phi(x) \quad , \quad (3.54)$$

it follows from (3.53) that

$$e^{-idx^\mu A_\mu} \to U(x+dx)e^{-idx^\mu A_\mu(x)}U^\dagger(x) \quad , \quad (3.55)$$

which is the fundamental relation we sought to obtain. Now, (3.51) can be integrated by iterating on the displacement: $\phi(y)$ can be obtained from $\phi(x)$ by taking small displacements along a curve that begins at x and ends at y, thus obtaining

$$\phi(y) = (Pe^{-i\oint_x^y dx\cdot A})\phi(x) \quad , \quad (3.56)$$

where the path ordered exponential is defined by

$$Pe^{-i\int dx\cdot A} \equiv \prod_k (i-idx_k^c\cdot A(x_k)) \quad , \quad (3.57)$$

$dx_k$ being the displacement centered around $x_k$ on the curve C:

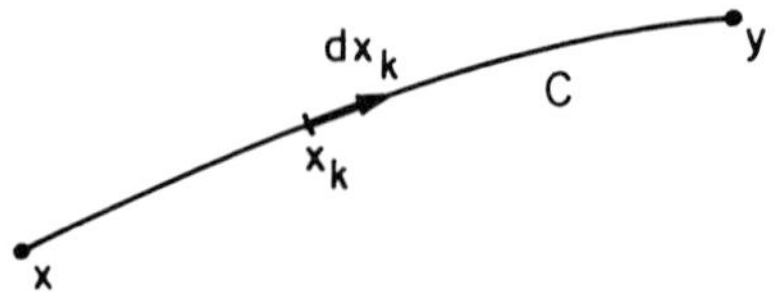


From (3.55) it follows that

$$Pe^{-i\oint_x^y dx\cdot A} \to U(y)Pe^{-i\oint_x^y dx\cdot A}U^\dagger(x) \quad , \quad (3.58)$$

and in particular the path ordered exponential along a closed path transforms like a local covariant quantity:

$$Pe^{-i\oint dx\cdot A} \rightarrow U(x)Pe^{-i\oint dx\cdot A}U^{\dagger}(x) \qquad , \qquad (3.59)$$

so that its trace is gauge invariant. It is a functional of the path.

There are many more aspects of the classical Yang-Mills theory we have not touched on, such as monopole solutions, generalization of instanton solutions, meron solutions with infinite Euclidean action (but finite Minkowski action and singular sources), etc. Alas it is time to go on and start thinking about how to define the quantum Yang-Mills theory.

Problems.

A. Show that the field configuration [S. Coleman, Phys. Lett. 70B, 59 (1977)]

$$A_1^B = A_2^B = 0$$

$$A_0^B = A_3^B = x_1 F_1^B(x^0+x^3) + x_2 F_2^B(x^0+x^3)$$

is a solution of the Yang-Mills equations of motion, where $F_{1,2}^B$ are arbitrary functions. Compare these solutions with the plane wave solutions of Maxwell's theory.

*B. Analyze the Wu-Yang Ansatz for SU(2) Yang-Mills

$$A_0^C = x^C \frac{g(r)}{r^2} ; \qquad A_i^C = \varepsilon_i^{cj} x^j \frac{f(r)}{r^2} ,$$

where C is the SU(2) index C = 1,2,3, r is the length of the position vector x. [Recall that for SU(2) $f^{ABC} = \varepsilon^{ABC}$ the Levi-Civitá tensor.] Derive the equations that f and g must satisfy. Show that they are satisfied by f = 1, g = constant. For this solution describe the potential and field configurations and find the energy density and energy.

C. For an SU(2) gauge theory, show that the 't Hooft-Corrigan-Fairlie-Wilczek Ansatz for the potentials in terms of one scalar field $\phi$

$$A_0^C = -\frac{1}{\phi} \partial^C \phi ; \qquad A_i^C = \frac{1}{\phi} [\delta_i^C \partial_0 \phi - \varepsilon_i^{\ Cj} \partial_j \phi] ,$$

implies that $\phi$ obeys the equation of motion for the $\lambda\phi^4$ theory where $\lambda$ is an arbitrary constant.

D. Show that the Noether energy momentum tensor for the Euclidean Yang-Mills theory can be written in the form

$$\theta_{\mu\nu} = \frac{1}{2g^2} (F^B_{\mu\rho} + \tilde{F}^B_{\mu\rho})(F^B_{\nu\rho} - \tilde{F}^B_{\nu\rho}) \quad .$$

E. Find the change in $W_\mu$ under a gauge transformation, and verify that $\partial_\mu W_\mu$ is gauge invariant.

F. Evaluate the trace of the path ordered exponential around a closed loop for the instanton solution described in the text. Choose a simple path at your convenience.

# VII. PATH INTEGRAL FORMULATION OF GAUGE THEORIES

Defining the Feynman Path Integral for gauge theories poses special problems. As we have just seen, these theories have Actions which are invariant under space-time dependent transformations

$$S[A_\mu] = S[A'_\mu] \quad .$$

Thus integrating blindly over all field configurations will lead to a tremendous amount of redundant infinite integrations and thus make the Path Integral more infinite than usual. We have treated in Appendix A the case of an integral whose integrand did not depend on all the variables it displayed. In that case we saw that the integral could be defined by clever alteration of the measure; the new measure limited itself to integrating over only nonredundant variables while keeping a covariant look. The application of this technique to the FPI for gauge theories leads to the Faddeev-Popov formulation.

Alternatively, we can require that the FPI really make sense only at the level of the Hamiltonian formalism, keeping in the spirit of the Quantum Mechanical correspondence of Chapter II. As it turns out both methods lead to the same answer. Still it is instructive to present them both.

## 1. Hamiltonian Formalism of Gauge Theories: Abelian Case

To start with, consider the case of an Abelian field (Maxwell's theory). There the Lagrangian is $[F_{\mu\nu} = \partial_\mu A_\nu - \partial_\nu A_\mu]$

$$\mathcal{L} = -\frac{1}{4} F_{\mu\nu}F^{\mu\nu} \tag{1.1}$$

$$= -\frac{1}{4} [2F_{0i}F^{0i} + F_{ij}F^{ij}] \quad . \tag{1.2}$$

Define the canonical momenta

$$\pi_\mu = \frac{\partial \mathcal{L}}{\partial[\partial_0 A^\mu]} \quad , \qquad (1.3)$$

and postulate the fundamental Poisson Bracket (PB) relations at equal times

$$\{A_\mu(\vec{x},t), \pi_\nu(\vec{y},t)\}_{PB} = -g_{\mu\nu}\delta(\vec{x}-\vec{y}) \quad , \qquad (1.4)$$

all other PB's vanishing at equal times. The Hamiltonian density is given by

$$\mathcal{H}(\vec{x},t) = \pi^\mu \partial_0 A_\mu - \mathcal{L} \quad , \qquad (1.5)$$

in terms of which the equations of motion are for any function f of the canonical variables $A_\mu$ and $\pi_\mu$

$$\dot{f} = \{f,H\}_{PB} \quad , \qquad (1.6)$$

where H is the energy

$$H = \int d^3x \; \mathcal{H}(\vec{x},t) \quad . \qquad (1.7)$$

The canonical procedure we have just outlined works well for many cases, such as the scalar field theory. However, when we apply it to gauge theories, it goes wrong right away. Indeed, for the Abelian theory, we find using (1.2) in (1.3)

$$\pi_\mu = F_{0\mu} \quad , \qquad (1.8)$$

which gives

$$\pi_0 = 0 \quad , \qquad (1.9)$$

using the antisymmetry of $F_{\mu\nu}$. This goes counter to the fundamental PB relation for $\mu = \nu = 0$. It then becomes evident that if we wish to keep the fundamental PB relation, we must treat (1.9) in a very special way.

Still let us continue for the moment, and compute the naive Hamiltonian which we call $H_0$. After integrating by parts we obtain from (1.5) and (1.7)

$$H_0 = \int d^3x[\frac{1}{4} F_{ij}F^{ij} - \frac{1}{2} \pi_i\pi^i + A_0\partial_i\pi^i] \quad , \qquad (1.10)$$

and we see that the velocities have disappeared: the raison d'être of the Hamiltonian formalism. However, the fact that $\pi^0$ vanishes means that the change of variables from velocities to momenta $\partial_0 A_\mu \rightarrow \pi_\mu$ is singular: you cannot map four things into three without paying a price. This means that the definition of $\mathcal{H}$ is not unique: we can add to it any arbitrary function proportional to $\pi_0$. So we write a new Hamiltonian

$$H = H_0 + \int d^3xF\pi_0 \quad , \qquad (1.11)$$

where F is arbitrary. In order to find the meaning of F, we apply the equation of motion (1.6) to $f = A_0$. We find

$$\dot{A}_0 = \{A_0,H\}_{PB} \qquad (1.12)$$

$$= F \quad , \qquad (1.13)$$

where we have used the fundamental PB (1.4). If F depends on canonical variables, there will be extra contributions to (1.12) but they will be multiplied by $\pi_0$ which is eventually set to zero. Hence we do not lose any generality by considering F to be independent of the canonical variables. It implies that if at a given time $t_0$ we start with a value $A_0$, its value at $t_0 + \delta t$ will be given by this totally arbitrary function. What does it mean? The extra term $\int d^3xF\pi^0$ has the effect of changing $A_0$ but leaves $A_i$ alone, so that it has the same effect as a gauge transformation

$$A_\mu \rightarrow A_\mu + \partial_\mu\lambda \quad , \qquad (1.14)$$

where $\lambda(\vec{x},t_0) = 0$ but with $\dot{\lambda}(\vec{x},t_0) \neq 0$. Therefore the extra term in (1.11) generates a special kind of gauge transformation.

But this is not the whole story: consider in turn the change in $\pi^0$

$$\dot{\pi}_0 = \{\pi_0, H\}_{PB} = -\partial_i \pi^i(\vec{x},t) \quad , \qquad (1.15)$$

using (1.4) and (1.10). But $\pi_0$ is zero for all times by the canonical procedure. Hence we obtain another constraint

$$\partial_i \pi^i = 0 \qquad (1.16)$$

which involves only the canonical momenta! Let us pause for a moment: first we had $\pi^0 = 0$ coming from the canonical procedure. Dirac calls this type of constraint a primary constraint. Then by using the equations of motion we find another constraint involving the $\pi$'s. Dirac calls this type a secondary constaint [Dirac, Can. J. Math., Vol. 2, 129 (1950)]. Thus we have even more relations between the $\pi$'s. It looks like we are mapping four velocities into two independent $\pi$'s. Thus we have to add yet another term to H to reflect this extra arbitrariness

$$H_{extra} = \int d^3x G(\vec{x},t) \partial_i \pi^i(\vec{x},t) \quad . \qquad (1.17)$$

Using (1.6) we can see what type of change it generates

$$\begin{aligned} \delta A_0 &= \{A_0, H_{extra}\}_{PB} \\ &= 0 \end{aligned} \quad , \qquad (1.18)$$

and

$$\begin{aligned} \delta A_j &= \{A_j, H_{extra}\} \\ &= \partial_j G \end{aligned} \quad , \qquad (1.19)$$

using (1.4). Therefore it generates a gauge transformation constant in time; G must not depend on t. So the alteration of the canonical formalism has led us to a Hamiltonian that does gauge transformations as it takes the system through time. By absorbing $A_0$ in G, we can rewrite our new and final Hamiltonian as

$$H_{new} = \int d^3x[\frac{1}{4} F_{ij}F^{ij} - \frac{1}{2} \pi^i\pi_i + G\partial_i\pi^i] \quad . \quad (1.20)$$

Note that we have dropped the $\pi^0$ term from (1.20) because $H_{new}$ no longer depends on $A_0$ in this form. Now we are ready to describe the resulting physical system. Let f be any function of $A_i$ and $\pi_i$. Its time variation is given by

$$\dot{f} = \{f,H_{new}\}_{PB} \quad , \quad (1.21)$$

and contains an arbitrary element due to the $\partial_i\pi^i$ term. This is not acceptable for a truly physical quantity whose time variation is not arbitrary. Hence we demand that

$$\{f_{phys.},\partial_i\pi^i\} = 0 \quad , \quad (1.22)$$

which means that $f_{phys}$ must not depend on the variable that is conjugate to $\partial_i\pi^i$. In other words a physical quantity must be defined only on some surface in the $(A_i,\pi_i)$ plane. We can always characterize such a surface by the equation

$$g(A_i,\pi_i) = 0 \quad , \quad (1.23)$$

provided that the change of variables between g and z, the variable conjugate to $\partial_i\pi^i$, is not singular, i.e.,

$$\det\left|\frac{\delta g}{\delta z}\right| = \det\left|\{g,\partial^i\pi_i\}_{PB}\right| \neq 0 \quad , \quad (1.24)$$

where we have used the definition of the PB's. Note that z is defined by

$$\frac{\delta}{\delta z(y)} = \int d^3x \, \frac{\delta(\partial_i \pi^i)(y)}{\delta \pi_j(x)} \, \frac{\delta}{\delta A^j(x)} = \{ \quad , \partial_i \pi_i \} \quad . \tag{1.25}$$

The physical meaning of (1.25) should be clear since $\partial_i \pi^i$ generates gauge transformations, and g must be able to fix the gauge.

Assuming that the condition (1.24) is satisfied, we can perform a canonical transformation

$$(A_i, \pi_i) \rightarrow (\tilde{A}_i, \tilde{\pi}_i) \quad ,$$

where we judiciously take

$$g(A_i, \pi_i) = \tilde{A}_3 \quad . \tag{1.26}$$

Now since $\tilde{A}_i$ and $\tilde{\pi}_i$ are conjugate variables, it follows that

$$\det\{g, \partial^i \pi_i\}_{PB} = \det\{ \frac{\delta g}{\delta \tilde{A}_j} \frac{\partial[\partial^i \pi_i]}{\partial \tilde{\pi}_j} - \frac{\delta g}{\delta \tilde{\pi}_j} \frac{\delta[\partial^i \pi_i]}{\delta \tilde{A}_j} \}_{PB}$$

$$= \det[ \frac{\delta[\partial^i \pi_i]}{\delta \tilde{\pi}_3} ] \tag{1.27}$$

just becomes the Jacobian for the transformation $\partial^i \pi_i \rightarrow \tilde{\pi}_3$. If it is not singular we can solve $\partial^i \pi_i = 0$ in order to express $\tilde{\pi}_3$ in terms of the remaining variables. Note that the Hamiltonian does not change under this transformation. Let us give several examples:

a) <u>The Coulomb Gauge</u>. It is defined by taking

$$g = \partial^i A_i \quad . \tag{1.28}$$

Then we form

$$\det\{\partial^i A_i, \partial_j \pi^j\}_{PB} = \det[\partial^i_x \partial_{iy} \delta(\vec{x}-\vec{y})] \quad , \qquad (1.29)$$

which is to be interpreted as the product of eigenvalues of the Laplace operator $\partial_i \partial^i$. It is well known not to have zero eigenvalues except for a constant solution which we eliminate by setting appropriate boundary conditions. Thus the Coulomb gauge satisfies our criterion to be a good gauge. It means that we can use $\partial^i \pi_i = 0$ to express the variable conjugate to (1.28) in terms of the remaining canonical variables. Dropping the twiddles we write

$$\pi_i = \pi^L_i + \pi^T_i \qquad (1.30)$$

$$A_i = A^L_i + A^T_i \quad , \qquad (1.31)$$

where by construction the transverse modes are divergenceless:

$$\partial^i \pi^T_i = \partial^i A^T_i = 0 \quad . \qquad (1.32)$$

Then the Coulomb gauge reads

$$A^L_i = 0 \qquad (1.33)$$

and the constraint (1.16) now reads

$$\partial^i \pi^L_i = \partial^i \partial_i \phi = 0 \quad , \qquad (1.34)$$

when expressing $\pi^L_i$ in terms of $\phi$. The invertibility of the Laplace operator is precisely what enables us to set $\phi = 0$. Then in this gauge we have the remaining canonical variables $\pi^T_i$ and $A^T_i$ both divergenceless. The Hamiltonian is now in terms of these variables

$$H = \frac{1}{2} \int d^3x (B_i B_i + \pi^T_i \pi^T_i) \quad , \qquad (1.35)$$

where

$$B_i = \frac{1}{2} \varepsilon_{ijk} \partial_j A_k^T \quad . \tag{1.36}$$

b) <u>The Arnowitt-Fickler or Axial Gauge</u>, characterized by

$$g = A_3 = 0 \quad . \tag{1.37}$$

In this case, the determinant condition is

$$\det[\frac{\partial}{\partial x_3} \delta(\vec{x}-\vec{y})] \neq 0 \quad , \tag{1.38}$$

which is satisfied since the operator $\frac{\partial}{\partial x_3}$ is invertible. It means that we can solve for $\pi_3$ by using (1.16). The result is

$$\pi_3(x,y,z,t) = -\int_{-\infty}^{z} dz'(\partial^1\pi_1+\partial^2\pi_2)(x,y,z',t) \quad , \tag{1.39}$$

where we have (arbitrarily) set a boundary condition on $\pi_3$. The system is now described in terms of the canonical variables $A_1$, $A_2$, $\pi_1$, $\pi_2$ and the (nonlocal) Hamiltonian

$$H = \frac{1}{2} \int d^3x[B_3^2+B_1^2+B_2^2+\pi_1^2+\pi_2^2+\pi_3^2(\pi_1,\pi_2)] \tag{1.40}$$

where $\pi_3$ is given by (1.39) and

$$B_1 = -\partial_3A_2,\ B_2 = \partial_3A_1,\ B_3 = (\partial_1A_2-\partial_2A_1) \quad . \tag{1.41}$$

It is now easy to write the FPI. Let $\tilde{A}_\perp$ and $\tilde{\pi}_\perp$ be the independent variables. The Feynman Path Integral is now taken to be

$$\int \mathcal{D}\tilde{A}_\perp \mathcal{D}\tilde{\pi}_\perp \, e^{i\int[\tilde{\pi}_\perp \cdot \dot{\tilde{A}}_\perp - \mathcal{H}(\tilde{A}_\perp,\tilde{\pi}_\perp)]d^4x} \tag{1.42}$$

$$= \int \mathcal{D}\tilde{A}_\perp \mathcal{D}\tilde{\pi}_\perp \mathcal{D}\tilde{A}_3 \delta[\tilde{A}_3] \mathcal{D}\tilde{\pi}_3 \delta[\tilde{\pi}_3-\tilde{\pi}_3(\tilde{\pi}_\perp)] e^{i\int[\ldots]} \quad , \tag{1.43}$$

where $\tilde{\pi}_3(\tilde{\pi}_\perp)$ is the expression of $\tilde{\pi}_3$ in terms of the transverse variables obtained by inverting (1.16). Now

$$\delta[\tilde{\pi}_3-\tilde{\pi}_3(\tilde{\pi}_\perp)] = \delta[\partial^i\tilde{\pi}_i]\det|\{\partial^i\tilde{\pi}_i,\tilde{A}_3\}_{PB}| \quad , \tag{1.44}$$

and

$$\delta[\partial^i\tilde{\pi}_i] = \int \mathcal{D}A_0 \; e^{i\int A_0\partial^i\tilde{\pi}_i} \quad . \tag{1.45}$$

These enable us to rewrite (1.43) in the form

$$\int \mathcal{D}\tilde{A}_\perp \mathcal{D}\tilde{A}_3 \mathcal{D}A_0 \mathcal{D}\tilde{\pi}_\perp \mathcal{D}\tilde{\pi}_3 \delta[\tilde{A}_3]\det|\{\partial^i\tilde{\pi}_i,\tilde{A}_3\}_{PB}| \cdot$$

$$\cdot \; e^{i\int d^4x[\tilde{\pi}_\perp\dot{\tilde{A}}_\perp+\tilde{\pi}_3\dot{\tilde{A}}_3 - \mathcal{H}(\tilde{\pi},\tilde{A})+A_0\partial_i\pi^i]} \quad . \tag{1.46}$$

We have added (at no extra charge because of $\delta[\tilde{A}_3]$) the term $\tilde{\pi}_3\dot{\tilde{A}}_3$ in the exponential, and also

$$\mathcal{H} = \frac{1}{2}(\tilde{\pi}_\perp\tilde{\pi}_\perp+\tilde{\pi}_3\tilde{\pi}_3+\tilde{B}_i\tilde{B}_i) \quad , \tag{1.47}$$

now contains all the components of $\tilde{B}_i$. Let us perform the inverse canonical transformation from the twiddled to the untwiddled variables. The only effect (besides dropping the twiddles) will be to change $\tilde{A}_3$ into the gauge function g. Then the FPI becomes

$$\int \mathcal{D}A_\mu \mathcal{D}\pi_i \delta[g]\det|\{\partial^i\pi_i,g\}_{PB}| e^{i\int d^4x[\pi_i\dot{A}_i-\mathcal{H}+A_0\partial^i\pi_i]} \quad . \tag{1.48}$$

Finally we integrate over the $\pi_i$'s: noting that the exponent can be rewritten as (integrating by parts the $A_0\partial^i\pi_i$ term)

$$\pi_i(\partial_0A_i-\partial_iA_0) - \frac{1}{2}\pi_i\pi_i - \frac{1}{2}B_iB_i \quad . \tag{1.49}$$

By completing the squares we arrive at

$$\int \mathcal{D}A_\mu \; e^{iS[A]} \int \mathcal{D}\pi_i \, \delta[g] \det|\{\partial^i \pi_i, g\}_{PB}| \, e^{-\frac{1}{2}\int d^4x (\pi_i - \partial_0 A_i + \partial_i A_0)^2} \qquad (1.50)$$

where S[A] is the Maxwell action in terms of the potentials. Now we let

$$\pi_i' = \pi_i - \partial_0 A_i + \partial_i A_0 \qquad , \qquad (1.51)$$

and change variables. When g does not depend on $\pi$, this change of variable does not affect g or the Poisson bracket. Thus we can take these out of the $\pi$ integration, but keep the interpretation of the P.B. as the infinitesimal change of g under gauge transformations. The integration over $\pi'$ leaves an infinite constant which we ignore. The end result is

$$\int \mathcal{D}A_\mu \, \delta[g] \det\left|\frac{\delta g}{\delta \omega}\right| e^{iS[A]} \qquad , \qquad (1.52)$$

where S[A] is the Maxwell Action, g is the gauge function, and $\frac{\delta g}{\delta \omega}$ its change under an infinitesimal gauge transformation. Let us apply this formula to the Coulomb gauge; we find

$$\int \mathcal{D}A_\mu \, \delta[\partial^i A_i] \det|\partial^2| e^{iS} \qquad , \qquad (1.53)$$

and we see that the determinant does not contain any dependence on A; it can therefore be absorbed in the normalization. We note that the $A_0$ variable which appears linearly in S gives upon integration a functional $\delta$-function. This shows that $A_0$ is not a dynamical variable although one finds the condition $A_0 = 0$ often referred to as a gauge condition (it isn't!). However, if we insist on setting $A_0 = 0$ we lose the constraint $\partial^i \pi_i = 0$ (Gauss' law) which then must be restored in the problem.

Finally let us mention that one can define a covariant gauge

$$\partial^\mu A_\mu = 0 \qquad , \qquad (1.54)$$

which is truly a gauge condition since it involves the dynamical variable $A_i$.

Problems.

A. Apply the canonical formalism to the Action

$$S = \int d^4x[\frac{1}{2}\,\partial_\mu\phi_1\partial^\mu\phi_1 + \frac{1}{2}\,\partial_\mu\phi_2\partial^\mu\phi_2 + m\chi(\phi_1^2+\phi_2^2)]$$

and define the corresponding Path Integral, treating $\phi_1$, $\phi_2$ and $\chi$ as canonical fields.

B. Consider the gauge condition

$$A_iA_i = m^2 \quad .$$

Discuss its validity as a gauge condition, and write the corresponding Path Integral for Electrodynamics in this gauge.

C. Repeat the above for the gauge condition

$$(\partial_i A_3)\partial^i A_3 = 0 \quad .$$

D. Consider the condition

$$0 = \int_{-\infty}^{\vec{x}} d\vec{z}\cdot\vec{A}(\vec{z},t) \quad ,$$

where the line integral is taken along a curve C. Can it be taken to be a gauge condition?

## 2. Hamiltonian Formalism for Gauge Theories: Non-Abelian Case

Starting from the Yang-Mills Lagrangian we could duplicate the procedure of the previous section with very similar results. However, let us start from the first order formalism where $F_{\mu\nu}$ and $A_\mu$ are taken to be independent variables and where S[F,A] is arranged so as to give $F_{\mu\nu}$ in terms of $A_\mu$ from an equation of motion. So we start from

$$S = -\frac{1}{g^2}\int d^4x \mathrm{Tr}[\frac{1}{2} F_{\mu\nu}F^{\mu\nu} - F^{\mu\nu}(\partial_\mu A_\nu - \partial_\nu A_\mu + i[A_\mu A_\nu])] \quad . \qquad (2.1)$$

It is clear that its variation with respect to $F_{\mu\nu}$ gives eq. (3.2) of Chapter VI, but $F_{\mu\nu}$ has no dynamical meaning since it has no time derivative: it is just an auxiliary field. However, in this form it is easier to rewrite S in a way that has no terms quadratic in time derivatives. Let us introduce the "electric" and "magnetic" fields

$$E_i = F_{0i} \qquad (2.2)$$

$$B_i = \frac{1}{2}\varepsilon_{ijk}F_{jk} \quad , \qquad (2.3)$$

or alternatively

$$F_{ij} = \varepsilon_{ijk}B_k \quad . \qquad (2.4)$$

Using the equation of motion

$$F_{ij} = \partial_i A_j - \partial_j A_i + i[A_i, A_j] \quad , \qquad (2.5)$$

we can rewrite S in the form

$$S = -\frac{1}{g^2}\int d^4x \mathrm{Tr}[B_iB_i + E_iE_i - 2E_i(\partial_0 A_i - \partial_i A_0 + i[A_0, A_i])] \qquad (2.6)$$

$$= -\frac{1}{g^2}\int d^4x \mathrm{Tr}[B_iB_i + E_iE_i - 2E_i\dot{A}_i - 2A_0(\partial_i E_i + i[A_i, E_i])] \quad , \qquad (2.7)$$

where we have used the cyclic property of the trace and integration by parts.

In this way S is rewritten in a way that translates easily to Hamiltonian form. Taking the trace we obtain

$$S = \frac{1}{g^2} \int d^4x[E_i^B \dot{A}_i^B - \frac{1}{2}(E_i^A E_i^A + B_i^A B_i^A) + A_0^B (\mathcal{D}_i E_i)^B] \quad , \quad (2.8)$$

where $E_i^B$ appears as the canonical momentum conjugate to $A_i$ (dot means time derivative); $A_0^B$ plays the role of a Lagrange multiplier, and

$$(\mathcal{D}_i E_i)^B = \partial_i E_i^B + f^{BCD} A_i^C E_i^D \quad . \quad (2.9)$$

Here the dynamical variables are $E_i^B$ and $A_i^B$ for which we postulate the fundamental Poisson Bracket relation at equal times

$$\{A_i^B(\vec{x},t), E_j^C(\vec{y},t)\}_{PB} = \delta^{BC}\delta_{ij}\delta(\vec{x}-\vec{y}) \quad , \quad (2.10)$$

all other PB's being zero at equal times. These variables are not all independent since they must satisfy the constraint

$$(\mathcal{D}_i E_i)^B \equiv \partial_i E_i^B + f^{BCD} A_i^C E_i^D = 0 \quad , \quad (2.11)$$

obtained by varying with respect to $A_0^B$. The equations of motion are given by

$$\frac{df}{dt} = \{f, H_0\}_{PB} + \{f, \int d^3x A_0^B(\vec{x},t)(\mathcal{D}_i E_i)^B(\vec{x},t)\}_{PB} \quad , \quad (2.12)$$

where

$$H_0 = \frac{1}{2} \int d^3x[E_i^A E_i^A + B_i^A B_i^A] \quad . \quad (2.13)$$

Thus, as in the previous section the time variation contains an extra term due to the Lagrange multiplier term. To see its meaning, let us calculate the Poisson Bracket

$$\delta A_i^B(\vec{x},t) = \{A_i^B(\vec{x},t), \int d^3y A_0^C(\vec{y},t)(\mathcal{D}_i E_i)^C(\vec{y},t)\}_{PB} \quad , \quad (2.14)$$

which is to be interpreted as the change in $A_i^B$ under an infinitesimal transformation generated by the extra term. Using the relation (2.10), we find

$$\delta A_i^B(\vec{x},t) = -\partial_i A_0^B(\vec{x},t) - f^{BCD} A_i^C(\vec{x},t) A_0^D(\vec{x},t) \qquad , \qquad (2.15)$$

which is a gauge transformation. Hence, as in the Abelian case, the extra term in H generates gauge transformations, with gauge parameter $A_0^B$.

It is easy to see that the time derivative of the constraints (2.11) is itself proportional to the constraints themselves (see problem). Hence we can generate no further constraints.

As before we take a function of $A_i^B$ and $E_i^B$ to be physical if its change under an infinitesimal time translation is not arbitrary, that is if

$$\{f, (\mathcal{D}_i E_i)^B\}_{PB} = 0 \quad \text{when } (\mathcal{D}_i E_i)^B = 0 \qquad . \qquad (2.16)$$

Now the PB can always be regarded as an integral operator

$$\frac{\delta}{\delta z^B(\vec{x},t)} = \{ \quad , (\mathcal{D}_i E_i)^B(\vec{x},t)\} \qquad , \qquad (2.17)$$

provided that the integrability condition

$$\frac{\delta}{\delta z^B(\vec{x},t)} \frac{\delta}{\delta z^C(\vec{y},t)} - \frac{\delta}{\delta z^C(\vec{y},t)} \frac{\delta}{\delta z^B(\vec{x},t)} = 0 \qquad (2.18)$$

is satisfied. It is not hard to show using the Jacobi identity and the PB between two $\mathcal{D}_i E_i$ (at equal times) that this condition is indeed satisfied (see problem). Thus our physical subspace can be defined by two conditions:

$$(\mathcal{D}_i E_i)^B = 0 \qquad \text{and } \{ \quad , (\mathcal{D}_i E_i)^B\} = 0 \qquad . \qquad (2.19)$$

These conditions restrict us from the functional space spanned by $A_i^B$ and $E_i^B$ i = 1,2,3 to a functional space spanned by $\tilde{A}_1^B$, $\tilde{A}_2^B$ and $\tilde{E}_1^B$, $\tilde{E}_2^B$ in an appropriately chosen basis.

Alternatively we can describe this subspace in another way by replacing the awkward PB condition by another set

$$g^B(A_i^C(\vec{x},t),E_i^C(\vec{x},t)) = 0 \qquad (2.20)$$

which we call the gauge conditions. This alternate definition must not involve any singular change of variables between the $z^C$ functions and the $g^C$ function, that is (functionally)

$$\det\left|\frac{\delta g^C}{\delta z^B}\right| = \det|\{g^C,(\mathcal{D}_iE_i)^B\}| \neq 0 \qquad . \qquad (2.21)$$

This is a necessary condition for $g^B$ to be a desirable gauge choice. Otherwise it would not fix the gauge. Assuming that (2.21) is satisfied, let us be a bit cleverer and restrict ourselves to gauge choices which satisfy (at equal times, again)

$$\{g^B,g^C\}_{PB} = 0 \qquad . \qquad (2.22)$$

Then we can regard $g^B$ as a canonical variable: consider the canonical transformation

$$(A_i^B,E_i^B) \rightarrow (\tilde{A}_j^B,\tilde{E}_j^B) \qquad , \qquad (2.23)$$

where the j indices are just used as labels and do not necessarily transform as vector indices under rotation, and where

$$\tilde{A}_3^B = g^B(A_i,E_i) \qquad . \qquad (2.24)$$

The constraint (2.21) now reads

$$\det\left|\frac{\delta(\mathcal{D}_i\tilde{E}_i)}{\delta\tilde{E}_3}\right| \neq 0 \qquad , \qquad (2.25)$$

since the twiddled variables are conjugate to one another. Now when we take $g^B = \tilde{A}_3^B = 0$, we can no longer make sense of the PB relation (2.10) involving $\tilde{E}_3^B$. This means that $\tilde{E}_3^B$ must now be expressed in terms of the remaining variables. But this is exactly what (2.25) enables us to do: solve the constraint (2.11) by expressing $\tilde{E}_3^B$ in terms of the remaining variables. Thus the Yang-Mills system is now defined in terms of the independent variables $\tilde{A}_\perp^B = (\tilde{A}_1^B, \tilde{A}_2^B)$ and $\tilde{E}_\perp^B = (\tilde{E}_1^B, \tilde{E}_2^B)$ with the Hamiltonian density

$$\mathcal{H} = \frac{1}{2}\,[\tilde{E}_\perp^A \tilde{E}_\perp^A + [\tilde{E}_3^A(\tilde{E}_\perp, \tilde{A}_\perp)]^2 + \tilde{B}_i^A \tilde{B}_i^A] \quad , \qquad (2.26)$$

where $\tilde{E}_3^A(\tilde{E}_\perp, \tilde{A}_\perp)$ is the function that solves the constraint (2.11), and the $\tilde{B}_i^A$ are given by (2.5) and (2.3). Needless to say, H is now very complicated for it involves cubic and quartic interaction terms, besides being nonlocal. Let us now give examples of popular gauge conditions:

a) the Coulomb gauge, defined as in the Abelian case

$$\partial^i A_i^B = 0 \quad . \qquad (2.27)$$

The requirement that (2.27) be a good gauge choice is easily seen to be that the operator

$$\partial^i \frac{\delta A_i^B}{\delta \omega^C} = \partial^i (\partial_i \delta^{BC} - f^{BDC} A_i^D) \qquad (2.28)$$

have no nontrivial zero eigenvalues. Using (2.27) we rewrite it as

$$\mathcal{O}^{BC} \equiv \partial^i \partial_i \delta^{BC} + f^{BCD} A_i^D \partial^i \quad . \qquad (2.29)$$

It has been recently pointed out by Gribov (Nucl. Phys. B139, 1 (1978)) that there exist nontrivial solutions to the equation

$$\mathcal{O}^{BC} f^C = 0 \quad , \qquad (2.30)$$

and that therefore the Coulomb gauge is not a well-defined gauge for Yang-Mills theories, in the sense that it does not allow for an unambiguous extraction of the independent canonical variables. However, there is some order in this madness because the potentials $A_i^B$ satisfying the Coulomb condition and for which the operator (2.28) has zero eigenvalues are not easy to come by. As an example of this problem, consider the potential (in matrix notation)

$$A_i = -iU^+ \partial_i U; \qquad \partial^i A_i = 0 \quad . \qquad (2.31)$$

If the condition (2.27) were sufficient to fix the gauge, then we should be able to derive that the only solution to the equation (2.31) is that $A_i = 0$. Let us specialize to SU(2) and write

$$U = \cos\frac{\omega}{2} + i\vec{\sigma}\cdot\vec{n}\sin\frac{\omega}{2} \quad , \qquad (2.32)$$

where $\vec{n}\cdot\vec{n} = 1$ and $\omega$ depends only on $\vec{x}$. Then it is straightforward to see that (a is the SU(2) index)

$$\partial^i A_i^a = (1+\cos\omega)(\partial_i\omega)\partial^i n^a + (1-\cos\omega)\varepsilon^{abc} n^b \partial^i \partial_i n^c + \sin\omega\,\partial_i\omega\,\varepsilon^{abc} n^b \partial^i n^c$$

$$+ \sin\omega\,\partial^i\partial_i n^a + n^a \partial^i \partial_i \omega \quad . \qquad (2.33)$$

This equation is clearly a mess, so we simplify it: following Gribov, we specialize to spherically symmetric solutions for which

$$n^a = \partial^a r = \frac{x^a}{r} \quad . \qquad (2.34)$$

Then we find that $\omega$ depends only on r and that $\omega(r)$, as a result of the Coulomb condition satisfies

$$\frac{d^2\omega}{dt^2} + \frac{d\omega}{dt} - \sin 2\omega = 0 \qquad , \qquad (2.35)$$

where $t = \ln r$. The nonsingular nature of U requires that

$$\omega(t=-\infty) = 0, 2\pi, 4\pi, \ldots \qquad . \qquad (2.36)$$

This equation is that of a damped pendulum in a constant gravitational field. The boundary condition (2.36) requires that it start at $\omega(t=-\infty)$ in a position of unstable equilibrium. Then, depending on the initial velocity of the bob, three things can happen:

1) it stays at $\omega = 0$ for all times,

2) it starts falling clockwise and then ends up at $t = +\infty$ in its position of stable equilibrium $\omega = -\pi$,

3) it starts falling counterclockwise and ends up the same way as in the previous case.

In addition, the pendulum could swing many times round and then fall in one of the three categories. The first solution corresponds to $A_i = 0$, what one would have expected, but the other two types of solution correspond to nontrivial $A_i$'s. Their existence leads to the Gribov ambiguity. If we let

$$A_i = -i\, e^{-i\ell\omega \frac{\vec{\sigma}\cdot\vec{x}}{r}}\, \partial_i\, e^{i\ell\omega \frac{\vec{\sigma}\cdot\vec{x}}{r}} \qquad , \qquad (2.37)$$

where $\ell = 0, \pm 1, \pm 2, \ldots$, the case $\ell = 0$ corresponds to case 1), while $\ell = \pm 1$ correspond to the other two cases. As $t \to +\infty$, we have the following boundary conditions

$$U \underset{t\to\infty}{\to} \begin{cases} 1 & \text{for } \ell = 0 \\ \pm i \dfrac{\vec{\sigma}\cdot\vec{x}}{r} & \text{for } \ell = \pm 1 \end{cases} \quad . \tag{2.38}$$

Further, if one computes the Pontryargin index for the various Gribov solutions, one finds that

$$n = -\frac{1}{24\pi^2}\int_V d^3x\varepsilon_{ijk}\mathrm{Tr}(A_iA_jA_k) = \begin{cases} 0 & \ell = 0 \\ \pm\frac{1}{2} & \ell = \pm 1 \end{cases} \quad . \tag{2.39}$$

Thus the nontrivial Gribov solutions have topological charge $\pm\frac{1}{2}$ (instantons have $\pm 1$). They certainly do not correspond to run-of-the-mill potential configurations!

Therefore it seems plausible to expect that the operator (2.28) has zero eigenvalue only for $A_i$'s that have nontrivial topological structure (I do not know of a proof of this), i.e., $n \neq 0$. Therefore if we limit ourselves to perturbations about zero potentials which have $n = 0$, we can ignore this problem. However, its resolution is still an open question whenever nonperturbative Yang-Mills phenomena are contemplated. All one can say is that the Coulomb gauge does restrict phase space as required but only modulo copies corresponding to $\ell = \pm 1, \pm 2, \ldots$ . We will come back to this problem after discussing the axial gauge.

b) the Arnowitt-Fickler or axial gauge characterized by

$$n^i A_i^B = 0 \qquad n^i n_i = 1 \quad , \tag{2.40}$$

where $\vec{n}$ is a constant vector. There the operator

$$n^i \frac{\delta A_i^B}{\delta\omega^C} = n^i[\partial_i \delta^{BC} + f^{BCD} A_i^D] \tag{2.41}$$

$$= n^i \partial_i \delta^{BC} \tag{2.42}$$

reduces to the same form as in the Abelian case and it is apparently invertible so that there does not seem to be any Gribov problem in this gauge [more on this later]. Thus we can invert the constraint (2.11) and solve for $E_3^A$ with the result ($n_i = \delta_{i3}$)

$$E_3^A(x,y,z,t) = -\int_{-\infty}^{z} dz' (\mathcal{D}_\perp E_\perp)^A (x,y,z',t) \quad , \tag{2.43}$$

the only tricky thing being the boundary condition at $z = -\infty$ (the soft underbelly of this gauge). Then the Hamiltonian is readily worked out - it is a messy expression not worth expressing here.

Let us for a moment return to the Gribov problem. On the surface, the axial gauge does not seem to be afflicted by the ambiguities we found in the Coulomb gauge. However, they could hide in the boundary condition on $E_3$ necessary to invert $n^i \partial_i$. Thus one could speculate that 1) the problem is not endemic to Yang-Mills and only reflects that the Coulomb gauge is an unfortunate choice or 2) the problem is really there, in which case it must show up in the axial gauge, and then the only place is at spatial infinity. No one knows the answer, but I. Singer has shown that if one defines the Path Integral on the surface of a sphere in Euclidean space, $S_4$, the Gribov problem is endemic and its cause lies in the fact that it is not possible to get away with the same gauge condition over all of space-time. Since we are mainly interested in the perturbative evaluation of the Yang-Mills Path Integral we shall ignore the Gribov problem in the sequel.

The Feynman Path Integral for Yang-Mills theory can now be set up in exactly the same way as in the Abelian case, with only a slight complication due to the indices. Consequently we just state the result

$$\int \mathcal{D}A_{\mu}^{B}\ e^{iS^{YM}[A]}\delta[g^{A}]\det\left|\frac{\delta g^{A}}{\delta\omega^{B}}\right| \quad , \qquad (2.44)$$

where $g^{A}$ is the gauge condition and $S^{YM}[A]$ is the Yang-Mills action in terms of potentials only.

Problems.

A. Evaluate the Poisson brackets

$$\{(\mathcal{D}_i E_i)^A(\vec{x},t), H\} \qquad \text{and} \qquad \{(\mathcal{D}_i E_i)^A(\vec{x},t), (\mathcal{D}_i E_i)^B(\vec{y},t)\} \quad ,$$

and show that the time change of $(\mathcal{D}_i E_i)^A$ vanishes when the constraints are satisfied.

B. Show that the integrability conditions for the z functions defined by

$$\frac{\delta}{\delta z^A(\vec{x},t)} = \{ \qquad , (\mathcal{D}_i E_i)^A(\vec{x},t)\}$$

are satisfied, using some results from problem A.

C. Express the Hamiltonian for the Yang-Mills system in the Coulomb and axial gauges, in terms of the relevant independent canonical variables.

D. Derive the equation of motion for a damped pendulum in a constant gravitational field, and compare with Gribov's equation. For its non-trivial solutions, compute the topological charge (2.39).

E. Show how, in the Coulomb gauge, the ability to use the constraint $(\mathcal{D}_i E_i) = 0$ to get rid of the longitudinal $E_i$ depends crucially on the invertibility of the operator $\partial^i \mathcal{D}_i$.

## 3. Direct Determination of the Yang-Mills FPI: the Faddeev-Popov Procedure

In the last two sections we saw how the classical Hamiltonian formalism could be used to derive the Yang-Mills FPI. The end result was a complicated expression coming from the constraints encountered in the Hamiltonian formalism; these constraints are due to the gauge invariance of the initial Action. There is another, more direct way to see this, by using a procedure pioneered by Faddeev and Popov [Physics Lett. 25B, 29 (1967)].

The Yang-Mills Action is by construction gauge invariant, that is

$$S^{YM}[A_\mu] = S^{YM}[A_\mu^U] \quad , \tag{3.1}$$

where

$$A_\mu^U = UA_\mu U^\dagger - iU\partial_\mu U^\dagger , \quad U(x) = e^{i\vec{\omega}(x)\cdot\vec{T}} \quad . \tag{3.2}$$

This means that the naive expression (in Euclidean space)

$$\int \mathcal{D}A_\mu \; e^{-S}$$

is not well defined if $\mathcal{D}A_\mu$ means summation over all $A_\mu$'s, even those related by gauge transformations. In Appendix A, we have seen how to handle this problem: we have to define a new measure which does not overcount, that is, a measure that sums over a gauge family only once. Roughly speaking it means that we must divide out the redundant integrations [a problem known to mathematicians as the determination of the Haar measure].

Consider the quantity

$$\Delta_g^{-1}[A_\mu] = \int \mathcal{D}U \; \delta[g^B(A_\mu^U)] \quad , \tag{3.3}$$

where $A_\mu^U$ is defined in (3.2). $\mathcal{D}U$ stands for the sum over all group elements, and the $g^B$ are functions that vanish for some $A_\mu^U$. $\Delta_g^{-1}$ is invariant [we neglect nontrivial homotopy classes and the Gribov problem]. Indeed, since

$$\Delta_g^{-1}[A_\mu^{U'}] = \int \mathcal{D}U \; \delta[g^B(A_\mu^{U'U})] \quad , \quad (3.4)$$

change the variables of integration from U to U' where

$$U'' = U'U \qquad \mathcal{D}U'' = \mathcal{D}U \quad . \quad (3.5)$$

The result is

$$\Delta_g^{-1}[A_\mu^{U'}] = \int \mathcal{D}U'' \; \delta(g^B[A_\mu^{U''}]) = \Delta_g^{-1}[A_\mu] \quad , \quad (3.6)$$

since U'' is an integration variable. Thus by cleverly inserting 1 into the naive sum over paths, we obtain

$$\int \mathcal{D}A_\mu \; e^{-S[A]} = \int \mathcal{D}A_\mu \Delta_g[A] \int \mathcal{D}U \; \delta[g^B[A^U]] e^{-S[A]} \quad . \quad (3.7)$$

Perform in the integrand a gauge transformation from $A_\mu^U$ to $A_\mu$, obtaining

$$\int \mathcal{D}A_\mu \Delta_g[A] \int \mathcal{D}U \; \delta[g^B[A_\mu]] e^{-S[A]} \quad , \quad (3.8)$$

where we have used (3.1) and (3.6) and the fact that $\mathcal{D}A_\mu$ is the same as $\mathcal{D}A_\mu^U$. But now nothing depends on U in the integrand and we can take out $\mathcal{D}U$ at the cost of a multiplicative infinity, which is the infinity we wanted to take out in the first place. Hence we define the correct FPI for Yang-Mills to be

$$\int \mathcal{D}A_\mu \Delta_g[A_\mu] \delta[g(A_\mu)] e^{-S[A_\mu]} \quad . \quad (3.9)$$

There remains to evaluate $\Delta_g[A]$. The trick is to regard $g^A(A^U)$ as a function of the group element U(x). Then we can change variables from U(x) to $g^A$. Writing symbolically

$$\mathcal{D}U = \mathcal{D}g \; \det\left|\frac{\delta U}{\delta g}\right| \quad , \quad (3.10)$$

we arrive at

$$\Delta_g^{-1}[A] = \int \mathcal{D}g \ \det\left|\frac{\delta U}{\delta g}\right| \delta[g] \tag{3.11}$$

or

$$\Delta_g[A] = \det\left|\frac{\delta g}{\delta U}\right| \Bigg|_{g=0} \quad . \tag{3.12}$$

These manipulations can be performed if the change of variables from U to g is well defined and not singular: to one group element U(x) corresponds only one g and vice versa. As we have seen this is not true in the Coulomb gauge [Gribov problem]. Furthermore, U(x) is labeled by the same number of parameters as g, but U(x) may have nontrivial boundary conditions and belong to a nonzero homotopy class. In the following we ignore such problems as long as we deal with the perturbative evaluation of the quantum field theory away from $A_\mu = 0$.

If we parametrize U(x) by the functions $\omega^A(x)$, we write $\Delta_g$ in the form

$$\Delta_g[A] = \det\left|\frac{\delta g^A(x)}{\delta \omega^B(y)}\right| \quad . \tag{3.13}$$

Putting it all together we arrive at the final expression for the gauge theory FPI (suppressing indices)

$$\int \mathcal{D}A_\mu \, \delta[g(A)] \det\left|\frac{\delta g}{\delta \omega}\right| e^{-S[A_\mu]} \quad , \tag{3.14}$$

which is the same as that obtained by the Hamiltonian formalism. It is gauge invariant [see Appendix A and problem].

Problems.

A. Show directly from the result

$$\Delta_g[A] = \det\left|\frac{\delta g^A}{\delta \omega^B}\right| \quad ,$$

that $\Delta_g$ is gauge invariant.

B. Show that the final expression

$$\int \mathcal{D}A_\mu \, \delta[g] \det\left|\frac{\delta g}{\delta \omega}\right| e^{-S_E^{YM}}$$

does not depend on g the gauge choice [see Appendix A for hints].

**C. When the Gribov problem is present, $\Delta_g^{-1}$ does not exist. Try to generalize the Faddeev-Popov procedure, i.e., define a new $\Delta_g^{-1}$.

## VIII. PERTURBATIVE EVALUATION OF GAUGE THEORIES

### 1. Feynman Rules for Gauge Theories in Euclidean Space

Our starting point will be the path integral for gauge theories derived in the previous chapter, to which we will add source terms in order to extract Green's functions. Thus consider in Euclidean space

$$W[J_\mu^A] \sim \int \mathcal{D}A_\mu^A \delta[g^A] \det\left|\frac{\delta g^A}{\delta\omega^B}\right| e^{-S_E^{YM}[A]+\int d^4x J_\mu^A A_\mu^A} \quad . \qquad (1.1)$$

In order to derive Feynman rules, we must rewrite the extra factors in the measure. First of all we note that the expression (1.1) does not depend on the gauge functions $g^A$. Hence we can just consider a new gauge choice

$$g'^A = g^A - c^A \quad , \qquad (1.2)$$

where $c^A(x)$ is some function independent of $A_\mu^A$. In addition, we can integrate functionally the expression (1.1) over $c^A$ with any weight; it would just result in a change in the normalization of W[J]. Usually the gauge function is linear in $A_\mu$ so if we want to introduce the gauge choice as a quadratic expression in the exponential, we consider

$$\int \mathcal{D}c^A \, e^{-\frac{1}{2\alpha}\int d^4x c^A c^A} \delta[g^A - c^A] = e^{-\frac{1}{2\alpha}\int d^4x g^A g^A} \quad , \qquad (1.3)$$

where $\alpha$ is an arbitrary coefficient. Then our starting point becomes

$$W[J] = \int \mathcal{D}A_\mu^A \det\left|\frac{\delta g^A}{\delta\omega^B}\right| e^{-S_E^{YM} - \frac{1}{2\alpha}\langle g^A g^A\rangle + \langle J_\mu^A A_\mu^A\rangle} \quad . \qquad (1.4)$$

The next step involves the rewriting of the determinant as a path integral. Since the determinant appears in the numerator it corresponds to a path integral over Grassmann numbers,

$$\det\left|\frac{\delta g^A(x)}{\delta\omega^B(y)}\right| = \int \mathcal{D}\eta^* \mathcal{D}\eta \; e^{i\int d^4x d^4y \, \eta^{*A}(x) \frac{\delta g^A(x)}{\delta\omega^B(y)} \eta^B(y)} \quad . \qquad (1.5)$$

These Grassmann fields are the famous Feynman and Faddeev-Popov ghosts, which transform as members of the adjoint representation of the group. We have now succeeded in writing the FPI in the form

$$\int \mathcal{D}\eta^* \mathcal{D}\eta \int \mathcal{D}A_\mu \; e^{-S_E^{eff}} \quad , \qquad (1.6)$$

where

$$S_E^{eff} = S_E^{YM} + \frac{1}{2\alpha} \langle g^A g^A \rangle - i \langle \eta^{*A}(x) \frac{\delta g^A(x)}{\delta\omega^B(y)} \eta^B(y) \rangle + \langle J_\mu^A A_\mu^A \rangle . \qquad (1.7)$$

This enables us to read off the Feynman rules, which obviously depend on the gauge choice. Let us remark that

$$\frac{\delta g^A(x)}{\delta\omega^B(y)} = \frac{\partial g^A(x)}{\partial A_\mu^C(x)} \frac{\delta A_\mu^C(x)}{\delta\omega^B(y)}$$

$$= \frac{\partial g^A(x)}{\partial A_\mu^C(x)} (\mathcal{D}_\mu)^{CB} \delta(x-y) \quad , \qquad (1.8)$$

leading to a local interaction for the $\eta$ fields. The same expression works in the Abelian case, the only difference being the absence of group indices. In the following we will treat both Abelian and non-Abelian cases. We first consider the covariant Lorentz gauge

$$g^A = \partial_\mu A_\mu^A = 0 \quad . \qquad (1.9)$$

Then it is easy to see that

$$\frac{\delta g^A(x)}{\delta\omega^B(y)} = \partial_\mu (\partial_\mu \delta^{AB} + f^{ABC} A_\mu^C) \delta(x-y) \qquad (1.10)$$

for the non-Abelian case and

$$\frac{\delta g(x)}{\delta\omega(y)} = \partial_\mu\partial_\mu\delta(x-y) \qquad (1.11)$$

for the Abelian case. Comparing with (1.7), we immediately note that in this gauge the ghosts do not interact with the gauge fields in the Abelian case. Hence they can be integrated out of the path integral: ghosts are not necessary in the covariant gauge (1.9) for QED.

However, in the non-Abelian case (1.10) indicates that we have a non-trivial interaction term. In this case the ghost part of the Action becomes after integration by parts and use of (1.9)

$$i\int d^4x[\partial_\mu\eta^{*A}(x)\partial_\mu\eta^A(x) - \frac{1}{2}gf^{ABC}\partial\cdot A^C\eta^{*A}\eta^B - \frac{1}{2}gf^{ABC}A^C_\rho\eta^{*A}\partial_\rho\eta^B]. \qquad (1.12)$$

This form suggests that the $\eta$ fields be interpreted as scalar-like fields in interaction with the gauge fields through their current. However, remember that they are Grassmann numbers, which means that their Feynman rules are crucially different: a minus sign must be added to a closed loop made up of $\eta$'s (see Chapter V). The current coupling to the divergence of the gauge field is antihermitean while the rest of the ghost Lagrangian is hermitean. However, the antihermitean part can be neglected since the determinant is to be evaluated when $\partial\cdot A = 0$. As we shall see later, the consequent alteration of the ghost Feynman rules gives results which only affect the propagation of the longitudinal part of the gauge field, which can always be reabsorbed by properly renormalizing the gauge parameter $\alpha$. We can read off the Feynman rules involving the ghosts:

ghost propagator:

```
A            B
------------
     p →
```

$$\longleftrightarrow \quad -i\,\frac{\delta^{AB}}{p^2} \qquad (1.13)$$

ghost-ghost-gauge vertex:

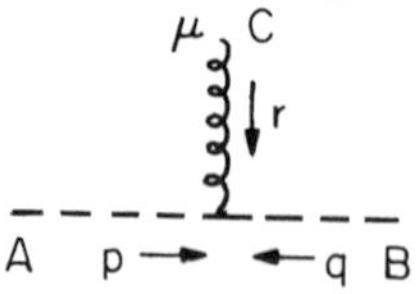


$$\longleftrightarrow \quad \frac{1}{2}\, g f^{ABC}(r_\mu+p_\mu-q_\mu) = -g f^{ABC} q_\mu \qquad (1.14)$$

where the wavy line corresponds to the gauge field $A^C_\mu$. We have also reinstated the factor g for clarity, and p,q,r are the momenta coming into the vertex. They, of course, satisfy the conservation equation

$$(p+q+r)_\mu = 0 \quad . \qquad (1.15)$$

The quadratic part of the effective Action for the gauge fields is given by

$$\int d^4x[\frac{1}{4}\,(\partial_\mu A^B_\nu-\partial_\nu A^B_\mu)(\partial_\mu A^B_\nu-\partial_\nu A^B_\mu) + \frac{1}{2\alpha}\,\partial_\mu A^B_\mu \partial_\rho A^B_\rho] \qquad (1.16)$$

where we have redefined $A \rightarrow gA$ and $\alpha \rightarrow g^2\alpha$; furthermore

$$\int d^4x[\frac{1}{2}\,\partial_\mu A^B_\nu \partial_\mu A^B_\nu - \frac{1}{2}\,\partial_\nu A^B_\mu \partial_\mu A^B_\nu + \frac{1}{2\alpha}\,\partial_\mu A^B_\mu \partial_\rho A^B_\rho]$$

$$= \frac{1}{2}\int d^4x\; A^B_\rho[-\partial_\mu\partial_\mu\delta_{\rho\nu}+(1-\frac{1}{\alpha})\partial_\rho\partial_\nu]A^B_\nu \quad , \qquad (1.17)$$

where each term has been integrated by parts. The propagator is the inverse of the operator in square brackets. [Note that in the absence of the gauge term the quantity in square brackets is a projection operator and has no inverse; that is the whole point for adding the gauge term, which is what Fermi did for QED.] Write it in momentum space in the form $X(p)\delta_{\mu\nu}+Y(p)p_\mu p_\nu$ without loss of generality; then by requiring

$$\delta_{\mu\rho} = [X(p)\delta_{\mu\nu}+Y(p)p_\mu p_\nu][p^2\delta_{\nu\rho}-(1-\frac{1}{\alpha})p_\nu p_\rho] \quad , \qquad (1.18)$$

we obtain for the Feynman propagator, represented pictorially by a pigtail (or spring)

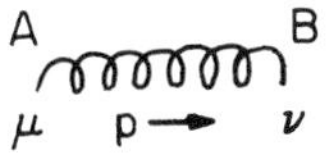


$$\longleftrightarrow \quad \frac{\delta^{AB}}{p^2}[\delta_{\mu\nu}-(1-\alpha)\frac{p_\mu p_\nu}{p^2}] \quad . \qquad (1.19)$$

We have up to this point kept the parameter $\alpha$ arbitrary. Let us mention that the simplest gauge to calculate in is the Feynman gauge where we take $\alpha = 1$

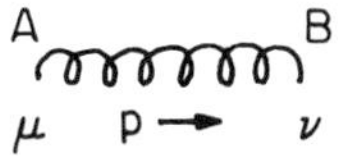


$$\longleftrightarrow \quad \frac{1}{p^2}\delta^{AB}\delta_{\mu\nu} \text{ (Feynman gauge)} \quad . \qquad (1.20)$$

When we take $\alpha = 0$, the numerator of the propagator (1.19) becomes just the projection operator needed to forbid one mode from propagating. This gauge is called the Landau gauge. Although it is not convenient for calculating Feynman diagrams it is useful when checking the unitarity of the Minkowski space amplitudes.

Next the effective Action contains a term cubic in the gauge fields

$$-g\int d^4x f^{ABC}A^A_\mu A^B_\nu \partial_\mu A^C_\nu \quad . \qquad (1.21)$$

In order to obtain the corresponding Feynman rule, we have to rewrite this term in momentum space in the form

$$\frac{1}{3!}\tilde{A}^A_\mu(p)\tilde{A}^B_\nu(q)\tilde{A}^C_\rho(r)V^{ABC}_{\mu\nu\rho}(p,q,r) \quad , \qquad (1.22)$$

with $-V$ being the Feynman rule, totally symmetric under the interchange of the A's. In particular we already know the index structure of V, it is just $f^{ABC}$. Thus we can write

$$V^{ABC}_{\mu\nu\rho}(p,q,r) = f^{ABC}V_{\mu\nu\rho}(p,q,r) \qquad , \qquad (1.23)$$

where $V_{\mu\nu\rho}(p,q,r)$ must be antisymmetric under the interchange of the pairs $(\mu,p)$, $(\nu,q)$, $(\rho,r)$, since the $f^{ABC}$ are themselves totally antisymmetric. From (1.21) we see that $V_{\mu\nu\rho}$ must contain $ir_\mu\delta_{\nu\rho}$. This is enough to generate all the other terms by symmetry. The result is

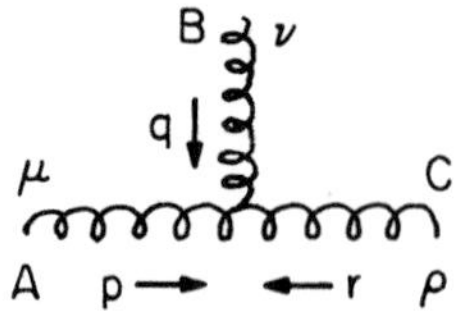


$$\longleftrightarrow \qquad -igf^{ABC}[r_\mu - q_\mu)\delta_{\nu\rho} + (q_\rho - p_\rho)\delta_{\mu\nu} + (p_\nu - r_\nu)\delta_{\rho\mu}]$$

(1.24)

with, of course

$$(p+q+r)_\mu = 0 \qquad .$$

Similarly, the effective Action contains a quartic term

$$\frac{1}{4} g^2 f^{ABE} f^{CDE} A^A_\mu A^B_\nu A^C_\mu A^D_\nu \qquad , \qquad (1.25)$$

which we must rewrite in the form

$$\frac{1}{4!} \tilde{A}^A_\mu(p)\tilde{A}^B_\nu(q)\tilde{A}^C_\rho(r)\tilde{A}^D_\sigma(s)V^{ABCD}_{\mu\nu\rho\sigma}(p,q,r,s) \qquad , \qquad (1.26)$$

where V is totally symmetric under the interchange of the triples $(A,\mu,p)$, $(B,\nu,q)$, $(C,\rho,r)$, $(D,\sigma,s)$. From (1.25) it is evident it does not contain any momenta, giving at least

$$\frac{1}{4} g^2 f^{ABE} f^{CDE} \delta_{\mu\rho}\delta_{\nu\sigma} \qquad .$$

From this term we must build symmetry in both $(A,\mu) \to (B,\nu)$ and $(C,\rho) \to (D,\sigma)$. Since the f's are antisymmetric, this means we must antisymmetrize under $\mu \to \nu$ and $\rho \to \sigma$ individually, i.e., let

$$\delta_{\mu\rho}\delta_{\nu\sigma} \to \frac{1}{2}(\delta_{\mu\rho}\delta_{\nu\sigma}-\delta_{\nu\rho}\delta_{\mu\sigma}) \quad ,$$

which takes care of both requirements. Next we must build in the remaining two types of symmetries $(A,\mu) \to (C,\rho)$ and $(A,\mu) \to (D,\sigma)$. This is done by adding these terms and dividing by 3. The result gives the last Feynman rules

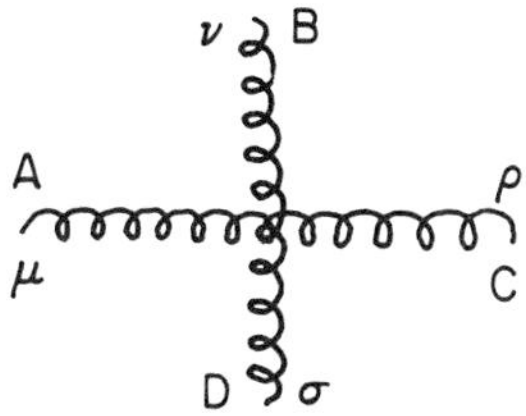


$$-g^2[f^{ABE}f^{CDE}(\delta_{\mu\rho}\delta_{\nu\sigma}-\delta_{\nu\rho}\delta_{\mu\sigma})+f^{CBE}f^{ADE}(\delta_{\mu\rho}\delta_{\nu\sigma}-\delta_{\nu\mu}\delta_{\rho\sigma})$$

$$+ f^{DBE}f^{CAE}(\delta_{\sigma\rho}\delta_{\nu\mu}-\delta_{\nu\rho}\delta_{\mu\sigma})] \quad . \tag{1.27}$$

These last two Feynman rules are unaffected by our gauge, and are absent in the Abelian case. These Euclidean space Feynman rules in the covariant gauge are summarized in Appendix C.

Let us now change gauge and take the Arnowitt-Fickler gauge which we write in a fancy way in Euclidean space:

$$n_\mu A_\mu^B = 0 \qquad n_\mu n_\mu = 1 \quad . \tag{1.28}$$

In this gauge we easily find that

$$\frac{\delta g^A(x)}{\delta\omega^B(y)} = n_\mu(\partial_\mu\delta^{AB}+f^{ABC}A_\mu^C)\delta(x-y) \tag{1.29}$$

$$= \delta^{AB}n_\mu\partial_\mu\delta(x-y) \tag{1.30}$$

is independent of $A_\mu$. Hence for both Abelian and non-Abelian theories, the ghosts do not couple to the gauge fields, thus obviating their use. It is in this type of gauge that the structure of the Abelian and non-Abelian theories are closest. Thus we only have to worry about the gauge field propagator. The term quadratic in $A_\mu$ now reads

$$\int d^4x[\frac{1}{4}(\partial_\mu A^B_\nu-\partial_\nu A^B_\mu)(\partial_\mu A^B_\nu-\partial_\nu A^B_\mu) - \frac{1}{2\alpha} n_\mu A^B_\mu n_\rho A^B_\rho] \quad ,$$

or after integration by parts

$$\frac{1}{2}\int d^4x\ A^B_\mu[-\partial_\rho\partial_\rho\delta_{\mu\nu}-\partial_\mu\partial_\nu - \frac{1}{\alpha}n_\mu n_\nu]A^B_\nu \quad . \quad (1.31)$$

The Feynman propagator is the inverse of the quantity in square brackets. In momentum space it gives (see problem)

$$= \frac{\delta^{AB}}{p^2}(\delta_{\mu\nu}-\frac{1}{n\cdot p}(n_\mu p_\nu+n_\nu p_\mu) - \frac{p_\mu p_\nu}{(n\cdot p)^2}(\alpha p^2-n^2)). \quad (1.32)$$

We see that this gauge brings a mixed blessing, giving no ghosts but a very complicated propagator structure.

We remark in passing that there is some arbitrariness in the Feynman rules we have just derived: the sign in front of the ghost propagator and of the ghost-ghost-gauge interaction do not matter since we will always be dealing with an even number of ghost lines.

Finally, let us mention the extra Feynman rules that come from coupling gauge fields to fermions. Although one can couple gauge fields to left-handed and right-handed fermion fields independently, let us concentrate on the pure vector coupling where left- and right-handed fermions couple in the same way. In that case the addition to the gauge Lagrangian is just

$$\mathcal{L}^f = \bar{\Psi}\gamma\cdot\mathcal{D}\Psi + im\bar{\Psi}\Psi \quad , \quad (1.33)$$

where the $\Psi(x)$ are Dirac spinors of mass m and $\not{D}_\mu$ is the relevant covariant derivative. We have suppressed all indices. The additional Feynman rules are then:

fermion line:

$$\underline{a \qquad b} \quad \longleftrightarrow \quad \frac{-i\delta^{ab}}{\not{p} + m} \quad , \qquad (1.34)$$

where $\not{p} = p_\mu \gamma_\mu$, and a,b are the indices of the fermion representation.

fermion-fermion-gauge vertex:

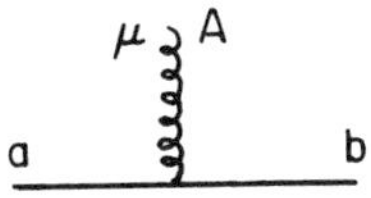


$$\longleftrightarrow \quad -ig\gamma_\mu (T^A)^a_b \quad , \qquad (1.35)$$

where $(T^A)^a_b$ are the matrix elements of the group generators in the appropriate fermion representation. As long as we deal with Dirac fermions, there is no essential difference between their Euclidean and Minkowski space treatments except for replacing $\bar{\psi}$ by $\psi^+$. In the following we keep the more relevant Minkowski notation although we write the Feynman rules in Euclidean space.

Problems.

A. Derive the expression for the gauge field propagator in the gauge $n_\mu A_\mu = 0$; $n_\mu n_\mu = 1$ and $n_\mu$ fixed.

B. Derive the Feynman rules for a complex scalar field coupling to a Yang-Mills Lagrangian. For definiteness, take SU(N) for the local invariance and assume the field transforms as the N-dimensional representation.

**C. For SU(N) gauge theories, consider the gauge condition

$$\partial_\mu A_\mu + a\{A_\mu, A_\mu\}$$

(here expressed in matrix form). a is an arbitrary coefficient. Derive the Feynman rules. Discuss the effect of the gauge conditions on the vertices. Note that this weird gauge condition is only possible when $\{A_\mu, A_\mu\}$ has the same group properties as $\partial \cdot A$.

## 2. QED: One-loop Structure

We now proceed to examine the perturbative treatment of the simplest of gauge theories which describes the interaction of the photon with charged particles. Its defining classical Lagrangian is

$$\mathcal{L}^{QED}_{c\ell} = \frac{1}{4}(\partial_\mu A_\nu - \partial_\nu A_\mu)(\partial_\mu A_\nu - \partial_\nu A_\mu) + \bar{\Psi}\gamma_\mu \partial_\mu \Psi + im\bar{\Psi}\Psi$$

$$+ ieA_\mu \bar{\Psi}\gamma_\mu \Psi + \frac{1}{2\alpha}(\partial_\mu A_\mu)(\partial_\rho A_\rho) \quad . \qquad (2.1)$$

Here $\Psi$ is a four-component Dirac field and e is its electric charge. In nature there are many charged fields; the leptons $e^-$, $\mu^-$, $\tau^-$ with charge -e, the "up" quarks u, c, (and perhaps t) with charge $\frac{2}{3}$ e, the "down" quarks d, s, b with charge $-\frac{1}{3}$ e, the intermediate vector beam of weak interaction $W^{\pm}_\mu$, and probably many others yet to be discovered. We restrict ourselves to one spin 1/2 field.

Since we are going to calculate in $2\omega$ dimensions, we replace the dimensionful coupling constant e with a dimensionless one

$$e \rightarrow e\,\mu^{2-\omega} \quad , \qquad (2.2)$$

where $\mu$ is the traditional mass parameter of dimensional regularization. [Recall that in $2\omega$ dimensions, spin 1/2 fields have dimension $-\omega + 1/2$, spin 1 fields have dimension $-\omega + 1$.] Thus the Euclidean space Feynman rules are (in the Feynman gauge $\alpha = 1$)

 $\longleftrightarrow$ $\dfrac{\delta_{\mu\nu}}{p^2}$

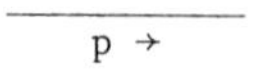 $\longleftrightarrow$ $\dfrac{-i}{\not{p} + m}$

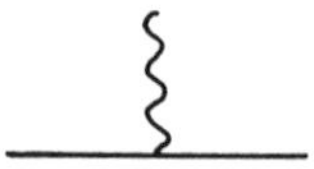 $\longleftrightarrow$ $-ie\mu^{2-\omega}\gamma_\rho \quad , \qquad (2.3)$

where we have suppressed all the spinor indices, and each fermion loop acquires a minus sign.

With these rules, we are led to the following one loop diagrams

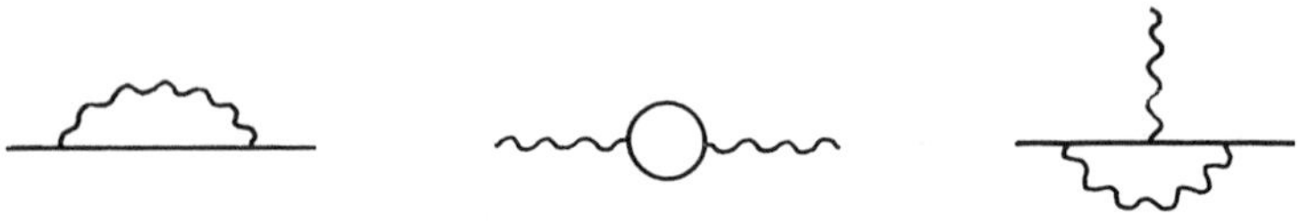 (2.4)

which correct the fundamental parameters and fields of the theory, and

 (2.5)

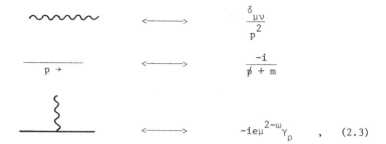

which seem to produce new interactions. Let us first deal with the diagrams of (2.5).

It is well known that the Dirac kinetic term is invariant under the charge conjugation operation

$$\Psi \rightarrow \Psi^C = C\bar{\Psi}^T \qquad . \qquad (2.6)$$

Under this discrete transformation the Dirac covariants $\bar{\Psi}\Psi$, $\bar{\Psi}\gamma_5\Psi$ and $\bar{\Psi}\gamma_5\gamma_\mu\Psi$ are even while the vector and tensor ones $\bar{\Psi}\gamma_\mu\Psi$ and $\bar{\Psi}\sigma_{\mu\nu}\Psi$ are odd. Hence it follows that the Lagrangian (2.1) is invariant under the combined discrete transformations

$$\Psi \to \Psi^C, \qquad A_\mu \to -A_\mu \qquad . \tag{2.7}$$

Therefore $\mathcal{L}_{QED}$ cannot generate interactions which are odd in the number of photon lines (recall the analogous case of $\lambda\phi^4$ theory which is symmetric under $\phi \to -\phi$ and thus does not have Green's functions with odd number of lines). This fact, known as Furry's theorem, disposes of the diagrams (2.5).

We have purposefully neglected the diagram

(2.8)

which describes scattering of light by light. By naive power counting, it is logarithmically divergent (in four dimensions) since each fermion propagator behaves as $(\not{p})^{-1}$. On the other hand since it involves four photon lines and comes from a gauge invariant theory, it must be proportional to $(F_{\mu\nu})^4$, and thus have dimension eight (when $\omega = 2$). Thus it would seem that we have found a diagram that is divergent and does not correspond to the fundamental interactions appearing in $\mathcal{L}$. Does it follow that QED is not renormalizable? On the surface it would seem impossible to stuff this divergence into the redefinition of the input parameters. Fortunately, contrary to the naive power counting, the box diagram (2.8) is UV convergent, thus obviating the question. Hence lesson No. 1: in gauge theories, do not trust the naive power counting, for divergent diagrams may turn out to be finite or at least not as badly divergent as naively believed.

After these words of wisdom, let us calculate the one-loop diagrams (2.4). We start with the correction to the fermion line (suppressing spinor indices)

$\Sigma(p) =$ 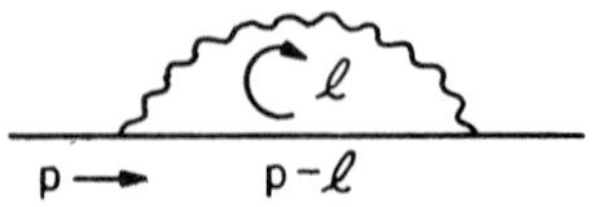 (2.9)



$$= -(e\mu^{2-\omega})^2 \int \frac{d^{2\omega}\ell}{(2\pi)^{2\omega}} \gamma_\mu \frac{(-i)}{\not{p}-\not{\ell}+m} \gamma_\nu \frac{\delta_{\mu\nu}}{\ell^2} \quad . \qquad (2.10)$$

Using the Euclidean space property of the $\gamma$-matrices

$$\{\gamma_\mu, \gamma_\nu\} = -2\delta_{\mu\nu} \quad , \qquad (2.11)$$

we rewrite the fermion propagator as

$$\frac{-i}{\not{q} + m} = i \frac{\not{q} - m}{q^2+m^2} \quad . \qquad (2.12)$$

Introducing Feynman parameter integration, we obtain

$$\Sigma(p) = -i(e\mu^{2-\omega})^2 \int_0^1 dx \int \frac{d^{2\omega}\ell}{(2\pi)^{2\omega}} \frac{\gamma_\mu(\not{p}-\not{\ell}-m)\gamma_\mu}{[\ell^2(1-x)+(p-\ell)^2x+m^2x]^2} \quad . \qquad (2.13)$$

Define the new variable of integration

$$\ell' = \ell - px \quad , \qquad (2.14)$$

in terms of which

$$\Sigma(p) = -ie^2\mu^{4-2\omega} \int_0^1 dx \int \frac{d^{2\omega}\ell'}{(2\pi)^{2\omega}} \frac{\gamma_\mu[\not{p}(1-x)-m-\not{\ell}']\gamma_\mu}{[\ell'^2+m^2x+p^2x(1-x)]^2} \quad . \qquad (2.15)$$

The term linear in $\ell'$ in the numerator vanishes upon integration while the other terms yield (using eq. B-16)

$$\Sigma(p) = -ie^2\mu^{4-2\omega} \int_0^1 dx\ \gamma_\mu[\not{p}(1-x)-m]\gamma_\mu \frac{\Gamma(2-\omega)}{(4\pi)^\omega} [p^2x(1-x)+m^2x]^{\omega-2} . \tag{2.16}$$

Before expanding about $\omega = 2$, we have to perform the $\gamma$-matrix algebra which is dimension dependent. Indeed from (2.11), we find that

$$\gamma_\mu\gamma_\mu = -2\omega \quad , \tag{2.17}$$

and

$$\gamma_\mu\gamma_\rho\gamma_\mu = [2-2(2-\omega)]\gamma_\rho \quad . \tag{2.18}$$

Letting $\varepsilon = 2-\omega$, these enable us to write

$$\Sigma(p) = -2i \frac{e^2}{16\pi^2} \Gamma(\varepsilon) \int_0^1 dx \left(\frac{p^2x(1-x)+m^2x}{4\pi\mu^2}\right)^{-\varepsilon} [\not{p}(1-x)+2m-\varepsilon(\not{p}(1-x)+m)] . \tag{2.19}$$

After expanding about $\varepsilon = 0$, we find

$$\Sigma(p) = \frac{-i}{\varepsilon}\cdot\frac{e^2}{16\pi^2}[\not{p}+4m]+i \frac{e^2}{8\pi^2}[\frac{1}{2} \not{p}(1+\gamma)+m(1+2\gamma)+\int_0^1 dx[\not{p}(1-x)+2m]\ln(\frac{p^2x(1-x)+m^2x}{4\pi\mu^2})]+\mathcal{O}(\varepsilon) , \tag{2.20}$$

where $\gamma$ is the Euler-Mascheroni constant. We store this result for future use.

Next we consider the correction to the photon line, also known as the vacuum polarization diagram

$$\Pi_{\mu\nu}(p) =$$ 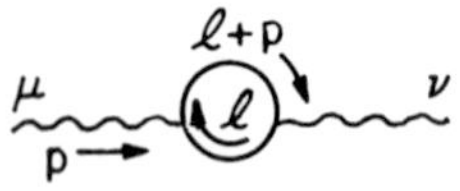

(2.21)

$$= -(e\mu^{2-\omega})^2 \int \frac{d^{2\omega}\ell}{(2\pi)^{2\omega}} \mathrm{Tr}[\gamma_\mu \frac{1}{\not\ell+\not p+m} \gamma_\nu \frac{1}{\not\ell + m}] \quad , \qquad (2.22)$$

where the (-) sign is present because of the fermion loop, and the trace is over the spinor indices, i.e., over the $\gamma$-matrices. This is rewritten as

$$\Pi_{\mu\nu}(p) = -(e\mu^{2-\omega})^2 \int \frac{d^{2\omega}\ell}{(2\pi)^{2\omega}} \frac{1}{[\ell^2+m^2][(p+\ell)^2+m^2]} \mathrm{Tr}[\gamma_\mu(\not\ell+\not p-m)\gamma_\nu(\not\ell-m)]. \qquad (2.23)$$

Introduce a Feynman parameter and the new loop momentum

$$\ell' = \ell + px \quad , \qquad (2.24)$$

in terms of which we obtain

$$\Pi_{\mu\nu}(p) = -(e\mu^{2-\omega})^2 \int_0^1 dx \int \frac{d^{2\omega}\ell'}{(2\pi)^{2\omega}} \frac{\mathrm{Tr}[\gamma_\mu(\not\ell'+\not p(1-x)-m)\gamma_\nu(\not\ell'-\not p x-m)]}{[\ell'^2+m^2+p^2x(1-x)]^2} . \qquad (2.25)$$

As usual terms odd in $\ell'$ drop out from the loop integration. In $2\omega$ dimensions, if we take the $\gamma$-matrices to be $2^\omega \times 2^\omega$ dimensional, we have the following trace formulae

$$\mathrm{Tr}(\gamma_\mu\gamma_\nu) = -2^\omega \delta_{\mu\nu} \quad , \qquad (2.26)$$

and

$$\mathrm{Tr}(\gamma_\mu\gamma_\rho\gamma_\nu\gamma_\sigma) = 2^\omega[\delta_{\mu\rho}\delta_{\nu\sigma}+\delta_{\mu\sigma}\delta_{\rho\nu}-\delta_{\mu\nu}\delta_{\rho\sigma}] \quad . \qquad (2.27)$$

Hence we rewrite the trace appearing in the numerator of (2.25) as

$$[\ell'_\rho\ell'_\sigma-p_\rho p_\sigma x(1-x)]\mathrm{Tr}(\gamma_\mu\gamma_\rho\gamma_\nu\gamma_\sigma) + m^2\mathrm{Tr}\,\gamma_\mu\gamma_\nu \quad , \qquad (2.28)$$

where we have used the fact that the trace of an odd number of matrices vanishes. Using (2.26) and (2.27), we arrive at the expression for the trace (2.28)

$$2^{\omega}[2\ell'_{\mu}\ell'_{\nu}-2x(1-x)[p_{\mu}p_{\nu}-\delta_{\mu\nu}p^2] - \delta_{\mu\nu}[\ell'^2+m^2+p^2x(1-x)]] \quad , \tag{2.29}$$

where we have added and subtracted $\delta_{\mu\nu}p^2x(1-x)$. Putting it all together, we obtain

$$\Pi_{\mu\nu}(p) = -(e\mu^{2-\omega})^2 \, 2^{\omega} \int_0^1 dx \int \frac{d^{2\omega}\ell}{(2\pi)^{2\omega}} \{\frac{2\ell_{\mu}\ell_{\nu}}{[\ell^2+m^2+p^2x(1-x)]^2}$$

$$- \frac{\delta_{\mu\nu}}{[\ell^2+m^2+p^2x(1-x)]} - \frac{2x(1-x)[p_{\mu}p_{\nu}-\delta_{\mu\nu}p^2]}{[\ell^2+m^2+p^2x(1-x)]^2}\}. \tag{2.30}$$

Integration over the loop momenta, using formulae B-16 and B-18 shows that the first two terms cancel one against the other, leaving us with

$$\Pi_{\mu\nu}(p) = \frac{e^2}{2\pi^2}\Gamma(\varepsilon)(p_{\mu}p_{\nu}-p^2\delta_{\mu\nu})\int_0^1 dx\; x(1-x)[\frac{m^2+p^2x(1-x)}{2\pi\mu^2}]^{-\varepsilon} \quad . \tag{2.31}$$

Expansion around $\varepsilon = 0$ yields

$$\Pi_{\mu\nu}(p) = \frac{e^2}{2\pi^2}(p_{\mu}p_{\nu}-\delta_{\mu\nu}p^2)[\frac{1}{6\varepsilon} - \frac{1}{6}\gamma - \int_0^1 dx\; x(1-x)\ln[\frac{m^2+p^2x(1-x)}{2\pi\mu^2}]]+\mathcal{O}(\varepsilon), \tag{2.32}$$

where we have used

$$\int_0^1 dx\; x(1-x) = \frac{1}{6} \quad . \tag{2.33}$$

The last one loop diagram is the vertex correction

$$\Gamma_\rho(p,q) \;=\; \tag{2.34}$$

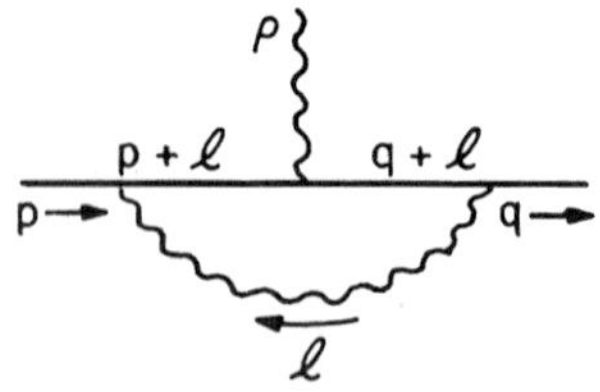


$$= -i[e\mu^{2-\omega}]^3 \int \frac{d^{2\omega}\ell}{(2\pi)^{2\omega}} \gamma_\tau \frac{1}{\not p+\not \ell+m} \gamma_\rho \frac{1}{\not \ell+\not q+m} \gamma_\sigma \frac{\delta_{\tau\sigma}}{\ell^2}. \tag{2.35}$$

It is more complicated than the previous two. We introduce two Feynman parameters and rewrite it as

$$\Gamma_\rho(p,q) = -2i(e\mu^{2-\omega})^3 \int_0^1 dx \int_0^{1-x} dy \int \frac{d^{2\omega}\ell}{(2\pi)^{2\omega}}$$

$$\gamma_\sigma(\not p+\not \ell-m)\gamma_\rho(\not \ell+\not q-m)\gamma_\sigma[\ell^2+m^2(x+y)+2\ell\cdot(px+qy)+p^2x+q^2y]^{-3}. \tag{2.36}$$

Introduce the new integration variable

$$\ell' = \ell + px + qy \tag{2.37}$$

in terms of which (2.36) becomes

$$\Gamma_\rho(p,q) = -2i(e\mu^{2-\omega})^3 \int_0^1 dx \int_0^{1-x} dy \int \frac{d^{2\omega}\ell}{(2\pi)^{2\omega}} \frac{\gamma_\sigma[\not \ell-\not q y+\not p(1-x)-m]\gamma_\rho[\not \ell-\not p x+\not q(1-y)-m]\gamma_\sigma}{[\ell^2+m^2(x+y)+p^2x(1-x)+q^2y(1-y)-2p\cdot qxy]^3}. \tag{2.38}$$

Only the piece of the numerator quadratic in $\ell$ gives a divergent loop integration. If we write

$$\Gamma_\rho(p,q) = \Gamma_\rho^{(1)}(p,q) + \Gamma_\rho^{(2)}(p,q) \quad , \tag{2.39}$$

where $\Gamma^{(1)}$ contains only the numerator part quadratic in the $\ell$'s, we find, using (B-18)

$$\Gamma_\rho^{(1)}(p,q) = -i\,\frac{(e\mu^{2-\omega})^3}{(4\pi)^\omega}\,\Gamma(2-\omega)\int_0^1 dx \int_0^{1-x} dy\,\frac{1/2\;\gamma_\sigma\gamma_\tau\gamma_\rho\gamma_\tau\gamma_\sigma}{[m^2(x+y)+p^2x(1-x)+q^2y(1-y)-2p\cdot qxy]^{2-\omega}}\,, \tag{2.40}$$

for the divergent part and using B-16

$$\Gamma_\rho^{(2)}(p,q) = -i\,\frac{(e\mu^{2-\omega})^3}{(4\pi)^\omega}\,\Gamma(3-\omega)\int_0^1 dx \int_0^{1-x} dy\,\frac{\gamma_\sigma[\not p(1-x)-\not q y-m]\gamma_\rho[\not q(1-y)-\not p x-m]\gamma_\sigma}{[m^2(x+y)+p^2x(1-x)+q^2y(1-y)-2p\cdot qxy]^{3-\omega}}\,. \tag{2.41}$$

In this last expression we can put $\omega = 2$ with impunity since it is convergent, obtaining

$$\Gamma_\rho^{(2)}(p,q) = -i(e\mu^{2-\omega})\,\frac{e^2}{16\pi^2}\int_0^1 dx \int_0^{1-x} dy\,\frac{\gamma_\sigma[\not p(1-x)-\not q y-m]\gamma_\rho[\not q(1-y)-\not p x-m]\gamma_\sigma}{[m^2(x+y)+p^2x(1-x)+q^2y(1-y)-2p\cdot qxy]}\,. \tag{2.42}$$

We will return to this expression later. The useful identity

$$\gamma_\sigma\gamma_\alpha\gamma_\rho\gamma_\beta\gamma_\sigma = 2\gamma_\beta\gamma_\rho\gamma_\alpha - 2(2-\omega)\gamma_\alpha\gamma_\rho\gamma_\beta \quad , \tag{2.43}$$

together with (2.18) enables us to rewrite (2.40) in the form

$$\Gamma_\rho^{(1)}(p,q) = -ie\mu^\varepsilon\gamma_\rho \cdot \frac{e^2}{8\pi^2}\,\Gamma(\varepsilon)(1-\varepsilon)^2\int_0^1 dx \int_0^{1-x} dy\left(\frac{m^2(x+y)+p^2x(1-x)+q^2y(1-y)-2p\cdot qxy}{4\pi\mu^2}\right)^{-\varepsilon} \tag{2.44}$$

$$= -ie\mu^\varepsilon\gamma_\rho\,\frac{e^2}{16\pi^2}\left[\frac{1}{\varepsilon} - \gamma - 1 - 2\int_0^1 dx \int_0^{1-x} dy\,\ell n\left[\frac{m^2(x+y)+p^2x(1-x)+q^2y(1-y)-2p\cdot qxy}{4\pi\mu^2}\right]\right]\,. \tag{2.45}$$

We shall return to these expressions later for they contain a lot of interesting physics. In order to properly analyze them, we will have to continue to Minkowski space and evaluate them on the fermion mass shell. We will notice that they are infrared divergent (all except $\Pi_{\mu\nu}$) and we will discuss how to circumvent this difficulty. For the moment, we concern ourselves with the structure of the field theory. The computation of the one loop diagrams enables us to build the counterterm structure necessary to renormalize QED.

It can be shown that in QED the number of primitively divergent diagrams is finite (see Chapter V). We have already noticed that a necessary condition for renormalizability is the ultraviolet finiteness of the light by light scattering diagram. Assume that it is indeed finite (see problem).

In the Feynman gauge, our starting Lagrangian was

$$\mathcal{L}_{c\ell} = \frac{1}{4}(\partial_\mu A_\nu - \partial_\nu A_\mu)^2 + \frac{1}{2}(\partial_\mu A_\mu)^2 + \bar{\Psi}\partial\!\!\!/\Psi + im\bar{\Psi}\Psi + ie\mu^{2-\omega}\bar{\Psi}A\!\!\!/\Psi \quad . \tag{2.46}$$

We try for a counterterm Lagrangian of the form

$$\mathcal{L}_{ct} = K_2\bar{\Psi}\partial\!\!\!/\Psi + imK_m\bar{\Psi}\Psi + ie\mu^{2-\omega}K_1\bar{\Psi}A\!\!\!/\Psi + \frac{1}{4}K_3F_{\mu\nu}F_{\mu\nu} + \frac{K_\alpha}{2}(\partial_\mu A_\mu)^2 \quad . \tag{2.47}$$

Then the renormalized Lagrangian

$$\mathcal{L}^{ren} = \mathcal{L}_{c\ell} + \mathcal{L}_{ct} \tag{2.48}$$

can be rewritten in terms of the bare quantities

$$\Psi_0 = (1+K_2)^{1/2}\Psi \equiv Z_2^{1/2}\Psi \tag{2.49}$$

$$A_0^\mu = (1+K_3)^{1/2}A^\mu \equiv Z_3^{1/2}A^\mu \tag{2.50}$$

$$e_0 = e\mu^{2-\omega}\frac{1+K_1}{(1+K_2)(1+K_3)^{1/2}} \equiv e\mu^{2-\omega}\frac{Z_1}{Z_2Z_3^{1/2}} \tag{2.51}$$

$$m_0 = m\,\frac{1+K_m}{1+K_2} \equiv m\,\frac{Z_m}{Z_2} \tag{2.52}$$

$$\alpha_0^{-1} = \frac{1+K_\alpha}{1+K_3} \equiv \frac{Z_\alpha}{Z_3} \tag{2.53}$$

as

$$\mathcal{L}^{ren} = \bar{\Psi}_0 \not{\partial} \Psi_0 + im_0 \bar{\Psi}_0 \Psi_0 + e_0 \bar{\Psi}_0 \not{A}_0 \Psi_0 + \frac{1}{4} (\partial_\mu A_{\nu 0} - \partial_\nu A_{\mu 0})^2 + \frac{1}{2\alpha_0} (\partial \cdot A_0)^2 \tag{2.54}$$

$$= Z_2 \bar{\Psi} \not{\partial} \Psi + imZ_m \bar{\Psi}\Psi + eZ_1 \bar{\Psi} \not{A} \Psi + \frac{Z_3}{4} (F_{\mu\nu})^2 + \frac{1}{2} Z_\alpha (\partial \cdot A)^2 \quad , \tag{2.55}$$

where we have introduced the Dyson Z-notation. In this form it is rather suggestive that the gauge invariance of $\mathcal{L}_{c\ell}$ will be preserved by $\mathcal{L}^{ren}$ when $Z_1 = Z_2$, in order to preserve the nature of the covariant derivative which for $\mathcal{L}^{ren}$ is seen to be

$$\mathcal{D}_\mu^{ren} = \partial_\mu + ie \frac{Z_1}{Z_2} A_\mu \quad . \tag{2.56}$$

However, it does not yet follow that $Z_1 = Z_2$ (it will!) because we have broken gauge invariance in our Lagrangian by putting in the gauge fixing term, so that the Z factors are gauge dependent.

The counterterms can be read off from the one loop calculations; firstly the fermion line calculation

$$\Sigma(p) = -i \frac{e^2}{16\pi^2} (\not{p} + 4m) \frac{1}{\varepsilon} + \text{finite terms} \tag{2.57}$$

yields the counterterms

$$K_2 = - \frac{e^2}{16\pi^2} [\frac{1}{\varepsilon} + F_2(\varepsilon, \frac{m}{\mu})] \tag{2.58}$$

$$K_m = - \frac{e^2}{4\pi^2} [\frac{1}{\varepsilon} + F_m(\varepsilon, \frac{m}{\mu})] \quad , \tag{2.59}$$

where $F_2$ and $F_m$ are the arbitrary finite parts which are analytic as $\varepsilon \to 0$ and depend on $\frac{m}{\mu}$. Secondly the photon line, being corrected by

$$\Pi_{\mu\nu}(p) = (p_\mu p_\nu - \delta_{\mu\nu} p^2)[\frac{e^2}{12\pi^2}\frac{1}{\varepsilon} + \text{finite part}] \quad , \quad (2.60)$$

corresponds to the new propagator

$$= \frac{\delta_{\mu\nu}}{p^2} + \frac{\delta_{\mu\rho}}{p^2}\Pi_{\rho\sigma}\frac{\delta_{\sigma\nu}}{p^2} + \ldots \quad (2.61)$$

$$= \frac{\delta_{\mu\nu}}{p^2}[1 - \frac{e^2}{12\pi^2}\frac{1}{\varepsilon}] + \frac{p_\mu p_\nu}{p^4}\frac{e^2}{12\pi^2}\frac{1}{\varepsilon} + \ldots \quad (2.62)$$

so that

$$K_3 = -\frac{e^2}{12\pi^2}[\frac{1}{\varepsilon} + F_3] \quad , \quad (2.63)$$

where $F_3$ is an arbitrary dimensionless function. The longitudinal part of the propagator gives

$$K_\alpha = \frac{e^2}{12\pi^2}[\frac{1}{\varepsilon} + F_\alpha] \quad , \quad (2.64)$$

which, of course, amounts to a renormalization of $\alpha$. Notice that in the Landau gauge where $\alpha = 1$, the correction to the gauge propagator contains the same projection operator as in the bare propagator, so that $\alpha$ does not get changed by corrections, but this is only true in the Landau gauge. Lastly, the vertex correction

$$\Gamma_\rho(p,q) = ie\mu^\varepsilon\gamma_\rho[\frac{e^2}{16\pi^2}\frac{1}{\varepsilon} + \text{finite}] \quad (2.65)$$

gives

$$K_1 = -\frac{e^2}{16\pi^2}[\frac{1}{\varepsilon} + F_1] \quad , \quad (2.66)$$

when $F_1$ is the finite part of the counterterm. Thus to summarize our results, we have

$$Z_1 = 1 - \frac{e^2}{16\pi^2}\left(\frac{1}{\varepsilon} + F_1\right) + \mathcal{O}(e^4) \tag{2.67}$$

$$Z_2 = 1 - \frac{e^2}{16\pi^2}\left(\frac{1}{\varepsilon} + F_2\right) + \ldots \tag{2.68}$$

$$Z_3 = 1 - \frac{e^2}{12\pi^2}\left(\frac{1}{\varepsilon} + F_3\right) + \ldots \tag{2.69}$$

$$Z_m = 1 - \frac{e^2}{4\pi^2}\left(\frac{1}{\varepsilon} + F_m\right) + \ldots \tag{2.70}$$

$$Z_\alpha = 1 + \frac{e^2}{12\pi^2}\left(\frac{1}{\varepsilon} + F_\alpha\right) \quad . \tag{2.71}$$

We remark that the suggestive relation $Z_1 = Z_2$ is satisfied to this order in perturbation theory, modulo the finite part of the counterterms. Thus, using (2.51) we can express the bare charge as

$$e_0 = e\mu^\varepsilon\left[1 + \frac{e^2}{24\pi^2}\frac{1}{\varepsilon} + \text{finite parts} + \mathcal{O}(e^3)\right] \quad . \tag{2.72}$$

Thus if we ignore the finite part of the counterterms by adopting a mass independent prescription, we can read off the scale variation of the gauge coupling constant, [Chapter III, eq. (6.13)]

$$\mu \frac{\partial e}{\partial \mu} = \frac{e^3}{12\pi^2} \quad , \tag{2.73}$$

which has the same sign as for the scalar theory. The solution of this equation is

$$\frac{1}{e^2(\mu)} - \frac{1}{e^2(\mu_0)} = -\frac{1}{6\pi^2}\ln\left(\frac{\mu}{\mu_0}\right) \tag{2.74}$$

where $\mu_0$ is an arbitrary scale, or in a perhaps more suggestive form

$$e^2(\mu) = \frac{e^2(\mu_0)}{1 - \frac{e^2(\mu_0)}{6\pi^2} \ln \frac{\mu}{\mu_0}} \quad , \qquad (2.75)$$

which has a singularity at

$$\mu = \mu_0 \exp[6\pi^2 e^{-2}(\mu_0)] \quad , \qquad (2.76)$$

the famous Landau singularity. However, well before we reach such a large scale, the perturbative equation (2.74) has to be amended by higher order effects which can no longer be neglected at large mass scales, because of the sign of the right-hand side of (2.74). When there are many charged fermions, each contributes according to its charge to (2.74) [see problem].

The fact that the electric charge grows weaker and weaker at large distances (i.e., small scales) means that the identification of the free Lagrangian ($e = 0$) in terms of physical photon and, say, electrons is perfectly justified. However, the long range nature of the electromagnetic field makes this identification a bit tricky, but the fact remains that electrons and photons can be directly recognized in their free states in the laboratory.

Problems.

A. Show that the apparently logarithmically divergent box diagram for light by light scattering

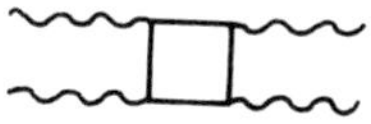

is, in fact, convergent.

B. Show by direct calculation that the diagrams

vanish.

C. Find the lowest order change of e with $\mu$ when $3n_u$ quarks of charge 2/3, $3n_d$ quarks of charge -1/3 and $n_\ell$ leptons of charge -1 are present. Assume that

$$\frac{e^2(\mu_0)}{4\pi} = \frac{1}{137} \text{ at } \mu_0 = 1 \text{ MeV} \quad ,$$

and find the location of the Landau singularity when the known charged fermions are included. How many species of fermions are needed for the Landau point to occur at the Planck mass?

*D. Consider electrodynamics coupled to a (charged) scalar field. Derive the Feynman rules and compute the vacuum polarization from a scalar loop and compare it with that already obtained from the fermion loop.

*E. Compute the counterterms $Z_1$, $Z_2$, $Z_3$, $Z_m$ and $Z_\alpha$ for an arbitrary covariant gauge, i.e., leave $\alpha$ arbitrary in the computations. Show that although the Z's are gauge ($\alpha$) dependent, the $\beta$-function is not.

## 3. QED: Ward Identities

Because of the local gauge invariance of QED, not all the Green's functions generated by

$$e^{-Z[J_\mu,\bar{\chi},\chi]} = N \int \mathcal{D}A_\mu \mathcal{D}\bar{\psi}\mathcal{D}\psi \; e^{-S_{eff}-<J_\mu A_\mu+i\bar{\chi}\psi+i\bar{\psi}\chi>} \quad , \qquad (3.1)$$

are independent. Here we have

$$S_{eff} = \int d^4x[\frac{1}{4}(F_{\mu\nu})^2 + \frac{1}{2\alpha}(\partial\cdot A)^2+\bar{\Psi}(\not{\partial}+ie\chi)\Psi+im\bar{\Psi}\Psi] \qquad (3.2)$$

and $J_\mu$, $\chi$, $\bar{\chi}$ are the sources, the last two being Grassmann sources. The generating functional (3.1) is not invariant under the gauge transformations

$$\delta A_\mu = \frac{1}{e}\partial_\mu \Lambda(x)$$

$$\delta\Psi = -i\Lambda(x)\Psi(x)$$

$$\delta\bar{\Psi} = i\Lambda(x)\bar{\Psi}(x) \quad , \qquad (3.3)$$

where $\Lambda(x)$ is an arbitrary function. This is caused by the gauge fixing term in $S_{eff}$ and the sources. In this section we derive a set of functional constraint equations on Z, from which we will extract relations between Green's functions known as Ward identities.

The technique we are going to use will readily generalize to the more complicated Yang-Mills theories; it is based on the work of Becchi, Rouet and Stora, Phys. Lett. 52B, 344 (1974). The first step consists in restoring some sort of invariance even in the presence of the gauge fixing term, neglecting for the moment the sources. This is done by reinstating the ghost Lagrangian which in this Abelian case amounts to no more than redefining the (infinite) normalization constant N. Then the new effective Action is given by

$$S'_{eff} = \int d^4x[\frac{1}{4}(F_{\mu\nu})^2+\bar{\Psi}(\not\partial+ie\not A)\Psi+im\bar{\Psi}\Psi + \frac{1}{2\alpha}(\partial\cdot A)^2+i\partial_\mu\eta^*\partial_\mu\eta] \tag{3.4}$$

where $\eta$ and $\eta^*$ are complex Grassmann fields. Then $S'_{eff}$ is invariant under the special gauge transformations

$$\delta A_\mu = \frac{1}{e}\partial_\mu(\zeta^*\eta+\zeta\eta^*) \tag{3.5}$$

$$\delta\Psi = -i(\zeta^*\eta+\zeta\eta^*)\Psi, \qquad \delta\bar{\Psi} = i\bar{\Psi}(\zeta^*\eta+\zeta\eta^*) \tag{3.6}$$

$$\delta\eta = -\frac{i}{\alpha e}(\partial\cdot A)\zeta, \qquad \delta\eta^* = \frac{i}{\alpha e}(\partial\cdot A)\zeta^* \quad , \tag{3.7}$$

where $\zeta$ and $\zeta^*$ are complex Grassmann variables which are independent of x. Under the above we see that

$$\delta S'_{eff} = \frac{1}{e}\int d^4x[\frac{1}{\alpha}(\partial\cdot A)\partial^2(\zeta^*\eta+\zeta\eta^*) - \frac{1}{\alpha}(\partial\cdot A)\partial^2\zeta^*\eta - \frac{1}{\alpha}\partial\cdot A\partial^2\zeta\eta^*], \quad , \tag{3.8}$$

after integrating by parts the variation of the last term in (3.4). In the above we have used the rule that for two Grassmann numbers $\omega$ and $\chi$ we have

$$(\omega\chi)^* = \omega^*\chi^* \quad , \tag{3.9}$$

so that $\omega\chi$ is real if $\omega$ and $\chi$ are real. Now, let us start from the new generating functional

$$e^{-Z[J_\mu,\chi,\bar{\chi},\sigma,\sigma^*]} \equiv N'\int \mathcal{D}A_\mu\mathcal{D}\bar{\Psi}\mathcal{D}\Psi\mathcal{D}\eta^*\mathcal{D}\eta\; e^{-S'_{eff}-\langle J\cdot A+i\bar{\Psi}\chi+i\bar{\chi}\Psi+\eta^*\sigma+\eta\sigma^*\rangle} \quad , \tag{3.10}$$

where $\sigma$ and $\sigma^*$ are the complex Grassmann sources for the ghost fields. In this expression, displace the fields by a BRS transformation (3.5) - (3.7): since $S'_{eff}$ is invariant under this transformation and since the Jacobian of the BRS transformation is unity (see problem), it follows that only the source terms get affected. Comparing both ways of writing the generating functional (3.10), we easily find that

$$e^{-Z} = N' \int \mathcal{D}A_\mu \ldots \mathcal{D}\eta \; e^{-S'_{eff} + \langle J\cdot A + \ldots\rangle - \langle J\cdot\delta A + i\bar{\chi}\delta\psi + i\delta\bar{\psi}\chi + \delta\eta^*\sigma + \delta\eta\sigma^*\rangle} \quad , \quad (3.11)$$

where the variations are given by (3.5) - (3.7). We note that if we specialize to BRS transformations for which $\zeta$ is real

$$\zeta^* = \zeta \quad , \quad (3.12)$$

we can easily expand the exponential since $\zeta^2 = 0$. The result leads to

$$0 = \int \mathcal{D}A_\mu \ldots \mathcal{D}\eta \; e^{-S'_{eff} + \langle J\cdot A + \ldots\rangle} \Big[\langle J_\mu \partial_\mu \frac{(\eta+\eta^*)}{e} - \bar{\chi}(\eta+\eta^*)\Psi + \bar{\Psi}(\eta+\eta^*)\chi$$

$$+ \frac{i}{\alpha e}(\partial\cdot A)\sigma - \frac{i}{\alpha e}\partial\cdot A\sigma^* \rangle\Big] \quad . \quad (3.13)$$

This is the desired statement, although it is stated in a somewhat awkward form. However, if we introduce the generating functional of one particle irreducible graphs

$$\Gamma[A_{\mu c\ell}, \Psi_{c\ell}, \bar{\Psi}_{c\ell}, \eta_{c\ell}, \eta^*_{c\ell}] \equiv Z[J_\mu, \chi, \bar{\chi}, \sigma, \sigma^*] - \langle J\cdot A_{c\ell} + i\bar{\chi}\psi_{c\ell} + i\bar{\psi}_{c\ell}\chi + \eta^*_{c\ell}\sigma + \eta_{c\ell}\sigma^* \rangle \; ,$$

$$(3.14)$$

where, $A_{\mu c\ell}$, etc., are the classical sources, we see that

$$J_\mu = - \frac{\delta\Gamma}{\delta A_{\mu c\ell}} \; , \qquad \text{etc.} \quad . \quad (3.15)$$

Thus in terms of $\Gamma$, we can immediately rewrite (3.13) in the form

$$\langle -\frac{1}{e}\frac{\delta\Gamma}{\delta A_{\mu c\ell}} \partial_\mu(\eta_{c\ell}+\eta^*_{c\ell}) + i\frac{\delta\Gamma}{\delta\Psi_{c\ell}}(\eta_{c\ell}+\eta^*_{c\ell})\Psi_{c\ell} - i\bar{\Psi}_{c\ell}(\eta_{c\ell}+\eta^*_{c\ell})\frac{\delta\Gamma}{\delta\bar{\Psi}_{c\ell}} +$$

$$+ \frac{i}{\alpha e}\partial\cdot A_{c\ell}\frac{\delta\Gamma}{\delta\eta_{c\ell}} - \frac{i}{\alpha e}\partial\cdot A_{c\ell}\frac{\delta\Gamma}{\delta\eta^*_{c\ell}} \rangle = 0 \quad . \quad (3.16)$$

This is the most manageable form of the Ward identities of QED. We now apply this formula to the simplest cases. The dependence of $\Gamma$ on $\eta_{c\ell}$ and $\eta^*_{c\ell}$ is very simple since these fields do not interact:

$$\Gamma = i \int d^4x d^4y \eta^*_{c\ell}(x)\Delta^{-1}(x-y)\eta_{c\ell}(y)+\Gamma'[A_{\mu c\ell},\Psi_{c\ell},\bar{\Psi}_{c\ell}] \quad , \qquad (3.17)$$

where $\Gamma'$ does not depend on $\eta_{c\ell}$ and $\eta^*_{c\ell}$, and $\Delta^{-1}$ is the inverse of the free massless propagator:

$$\Delta^{-1}(x-y) = -\partial^2\delta(x-y) \quad . \qquad (3.18)$$

The expression for $\Gamma'$ is more complicated and starts with

$$\Gamma' = \int d^4x d^4y [\tfrac{1}{2} A_{\mu c\ell}(x)\Delta^{-1}_{\mu\nu}(x-y)A_{\nu c\ell}(y)+\bar{\Psi}_{c\ell}(x)S^{-1}(x-y)\Psi_{c\ell}(y)]$$

$$+ \int d^4x d^4y d^4z \bar{\Psi}_{c\ell}(x)A_{\rho c\ell}(y)\Gamma_\rho(x,y,z)\Psi_{c\ell}(z)+\ldots \quad , \qquad (3.19)$$

where $\Delta^{-1}_{\mu\nu}$ is the full inverse photon propagator, $S^{-1}$ the full inverse fermion propagator and $\Gamma_\rho$ the three point function. Of course, $\Gamma'$ contains many more terms which correspond to induced interactions not present in the original Lagrangian.

We start by applying (3.16) to (3.17) and (3.19), just keeping the terms containing $A_\mu$ and $\eta+\eta^*$. The result is seen to be (in momentum space)

$$k_\mu \Delta^{-1}_{\mu\nu}(k) + \frac{1}{\alpha} k_\mu k^2 = 0 \quad . \qquad (3.20)$$

If we write

$$\Delta^{-1}_{\mu\nu} = A\delta_{\mu\nu} + Bk_\mu k_\nu \quad , \qquad (3.21)$$

the Ward identity (3.20) reduces to

$$A + Bk^2 + \frac{1}{\alpha} k^2 = 0 \quad . \qquad (3.22)$$

For example in the Feynman gauge $\alpha = 1$, it says that

$$\Delta^{-1}_{\mu\nu} = -\delta_{\mu\nu}k^2 + (\delta_{\mu\nu}k^2 - k_\mu k_\nu)F(k^2) \quad , \tag{3.23}$$

where $F(k^2)$ is at least of order $e^2$. We have already verified this result to second order in e.

Next, the terms that contain $\bar{\Psi}$, $\Psi$, $\eta$ (or any $\eta^*$) give another identity,

$$\frac{\partial}{\partial y_\mu}\Gamma_\mu(x,y,z) - iS^{-1}(x-z)\delta(z-y) + iS^{-1}(x-z)\delta(x-y) \ . \tag{3.24}$$

Alternatively, we can write this equation in momentum space

$$(p-q)_\mu\tilde{\Gamma}_\mu(p,p-q,q) = S^{-1}(p) - S^{-1}(q) \quad . \tag{3.25}$$

This is the original formulation of Ward identities. We can easily test it in perturbation theory

$$\tilde{\Gamma}_\mu = i\gamma_\mu + \ldots \tag{3.26}$$

$$S^{-1}(p) = i(\not{p} + m) + \ldots \quad . \tag{3.27}$$

Moreover, since $S^{-1}(p)$ is multiplicatively renormalized by $Z_2$, and $\Gamma_\mu$ by $Z_1$, it follows from this Ward identity that

$$Z_1 = Z_2 \quad , \tag{3.28}$$

as we promised earlier. It would be foolish to adopt a subtraction procedure that violates (3.25). Hence the finite part of $Z_1$ and $Z_2$ is always chosen to be equal - we have already seen by explicit calculation that their pole parts are equal. We note in passing that if we write

$$\Gamma_\mu(p,p-q,q) = C_1\gamma_\mu + C_2\sigma_{\mu\nu}(p-q)_\nu \quad , \tag{3.29}$$

the magnetic moment $C_2$ term decouples from (3.25) by antisymmetry.

We have remarked in several places that theories with massless particles are beset by infrared divergences, and QED because of the masslessness of the photon is no exception. One device for avoiding these divergences is the introduction of a small mass for the photon. Let us remark that in the QED case this does not spoil the Ward identities because we can still maintain BRS invariance even in the presence of a photon mass term provided that the ghost acquires the same mass. The reason is that the mass term varies as

$$A_\mu A_\mu = \lambda A_\mu \partial_\mu (\zeta^* \eta + \zeta \eta^*) \tag{3.30}$$

$$\delta \frac{1}{2} A_\mu A_\mu = -\lambda \partial \cdot A(\zeta^* \eta + \zeta \eta^*) + \text{surface term} \quad . \tag{3.31}$$

In the Feynman gauge $\alpha = 1$, this variation is seen to be equal to

$$-\lambda(\delta\eta^* \eta + \eta^* \delta\eta) = -\lambda\delta(\eta^* \eta) \quad . \tag{3.32}$$

This means that the BRS invariance can be maintained at the cost of having a massive ghost particle which decouples anyway! As we shall see this extra bonus does not seem to be generalizable to non-Abelian theories.

Problems.

A. Show that the Jacobian of the BRS transformation is equal to one.

B. Given the functional

$$e^{-Z[J]} = \int \mathcal{D}\phi \; e^{-S[\phi] - \langle J\phi \rangle} ,$$

show that

$$\int \mathcal{D}\phi \; e^{-S[\phi] - \langle J\phi \rangle} \langle J\phi \rangle = - \langle \frac{\delta\Gamma}{\delta\phi_{c\ell}} \phi_{c\ell} \rangle ,$$

where

$$\Gamma[\phi_{c\ell}] = Z[J] - \langle J\phi_{c\ell} \rangle .$$

*C. Derive the Ward identities for scalar electrodynamics (complex scalar field coupled to the photon) in the Feynman gauge.

*D. Derive the Ward identities for QED in the axial gauge.

**E. Derive the Ward identities for QED in the gauge

$$\partial_\mu A_\mu + a A_\mu A_\mu = 0 .$$

## 4. QED: Applications

Before applying our results concerning QED we have to continue the Green's functions into Minkowski space and choose a renormalization prescription.

In Euclidean space, the photon propagator is given by [to avoid confusion, we reinstate bars over Euclidean momenta]

$$\Delta_{\mu\nu}(\bar{p}) = \frac{\delta_{\mu\nu}}{\bar{p}^2}\left[1 + \frac{e^2}{2\pi^2}\left(-\frac{1}{6}\gamma - \int_0^1 dx\, x(1-x)\ln\frac{m^2+\bar{p}^2x(1-x)}{2\pi\mu^2} - \frac{1}{6}F_3\right)\right]$$

$$- \frac{\bar{p}_\mu\bar{p}_\nu}{\bar{p}^4}\frac{e^2}{2\pi^2}\left[-\frac{1}{6}\gamma - \int_0^1 dx\, x(1-x)\ln\frac{m^2+\bar{p}^2x(1-x)}{2\pi\mu^2} - \frac{1}{6}F_\alpha\right] + \mathcal{O}(e^4). \qquad (4.1)$$

The finite parts of the counterterms $Z_3$ and $Z_\alpha$ are fixed by requiring that $\Delta_{\mu\nu}$ look like the original propagator as $\bar{p}^2 \to 0$,

$$\Delta_{\mu\nu}(\bar{p}) = -\left.\frac{\delta_{\mu\nu}}{\bar{p}^2}\right|_{\bar{p}^2 = 0} , \qquad (4.2)$$

so that

$$F_3 = F_\alpha = -\gamma - \ln\frac{m^2}{2\pi\mu^2} . \qquad (4.3)$$

Similarly, to $\mathcal{O}(e^2)$ we have

$$S^{-1}(p) = i\bar{\not p}\left[1 + \frac{e^2}{8\pi^2}\left(\frac{1+\gamma}{2} + \int_0^1 dx(1-x)\ln\frac{\bar{p}^2x(1-x)+m^2x}{4\pi\mu^2} + \frac{1}{2}F_2\right)\right]$$

$$+ im\left[1 + \frac{e^2}{8\pi^2}\left(1+2\gamma+2\int_0^1 dx\,\ln\frac{\bar{p}^2x(1-x)+m^2x}{4\pi\mu^2} - 2F_m\right)\right] , \qquad (4.4)$$

and it would seem that a good subtraction procedure would be to continue to Minkowski space, i.e., $\bar{\not p} \to -\not p$ and $\bar{p}^2 \to -p^2$, and require that

$$S^{-1}(p) = i(m_e - \not{p}) \quad \text{at} \quad p^2 = m_e^2 \quad , \tag{4.5}$$

which fixes

$$F_2 = -\gamma + 2 - \ln \frac{m^2}{4\pi\mu^2} \tag{4.6}$$

$$F_m = \gamma - 1 + \frac{1}{2} \ln \frac{m^2}{4\pi\mu^2} \quad . \tag{4.7}$$

However, this choice of subtraction procedure is ambiguous. The reason is that our prescription assumes that the expansion of $\Sigma(p)$ about $\not{p} = m$ is well-defined. A simple argument shows that it is not. The prescription (4.5) states that

$$\Sigma(p) = 0 \quad \text{at} \quad \not{p} = m \quad , \tag{4.8}$$

so let us expand the original expression for $\Sigma(p)$ about $\not{p} = m$. This is most easily done by expanding the fermion propagator in (2.10) in powers of $\bar{\not{p}} + m$

$$\frac{1}{\not{p} - \not{\ell} + m} = \frac{-1}{\not{\ell}} \left[1 - \frac{1}{\not{\ell}} (\not{p} + m) + \ldots\right] \quad . \tag{4.9}$$

We immediately notice that the terms appearing in this expansion are not well defined. For instance, the term linear in $\not{p} + m$ gives

$$(e\mu^{2-\omega})^2 \int \frac{d^{2\omega}\ell}{(2\pi)^{2\omega}} \frac{1}{\ell^4} \quad , \tag{4.10}$$

which, besides diverging at the upper end of integration when $\omega = 2$, also diverges at the lower end. This type of divergence is called an infrared divergence, and must be dealt with separately from the previously considered ultraviolet divergences.

One way to deal with this problem is to give the photon a fictitious mass $\lambda$ and keep it in the calculation. Then the infrared divergence is avoided and all is well again. In QED this procedure is favored because it does not affect the gauge invariance of the calculation. Note that infrared divergences occur on the "mass-shell" of the external particles: (4.10) is $\frac{\partial\Sigma}{\partial \not{p}}$ at the Euclidean mass shell $\not{p} + m = 0$.

The fact that IR divergences in QED can be circumvented by adding a mass to the photon, and arise on mass-shell of external particles indicates that they are connected with the presence of long-range forces which make ambiguous the definition of asymptotic states.

Another way of remedying IR divergences is to use dimensional regularization. In this case the procedure consists of performing the parameter integrations in $2\omega$-dimensional space and then in expanding about $\omega = 2$. Infrared (as well as ultraviolet) divergences will appear as poles in the dimension plane. More on this later.

There remains to fix the vertex counterterm. Before doing so we must rewrite $\Gamma_\rho(\bar{p},\bar{q})$ as calculated earlier in a recognizable form. We are going to use a prescription for which the fermions are on their "mass-shell". This leads us to understand $\Gamma_\rho(\bar{p},\bar{q})$ as being sandwiched between spinors on which the (Euclidean) equations of motion are satisfied. Thus it is smart to rewrite $\Gamma_\rho$ in the form

$$\Gamma_\rho(\bar{p},\bar{q}) = \tilde{\Gamma}_\rho(\bar{p},\bar{q}) + (\bar{\not{p}}+m)A_\rho + B_\rho(\bar{\not{q}}+m) \qquad , \qquad (4.11)$$

where $\tilde{\Gamma}_\rho$ is what we are after because the A and B terms will give no contribution due to the equations of motion. In this reduction process the Gordon identities

$$\gamma_\rho \bar{\not{q}} = - \bar{q}_\rho - 2i\sigma_{\rho\tau}\bar{q}_\tau = - m\gamma_\rho + \gamma_\rho(\bar{\not{q}}+m) \qquad (4.12)$$

$$\bar{\not{p}}\gamma_\rho = - \bar{p}_\rho + 2i\sigma_{\rho\tau}\bar{p}_\tau = - m\gamma_\rho + (\bar{\not{p}}+m)\gamma_\rho \qquad (4.13)$$

where

$$\sigma_{\mu\nu} = \frac{1}{4i} [\gamma_\mu, \gamma_\nu] \tag{4.14}$$

will prove useful. They are easily derived by expressing the product of two $\gamma$-matrices as the sum of half the commutator and half the anticommutator. Thus equipped, we are ready for a nasty bit of Diracology.

Using the identities (valid only for $\tilde{\Gamma}_\rho$)

$$\bar{\not{q}}\gamma_\rho\bar{\not{p}} = m^2\gamma_\rho - \bar{k}^2\gamma_\rho - 4im\sigma_{\rho\tau}\bar{k}_\tau; \qquad \bar{k}_\tau = \bar{q}_\tau - \bar{p}_\tau \tag{4.15}$$

$$\bar{\not{q}}\gamma_\rho = -m\gamma_\rho + 4i\sigma_{\rho\tau}\bar{q}_\tau \tag{4.16}$$

$$\gamma_\rho\bar{\not{p}} = -m\gamma_\rho - 4i\sigma_{\rho\tau}\bar{p}_\tau \quad , \tag{4.17}$$

and (2.42), the numerator of the integrand of $\Gamma^{(2)}$ is reduced to

$$2m^2\gamma_\rho[(x+y)^2 - 2(1-x-y)] - 2\bar{k}^2(1-x)(1-y)\gamma_\rho$$
$$+ 8im\sigma_{\rho\tau}\bar{q}_\tau[x-y(y+x)] - 8im\sigma_{\rho\tau}\bar{p}_\tau[y-x(x+y)] \quad . \tag{4.18}$$

This result enables us to split up the contributions of $\Gamma^{(2)}$ into two parts, one proportional to $\gamma_\rho$ and the other proportional to $\sigma_{\rho\tau}$. At the moment, we are interested in the $\gamma_\rho$ part. It is given by

$$\Gamma_\rho(\bar{p},\bar{q}) = \Gamma_\rho^{(1)}(\bar{p},\bar{q}) - ie\mu^\varepsilon\gamma_\rho \frac{e^2}{16\pi^2}\int_0^1 dx \int_0^{1-x} dy \frac{(x+y)^2-2(1-x-y)-2\bar{k}^2(1-x)(1-y)}{m^2(x+y)+\bar{p}^2x(1-x)+\bar{q}^2y(1-y)-2\bar{p}\cdot\bar{q}xy} \quad . \tag{4.19}$$

When we evaluate this expression on the mass-shells

$$\bar{p}^2 = \bar{q}^2 = -m^2, \qquad \bar{k}^2 = 0 \quad , \tag{4.20}$$

the last integral reduces to

$$\int_0^1 dx \int_0^{1-x} dy \, \frac{[(x+y)^2-2(1-x-y)]}{(x+y)^2} \quad , \qquad (4.21)$$

and it is painfully apparent that it diverges. This is our old friend, the infrared divergence. As we mentioned above, we can get rid of it by giving the photon a fictitious mass $\lambda$, which amounts to replacing $m^2(x+y)$ by $m^2(x+y) + \lambda(1-x-y)$ in the denominator. Then the integral (4.21) is replaced by

$$\int_0^1 dx \int_0^{1-x} dy \, \frac{(x+y)^2-2(1-x-y)}{(x+y)^2 m^2+\lambda(1-x-y)} \quad , \qquad (4.22)$$

which nicely converges. This infrared divergence is not present in the $\sigma_{\mu\nu}$ terms in $\Gamma^{(2)}$ nor in $\Gamma^{(1)}$. Still it stands in the way of using an on-shell renormalization prescription, here shown in Minkowski space

$$\Gamma_\rho(p,q) = -e\mu^\varepsilon \gamma_\rho \qquad \text{at} \qquad p^2 = q^2 = m^2, \quad (p-q)^2 = 0 \ , \qquad (4.23)$$

since if we did not control the divergence in (4.21) it would mean that $F_1$ is (infrared) divergent. Note that we can carry out this procedure in Euclidean space except that $\bar{k}^2 = 0$ implies that $\bar{p} = \bar{q}$. From (4.22), we have

$$F_1 = 1+\gamma+ \int_0^1 dx \int_0^{1-x} dy[2 \,\ell n \frac{m^2(x+y)^2+\lambda(1-x-y)}{4\pi\mu^2} - \frac{(x+y)^2-2(1-x-y)}{(x+y)^2 m^2+\lambda(1-x-y)}] \quad . \qquad (4.24)$$

The evaluation of these integrals is left to the reader.

As we have mentioned before, another way to regulate infrared divergences is to do the parametric integrals before taking the limit $\omega \to 2$. For instance, let us start from the expression (2.40) for $\Gamma_\rho^{(2)}$. The numerator is evaluated in the same way except that there will be correction terms to (4.18) which vanish like $\omega - 2$. Neglecting those, we find for all particles on shell

$$\Gamma_\rho^{(2)}(\bar{p},\bar{q}) = -i\frac{(e\mu^\varepsilon)^3}{(4\pi)^\omega}\Gamma(3-\omega)\int_0^1 dx\int_0^{1-x} dy\,\frac{2m^2\gamma_\rho[(x+y)^2-2(1-x-y)]+8im\sigma_{\rho\tau}(\bar{q}-\bar{p})_\tau[x-y(y+x)]+\mathcal{O}(\varepsilon)}{[m^2(x+y)^2]^{3-\omega}}$$

(4.25)

where we have used the symmetry of the integrand as $x \to y$ to rewrite the $\sigma_{\mu\nu}$ term. Since we are only interested in the infinite part as $\varepsilon \to 0$, we concentrate on the integral

$$\int_0^1 dx \int_0^{1-x} dy\ (x+y)^{2(\omega-3)} = \frac{1}{2\omega-5}\left[1 - \frac{1}{2\omega-4}\right] ,$$

$$\to -\frac{1}{2\varepsilon} \quad \text{as} \quad \varepsilon \to 0 \quad . \tag{4.26}$$

Thus we explicitly see that the infrared divergence in $\Gamma_\rho$ now appears as a pole in the dimension plane, just like the UV divergence. Unlike it, however, it only appears when the particles are on their mass shells, and at least one particle is massless. It will turn out that this pole, present after regularization, will be canceled for quantities of physical interest by the contribution of other diagrams when integrated over $2\omega$-dimensional phase space. However, this trick provided us with a neat bookkeeping device and does not affect the gauge properties of the theory.

While we are at it, let us evaluate the $\sigma_{\mu\nu}\bar{k}_\nu$ contribution to $\Gamma^{(2)}$. Since it is both UV and IR convergent, there is no need to be fancy: set $\omega = 2$. The contribution is

$$\Gamma_\rho^{(2)}(\bar{p},\bar{q}) = \frac{e^3}{2m\pi^2}\sigma_{\rho\tau}(\bar{q}-\bar{p})_\tau \int_0^1 dx \int_0^{1-x} dy\ \frac{x-y(x+y)}{(x+y)^2} \tag{4.27}$$

$$= \frac{e^3}{8m\pi^2}\sigma_{\rho\tau}(\bar{q}-\bar{p})_\tau \quad . \tag{4.28}$$

The physical meaning of this induced interaction between the photon and the fermion is evident since it contributes to the effective action a term of the form

$$\frac{e}{m}\frac{e^2}{8\pi^2}\bar{\Psi}(\bar{p})\sigma_{\rho\tau}(\bar{q}-\bar{p})_\tau A_\rho(\bar{p}-\bar{q})\Psi(\bar{q}) \qquad (4.29)$$

or in position space

$$\frac{ie}{2m}\frac{e^2}{8\pi^2}\int \bar{\Psi}(\bar{x})\sigma_{\rho\tau}F_{\rho\tau}(\bar{x})\Psi(\bar{x})d^4\bar{x} \quad . \qquad (4.30)$$

It gives a correction to the intrinsic magnetic moment g of the fermion (in units of Bohr magnetons)

$$g = 2(1+\frac{\alpha}{2\pi}) \qquad \alpha = \frac{e^2}{4\pi} \quad . \qquad (4.31)$$

Thus we see how the field theory induces a correction to the Dirac magnetic moment of the fermion. Further we note that this correction is finite. This is because the induced interaction term (4.30) is not present in the input Lagrangian (it has dimension 5) and, since the theory is renormalizable, all counterterms come only in the form of the fundamental vertices and not with the induced new interactions. At this point let us remark that we have not proved renormalizability of QED, but the suspicious reader should consult one of the many excellent textbooks on the subject.

The rest of $\Gamma_\rho$ also gives a correction to the electromagnetic force between two charged fermions, contributing in the form

$$i(e\mu^\varepsilon)\bar{\Psi}(\bar{q})\gamma_\mu F[(\bar{q}-\bar{p})^2]A_\mu(\bar{q}-\bar{p})\Psi(\bar{q}) \quad . \qquad (4.32)$$

Using our prescription, it is straightforward to find $F(\bar{k}^2)$ to $\mathcal{O}(e^2)$. The result is in Euclidean space (see problem)

$$F(\bar{k}^2) = 1 + \frac{2}{3}\frac{\bar{k}^2}{m^2}\frac{\alpha}{2\pi}\left[\ln\frac{m}{\lambda} - \frac{3}{8}\right] + \mathcal{O}(\alpha^2) \qquad k^2 << m^2 \; . \quad (4.33)$$

We should emphasize that this correction depends on the way we have chosen $\alpha$, i.e., on the subtraction (4.23).

This particular subtraction procedure was chosen because it lends itself to direct comparison with experiment, and therefore a numerical evaluation of e. For instance, we can compare e defined by (4.23) with the nonrelativistic limit of the electron Coulomb scattering cross section formula. This is not exactly kosher because the exchanged photon is never really on its mass shell, but because of the smallness of the electron mass it nearly is.

The correction to the vertex (4.33) breaks the degeneracy between the $2S_{1/2}$ and $2P_{1/2}$ states of the (relativistic) hydrogen atom. Calculations show that it depresses the $2P_{1/2}$ level with respect to the $2S_{1/2}$ level by 1010 Mc/sec compared to the experimental value of 1057.77 Mc/sec for the Lamb shift. But, as in all correct theories, we are saved by including additional corrections. First of all, the anomalous magnetic moment also breaks the degeneracy and adds another 68 Mc/sec. Now we are too high, but the modification to the photon propagator due to vacuum polarization subtracts a healthy 27 Mc/sec, leaving us with 1052 Mc/sec, which to $\mathcal{O}(\alpha)$ is sufficiently accurate. In fact, $\mathcal{O}(\alpha^2)$ calculations bring perfect agreement between theory and experiment!!

We have not treated all applications of QED, but it should be clear that it is an unusually successful theory in spite of all the intermediate steps needed to regularize it.

Problems.

A. Evaluate the $\gamma_\rho$ part of $\Gamma_\rho$ with a small mass for the photon and find the expression for $F_\nu$ when the on-shell prescription is used for the vertex.

B. Using the result of the previous problem, calculate the vertex correction to the Lamb shift in first order perturbation theory, and then compute the contribution from the vacuum polarization.

C. Compute the contribution of the anomalous magnetic moment to the Lamb shift.

## 5. Yang-Mills Theory: Preliminaries

We now start our long-awaited journey into the perturbative evaluation of Yang-Mills theories. In particular, we study the Yang-Mills theory in interaction with Dirac spinors in Euclidean space.

In the covariant gauge $\partial \cdot A^B = 0$, the effective Lagrangian is given by

$$\mathscr{L}_{\text{eff}} = \frac{1}{4} F^B_{\mu\nu} F^B_{\mu\nu} + \frac{1}{2\alpha} \partial \cdot A^B \partial \cdot A^B + i\partial_\mu \eta^{*B} \partial_\mu \eta^B - \frac{i}{2} g\mu^{2-\omega} f^{ABC} A^C_\mu \eta^{*A} \overleftrightarrow{\partial}_\mu \eta^B$$

$$- \frac{i}{2} g\mu^{2-\omega} f^{ABC} \eta^{*A} \eta^B \partial \cdot A^C + \bar{\Psi}(\not{\partial}+im)\Psi + ig\mu^{2-\omega} A^B_\mu \bar{\Psi} \gamma_\mu T^B_f \Psi \quad , \qquad (5.1)$$

where

$$F^B_{\mu\nu} = \partial_\mu A^B_\nu - \partial_\nu A^B_\mu - g\mu^{2-\omega} f^{BCD} A^C_\mu A^D_\nu \quad . \qquad (5.2)$$

We have suppressed the group indices of the fermions and have retained their Minkowski space notation (using $\bar{\Psi}$ instead of $\Psi^\dagger$) although we are in Euclidean space; as discussed earlier this does not change the Green's functions as long as we deal with Dirac spinors. The spinor field $\Psi$ transforms according to the $d_f$-dimensional irreducible representation of the group G. The N $d_f \times d_f$ hermitean matrices $T^B_f$ obey the Lie algebra of G

$$[T^B_f, T^C_f] = if^{BCD} T^D_f \qquad B,C,D = 1,\ldots N \quad , \qquad (5.3)$$

where $f^{BCD}$ are the structure constants of G and N its dimension.

When G is taken to be the eight-dimensional unitary group SU(3), with each Dirac fermion transforming as its three-dimensional representation, this theory is believed to describe the interaction between the constituents of nuclear matter: the quarks. The fermions are identified with the quarks and the eight self-interacting vector particles, called gluons, generate interquark forces. The SU(3) degrees of freedom of each quark are called colors ("red, white and blue"). In nature, there are five (perhaps six) such varieties of quarks, called flavors; they are denoted by the symbols u (up), d (down), s (strange), c (charm), b (bottom) and possibly t (top, yet undiscovered, but expected). This theory of interquark interactions is called Quantum Chromodynamics (QCD).

In the following we keep our treatment general. The Feynman rules for the theory (5.1) have been derived in Section 1 of this chapter; they are summarized in Appendix C. It might be appropriate at this point to mention an ambiguity concerning the Feynman rules for the ghost part. The interaction Lagrangian contains a term of the form

$$-\frac{i}{2}\, g\mu^{2-\omega} f^{ABC} \eta^{*A} \eta^{B} \partial\cdot A^{C} \quad , \qquad (5.4)$$

which is proportional to the gauge function $\partial\cdot A^C = 0$. This term comes from the Faddeev-Popov determinant which is the determinant of the change of the gauge condition evaluated at the gauge condition, i.e., when $\partial\cdot A^C = 0$. Hence the term (5.4) can be dropped from the effective Lagrangian without affecting the physics since it involves the coupling of the spurious longitudinal part of the gauge field. The formal advantage is that the remaining ghost part of the Lagrangian is now real since the term (5.4) was pure imaginary. By not taking this term we alter the ghost-ghost-gauge vertex to yield the new Feynman rule

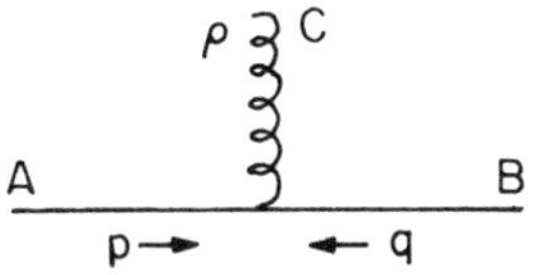


$$\leftrightarrow \frac{1}{2}\, g\mu^{2-\omega} f^{ABC}(p-q)_{\rho} \quad \text{[Altered]} \tag{5.5}$$

The alert reader may have noticed that the dimensionless coupling constant g appears in many different places in the Lagrangian (5.1). In the classical theory there was only one coupling constant because of the gauge invariance of the Lagrangian. However, our effective Lagrangian contains terms that break the gauge invariance, and it would seem we have no right to take these various coupling constants to be the same, unless there is a good reason which keeps the quantum corrections from messing up their equivalence. We shall see later that there exists such a good reason in the form of the BRS transformation which we analyzed in its simplest form for the Abelian case in Section 3 of this chapter.

As a first application, let us compute the lowest order contribution to the non-Abelian Compton amplitude, which will serve as a check on the sign of the Feynman rules. We have the diagrams

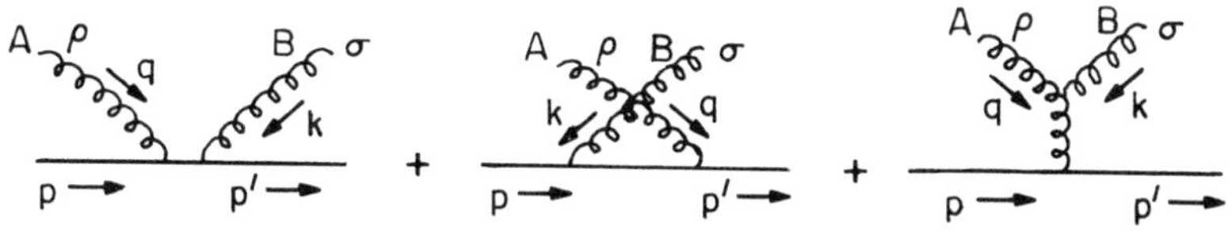


(5.6)

The first two diagrams give

$$-g^2\mu^{4-2\omega}[T_f^A T_f^B \gamma_\rho \frac{-i}{\not p+\not q+m}\gamma_\sigma + T_f^B T_f^A \gamma_\sigma \frac{-i}{\not p'-\not q+m}\gamma_\rho] \quad (5.7)$$

$$= -ig^2\mu^{4-2\omega}[T_f^A T_f^B \gamma_\rho \frac{\not p+\not q-m}{(p+q)^2+m^2}\gamma_\sigma + T_f^B T_f^A \gamma_\sigma \frac{\not p'-\not q-m}{(p'-q)^2+m^2}\gamma_\rho], \quad (5.8)$$

while the third diagram is represented by (in the Feynman gauge)

$$-g^2\mu^{4-2\omega}f^{ABC}T_f^C\gamma_\mu \frac{1}{(q+k)^2}\{\delta_{\sigma\mu}(-q-2k)_\rho + \delta_{\rho\sigma}(k-q)_\mu + \delta_{\mu\rho}(2q+k)_\sigma\} \quad . \quad (5.9)$$

The sum of these three diagrams must satisfy gauge invariance, which means that the longitudinal part of the gauge particles must not contribute since our gauge condition is $\partial\cdot A^B = 0$. Thus we contract with $q_\rho$ and see what happens. The first two diagrams give

$$-ig^2\mu^{4-2\omega}[T_f^A T_f^B \not q \frac{\not p+\not q-m}{(p+q)^2+m^2}\gamma_\sigma + T_f^B T_f^A \gamma_\sigma \frac{\not p'-\not q-m}{(p'-q)^2+m^2}\not q] \quad . \quad (5.10)$$

Using

$$\not q\not q = -q^2 \quad , \quad (5.11)$$

$$\not q(\not p-m) = -2q\cdot p - (\not p+m)\not q \quad , \quad (5.12)$$

$$(\not p'-m)\not q = -2p'\cdot q - \not q(\not p'+m) \quad , \quad (5.13)$$

and the fermion Euclidean mass shell conditions which say that $\not p+m$ on the left and $\not p'+m$ on the right of the amplitude give zero because of the external fermion wave functions that sandwich (5.10), as well as

$$p^2 = p'^2 = -m^2 \quad , \quad (5.14)$$

we rewrite (5.10) in the form

$$-ig^2\mu^{4-2\omega}[T_f^A T_f^B \frac{-2q\cdot p-q^2}{2q\cdot p+p^2}\gamma_\sigma+T_f^B T_f^A \frac{-2p'\cdot q+q^2}{-2p'\cdot q+q^2}\gamma_\sigma] \tag{5.15}$$

$$= ig^2\mu^{4-2\omega}[T_f^A,T_f^B]\gamma_\sigma$$

$$= -g^2\mu^{4-2\omega}f^{ABC}T_f^C\gamma_\sigma \quad . \tag{5.16}$$

In the Abelian case, these two diagrams would have sufficed and the contribution of the longitudinal gauge particle would have given zero (since $f^{ABC} = 0$) even for arbitrary values of $q^2$ and $k^2$. In Yang-Mills we have the third diagram to contend with. Its contribution to the longitudinal part is

$$-g^2\mu^{4-2\omega}f^{ABC}T_f^C \frac{1}{(q+k)^2}\{\gamma_\sigma(-q^2-2k\cdot q)+q_\sigma(\not{k}-\not{q})+\not{q}(2q+k)_\sigma\} \tag{5.17}$$

$$= -g^2\mu^{4-2\omega}f^{ABC}T_f^C \frac{1}{(q+k)^2}[-\gamma_\sigma(q^2+2k\cdot q)-\not{k}k_\sigma] \quad . \tag{5.18}$$

In the last equation we have used the fermion Dirac equations to eliminate $\not{k}+\not{q}$ and rewrite $\not{q}k_\sigma$ as $-\not{k}k_\sigma$. The end result is obtained by taking the sum of (5.16) and (5.18); it is

$$q_\rho T_{\rho\sigma} = -g^2\mu^{4-2\omega}f^{ABC}T_f^C \frac{1}{(q+k)^2}\gamma_\mu[k^2\delta_{\mu\sigma}-k_\mu k_\sigma] \quad , \tag{5.19}$$

and it does not vanish unlike the equivalent result in QED! However, it is proportional to the projection operator for the second gauge particle. Thus the longitudinal Compton amplitude contracted into the polarization vector of the second gauge field does vanish provided the gauge field satisfies the equation of motion (unlike QED where no constraint is put on the second photon line). The distinction arises because in Yang-Mills theories the gauge lines carry charge (unlike QED) and therefore their sources will have charge (see problem) and will interact with other gauge lines unless the gauge field is on mass shell.

Problems.

*A. Compute the tree level Compton amplitude in the axial gauge $n_\mu A_\mu = 0$ in Euclidean space. Contract one of its vector lines with $n_\mu$. Interpret the result.

*B. Invent a new Feynman rule for an off-shell source such that the longitudinal Compton amplitude of (5.19) when added to the additional relevant diagram vanishes.

*C. Repeat problem B in the axial gauge.

## 6. Yang-Mills Theory: One-loop Structure

In this section we examine the one-loop structure of Yang-Mills theories in abundant detail. Since the theory is very complicated, we need to introduce an organizing principle behind our computations: we will assume that the theory is renormalizable (thanks to 't Hooft we know this to be true). This means that we can absorb all the ultraviolet divergences by adding to the Lagrangian (5.1) a counterterm Lagrangian that looks exactly the same except for unknown coefficients for each term which diverge as $\varepsilon \to 0$; $\varepsilon = 2 - \frac{d}{2}$. We write it as

$$\begin{aligned}\mathcal{L}_{ct} = {} & \frac{1}{4} K_3 (\partial_\mu A^B_\nu - \partial_\nu A^B_\mu)^2 - K_4 g \mu^\varepsilon f^{ABC} A^B_\mu A^C_\nu \partial_\mu A^A_\nu \\ & + \frac{1}{4} K_5 g^2 \mu^{2\varepsilon} f^{ABC} f^{ADE} A^B_\mu A^C_\nu A^D_\mu A^E_\nu + \frac{1}{2\alpha} K_\alpha \partial \cdot A^B \partial \cdot A^B \\ & + i K_6 \partial_\mu \eta^{*B} \partial_\mu \eta^B - \frac{i}{2} K_7 g \mu^\varepsilon f^{ABC} A^C_\mu \eta^{*A} \overleftrightarrow{\partial}_\mu \eta^B - \frac{i}{2} K_8 g \mu^\varepsilon f^{ABC} \eta^{*A} \eta^B \partial_\mu A^C_\mu \\ & + K_2 \bar{\Psi} \not{\partial} \Psi + i K_1 g \mu^\varepsilon A^B_\mu \bar{\Psi} \gamma_\mu T^B_f \Psi + i m K_m \bar{\Psi} \Psi \quad . \end{aligned} \tag{6.1}$$

The renormalized Lagrangian

$$\mathcal{L}_{ren} = \mathcal{L}_{eff} + \mathcal{L}_{ct} \tag{6.2}$$

generates Green's functions which are ultraviolet finite when the limit $\varepsilon \to 0$ is taken. This is the essence of a renormalizable theory. The renormalizable Lagrangian can then be written in terms of bare fields and parameters in the form

$$\mathcal{L}_{\text{ren}} = \frac{1}{4}\left(\partial_\mu A^B_{\nu 0} - \partial_\nu A^B_{\mu 0}\right)^2 - g'_0 f^{ABC} A^B_{\mu 0} A^C_{\nu 0} \partial_\mu A^A_{\nu 0}$$

$$+ \frac{1}{4} g''^2_0 f^{ABC} f^{ADE} A^B_{\mu 0} A^C_{\nu 0} A^D_{\mu 0} A^E_{\nu 0} + \frac{1}{2\alpha_0} \partial \cdot A^B_0 \partial \cdot A^B_0$$

$$+ i\partial_\mu \eta^{*B}_0 \partial_\mu \eta^B_0 - \frac{i}{2} g'''_0 f^{ABC} A^C_{\mu 0} \eta^{*A}_0 \overleftrightarrow{\partial}_\mu \eta^B_0 - \frac{i}{2} g''''_0 f^{ABC} \eta^{*A}_0 \eta^B_0 \partial \cdot A^C_0$$

$$+ \bar{\Psi}_0 \not{\partial} \Psi_0 + i g_0 A^B_{\mu 0} \bar{\Psi}_0 \gamma_\mu T^B_f \Psi_0 + i m_0 \bar{\Psi}_0 \Psi_0 \quad . \qquad (6.3)$$

The bare fields, coupling constants and masses are related to their renormalized (finite) counterparts by the relations

$$\Psi_0 = (1+K_2)^{1/2} \Psi \equiv Z_2^{1/2} \Psi \qquad (6.4)$$

$$A^B_{\mu 0} = (1+K_3)^{1/2} A^B_\mu \equiv Z_3^{1/2} A^B_\mu \qquad (6.5)$$

$$\eta^B_0 = (1+K_6)^{1/2} \eta^B \equiv Z_6^{1/2} \eta^B \qquad (6.6)$$

$$m_0 = m \frac{1+K_m}{1+K_2} \equiv m \frac{Z_m}{Z_2} \qquad (6.7)$$

$$\alpha_0^{-1} = \alpha^{-1} \frac{1+K_\alpha}{1+K_3} \equiv \alpha^{-1} \frac{Z_\alpha}{Z_3} \qquad (6.8)$$

$$g_0 = g\mu^\varepsilon \frac{1+K_1}{Z_2 Z_3^{1/2}} \equiv g\mu^\varepsilon \frac{Z_1}{Z_2 Z_3^{1/2}} \qquad (6.9)$$

$$g'_0 = g\mu^\varepsilon \frac{1+K_4}{Z_3^{3/2}} \equiv g\mu^\varepsilon \frac{Z_4}{Z_3^{3/2}} \qquad (6.10)$$

$$g''_0 = g\mu^\varepsilon \frac{(1+K_5)^{1/2}}{Z_3} \equiv g\mu^\varepsilon \frac{Z_5^{1/2}}{Z_3} \qquad (6.11)$$

$$g_0''' = g\mu^\varepsilon \frac{1+K_7}{Z_3^{1/2} Z_6} \equiv g\mu^\varepsilon \frac{Z_7}{Z_3^{1/2} Z_6} \tag{6.12}$$

$$g_0'''' = g\mu^\varepsilon \frac{1+K_8}{Z_3^{1/2} Z_6} \equiv g\mu^\varepsilon \frac{Z_8}{Z_3^{1/2} Z_6} \quad . \tag{6.13}$$

This list of definitions has been lengthened by our introduction of five independent bare coupling constants. They will turn out to be all equal as a result of the Slavnov-Taylor identities which are the Ward identities of Yang-Mills theories. Their equality then implies many relations between the renormalization constants:

$$\frac{Z_1}{Z_2} = \frac{Z_4}{Z_3} = \frac{Z_5^{1/2}}{Z_3^{1/2}} = \frac{Z_7}{Z_6} = \frac{Z_8}{Z_6} \quad . \tag{6.14}$$

We now would like to compute these renormalization constants to $\mathcal{O}(\hbar)$ in the scheme of dimensional regularization.

Let us first indicate diagrammatically the contributions to the fundamental interactions of the Yang-Mills theory up to one loop ($\mathcal{O}(\hbar)$):

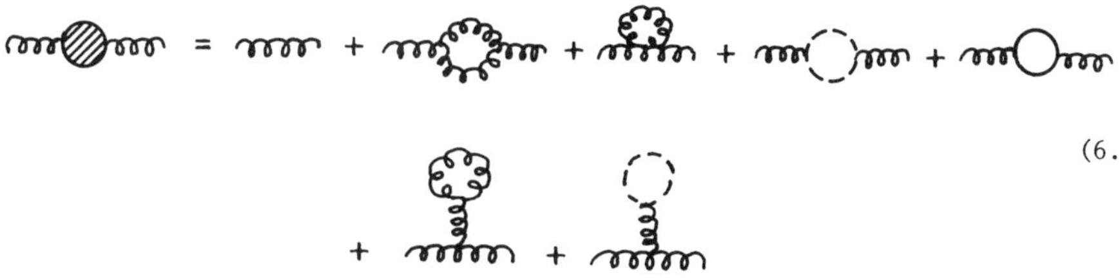

(6.15)

(6.16)

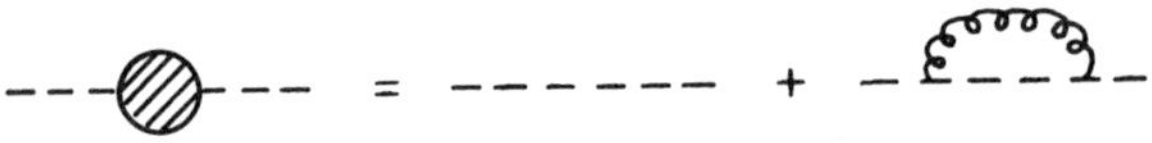

(6.17)

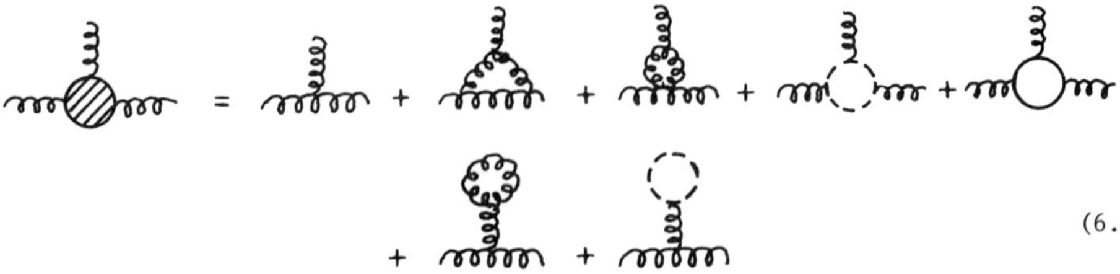

(6.18)

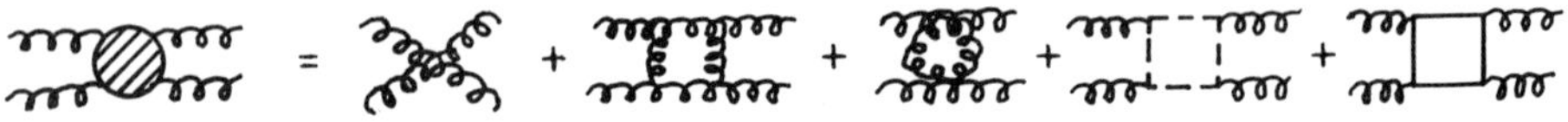

(6.19)

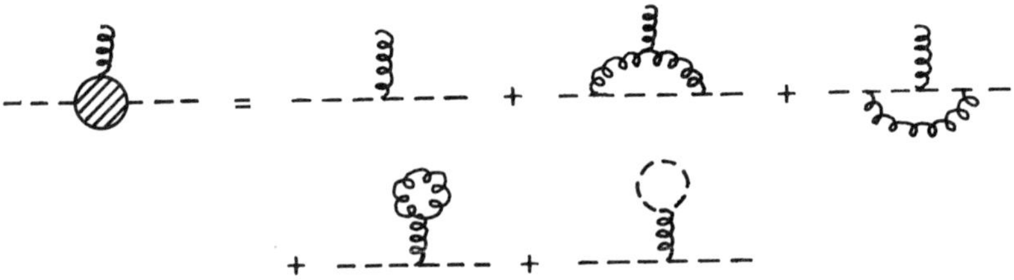

(6.20)

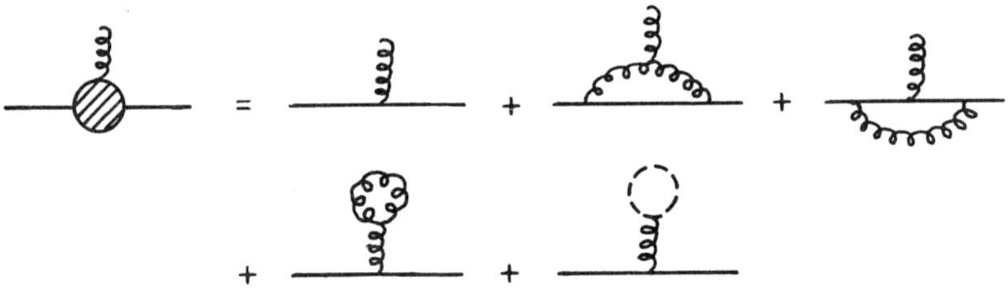

(6.21)

These pictures should convince the reader of the complexity of Yang-Mills theories and make him understand why high order calculations necessitate the use of sophisticated computer programs. The two ghost interactions with the gauge field have been represented only once by the diagrams (6.20).

We start by computing the one loop correction to the vacuum polarization. As shown in (6.15) it consists of six parts. We first compute

$\Pi^{(1)AB}_{\mu\nu}(p) =$ 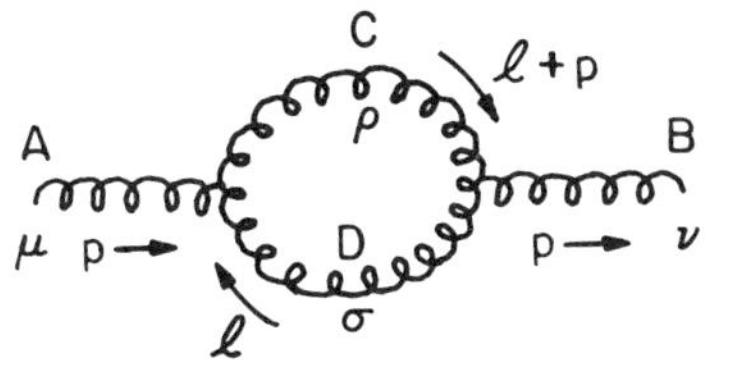


(6.22)

In the Feynman gauge, we find, not forgetting the symmetry factor,

$$\Pi^{(1)AB}_{\mu\nu}(p) = -\frac{1}{2} g^2 \mu^{2\varepsilon} f^{ACD} f^{BDC} \int \frac{d^{2\omega}\ell}{(2\pi)^{2\omega}} \frac{N_{\mu\nu}(\ell,p)}{\ell^2(\ell+p)^2} , \qquad (6.23)$$

where

$$N_{\mu\nu}(\ell,p) = [(2\ell+p)_\mu \delta_{\rho\sigma} - (\ell+2p)_\sigma \delta_{\mu\rho} + (p-\ell)_\rho \delta_{\mu\sigma}][(2\ell+p)_\nu \delta_{\rho\sigma} - (2p+\ell)_\sigma \delta_{\rho\nu} + (p-\ell)_\rho \delta_{\sigma\nu}] \qquad (6.24)$$

$$= (8\omega-6)\ell_\mu \ell_\nu + (4\omega-3)(\ell_\mu p_\nu + \ell_\nu p_\mu) + (2\omega-6)p_\mu p_\nu + [(p-\ell)^2 + (2p+\ell)^2]\delta_{\mu\nu} ; \qquad (6.25)$$

we have used $\delta_{\mu\mu} = 2\omega$. This expression is both infrared and ultraviolet divergent in four dimensions. Let us mention that the traditional remedy for infrared divergences of adding a small mass to the gauge particle leads to a lot of grief in this case because it does not respect gauge invariance. Thus in the following we will try to stay away from $p^2 = 0$ where the IR divergence occurs. When it becomes impossible to do so we will revert to the dimensional regularization method we outlined in Section 4 of this chapter.

Introduce a Feynman parameter and change the loop variable to obtain

$$\Pi^{(1)AB}_{\mu\nu}(p) = -\frac{1}{2} g^2 \mu^{2\varepsilon} f^{ACD} f^{BDC} \int_0^1 dx \int \frac{d^{2\omega}\ell}{(2\pi)^{2\omega}} \frac{N_{\mu\nu}(\ell-px,p)}{[\ell^2+p^2x(1-x)]^2} \quad . \tag{6.26}$$

A little bit of algebra yields

$$N_{\mu\nu}(\ell-px,p) = (8\omega-6)\ell_\mu\ell_\nu+[(8\omega-6)x(x-1)+2\omega-6]p_\mu p_\nu$$
$$+ [2\ell^2+p^2(2x(x-1)+5)]\delta_{\mu\nu} \quad , \tag{6.27}$$

where we have neglected terms linear in $\ell$ which will integrate to zero. Integration over the loop variable yields

$$\Pi^{(1)AB}_{\mu\nu}(p) = -\frac{g^2\mu^{2\varepsilon}}{2(4\pi)^\omega} f^{ACD}f^{BDC} \int_0^1 dx\{\frac{(6\omega-3)\delta_{\mu\nu}\Gamma(1-\omega)}{[p^2x(1-x)]^{1-\omega}} +$$
$$+ \frac{\Gamma(2-\omega)}{[p^2x(1-x)]^{2-\omega}}[\delta_{\mu\nu}p^2(5-2x(1-x))+p_\mu p_\nu(2\omega-6-(8\omega-6)x(1-x))]\} \tag{6.28}$$

Let us leave this result as it stands and compute the contribution from the ghost loop; it is given by

$\Pi^{(2)AB}_{\mu\nu}(p) =$ 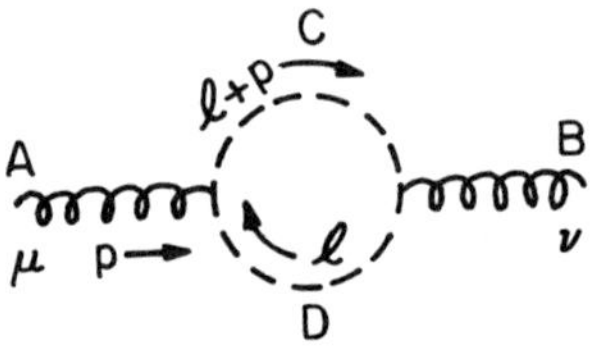 (6.29)



Remembering the minus sign for the ghost loop, the rules of Appendix C give

$$\Pi^{(2)AB}_{\mu\nu}(p) = g^2\mu^{2\varepsilon} f^{DCA} f^{CDB} \int \frac{d^{2\omega}\ell}{(2\pi)^{2\omega}} \frac{(\ell+p)_\mu \ell_\nu}{\ell^2(\ell+p)^2} \quad . \tag{6.30}$$

This expression is also infrared divergent. After introducing a Feynman parameter it becomes

$$\Pi^{(2)AB}_{\mu\nu}(p) = -g^2\mu^{2\varepsilon} f^{ACD} f^{BCD} \int_0^1 dx \int \frac{d^{2\omega}\ell}{(2\pi)^{2\omega}} \frac{(\ell-px)_\nu(\ell+p(1-x))_\mu}{[\ell^2+p^2x(1-x)]^2} \tag{6.31}$$

$$= -\frac{g^2\mu^{2\varepsilon}}{(4\pi)^\omega} f^{ACD} f^{BCD} \int_0^1 dx \{ \frac{\frac{1}{2}\delta_{\mu\nu}\Gamma(1-\omega)}{[p^2x(1-x)]^{1-\omega}} - \frac{p_\mu p_\nu x(1-x)\Gamma(2-\omega)}{[p^2x(1-x)]^{2-\omega}} \} \; . \tag{6.32}$$

Had we used the altered rule (5.5) for the ghost vertex, we would have only altered the $p_\mu p_\nu$ term (see problem). The third diagram is

$\Pi^{(3)AB}_{\mu\nu}(p) =$ 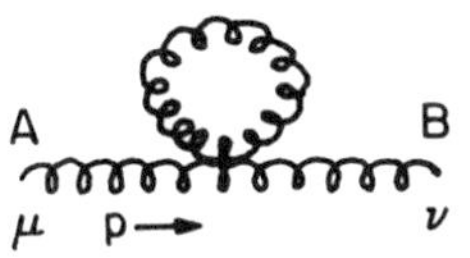 (6.33)


Fortunately this diagram vanishes in dimensional regularization along with the other two tadpole graphs of (6.15). It has the structure of a tadpole diagram and since no momentum dependence occurs at the vertices, it could only give a correction to a mass for the gauge particles. But we know from gauge invariance that gauge particles have no mass terms. However, if we adopt any regularization procedure that violates gauge invariance, this diagram will contribute a nonzero mass term. Since this would be an artifact, it cannot possibly affect any physical answer. For instance, if one adds a small mass to the gauge particles to regulate infrared divergences, the diagram (6.33) would contribute to this fake mass. This is another example of the power of dimensional regularization which controls infinities without violating gauge invariance.

We now proceed to extract the pole parts from $\Pi^{(1)}$ and $\Pi^{(2)}$. After a little bit of algebra, we find

$$\Pi^{(1)AB}_{\mu\nu}(p) = \frac{g^2}{32\pi^2} f^{ACD} f^{BCD} \{ (\frac{19}{6} \delta_{\mu\nu} p^2 - \frac{11}{3} p_\mu p_\nu)\frac{1}{\varepsilon}$$

$$- (\frac{19}{6} \gamma + \frac{1}{2})\delta_{\mu\nu} p^2 + (\frac{11}{3} \gamma - \frac{2}{3}) p_\mu p_\nu$$

$$+ \int_0^1 dx \, \ell n(\frac{p^2 x(1-x)}{4\pi\mu^2})[2p_\mu p_\nu (1+5x(1-x)) - p^2 \delta_{\mu\nu}(5-11x(1-x))] + \mathcal{O}(\varepsilon)\} \; . \tag{6.34}$$

Note the appearance of $\ell n\, p^2$ which indicates the presence of an infrared divergence on mass-shell at $p^2 = 0$ [in Minkowski space only, since $p^2 = 0$ in Euclidean space implies that $p_\mu$ itself is zero]. We also find that

$$\Pi^{(2)AB}_{\mu\nu}(p) = \frac{g^2}{32\pi^2} f^{ACD} f^{BCD} \{ (\frac{1}{6} \delta_{\mu\nu} p^2 + \frac{1}{3} p_\mu p_\nu)\frac{1}{\varepsilon} - \frac{1}{6}(\gamma - 1)\delta_{\mu\nu} p^2 - \frac{1}{3} \gamma p_\mu p_\nu$$

$$- \int_0^1 dx \, \ell n(\frac{p^2 x(1-x)}{4\pi\mu^2})[\delta_{\mu\nu} p^2 x(1-x) + 2p_\mu p_\nu x(1-x)] + \mathcal{O}(\varepsilon)\} \quad . \tag{6.35}$$

The sum of $\Pi^{(1)}$ and $\Pi^{(2)}$ gives the total pure Yang-Mills contribution to the vacuum polarization

$$\Pi^{AB}_{\mu\nu}(p) = \frac{g^2}{32\pi^2} f^{ACD} f^{BCD} (\delta_{\mu\nu} p^2 - p_\mu p_\nu)\{\frac{10}{3} \cdot \frac{1}{\varepsilon} + \frac{62}{9} - \frac{10}{3}\gamma - \frac{10}{3} \ell n \frac{p^2}{4\pi\mu^2}\} \tag{6.36}$$

and we recover the magic projection operator structure. The finite part shows a similar structure, and contains the $\ell n\, p^2/4\pi\mu^2$ factor, already noted to be infrared divergent. The elements of the matrices $T^A$ that represent the Lie algebra of G in the adjoint representation can be identified with the structure constants

$$(T^A)^{BC} = -if^{ABC} \quad . \qquad (6.37)$$

Now, for any representation R of G, with the representation matrices $T_R^A$, we can set

$$\mathrm{Tr}(T_R^A T_R^B) = C_R \delta^{AB} \quad , \qquad (6.38)$$

where the trace is taken over the representation indices. The number $C_R$ is called the Dynkin index of the representation R; it is equal to the quadratic Casimir operator multiplied by the dimension of the representation and divided by the dimension of the group. Its precise value depends on the normalization of the T-matrices, which is fixed once and for all by specifying the structure constants of the theory. The scale of the structure constants is in turn fixed by the definition of g since they always appear in the combination $gf^{ABC}$. It is very convenient to rewrite all our expressions in terms of the Dynkin index. Thus we rewrite

$$\Pi_{\mu\nu}^{AB}(p) = \delta^{AB} \frac{g^2}{16\pi^2} C_{ad.}(\delta_{\mu\nu}p^2 - p_\mu p_\nu) \cdot \frac{5}{3}\frac{1}{\varepsilon} + \ldots \quad . \qquad (6.39)$$

Let us remark that there is nothing sacred about the projection operator structure since it could have been changed had we adopted the altered rule (5.5) for the ghosts.

The fermion contribution to the vacuum polarization

$\Pi_{\mu\nu}^{fAB}(p) =$ 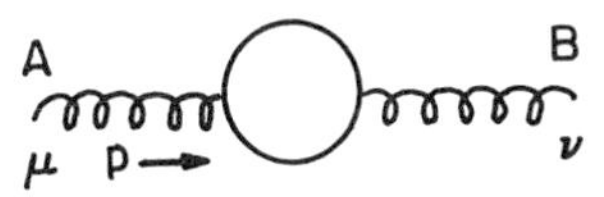 (6.40) need not be recalculated since the structure is the same as in QED, save for some group theoretical factors. Indeed we find

$$\Pi^{fAB}_{\mu\nu}(p) = \mathrm{Tr}(T^A_f T^B_f)\Pi^{QED}_{\mu\nu}(p) \tag{6.41}$$

$$= -C_f \delta^{AB} \frac{g^2}{16\pi^2}(\delta_{\mu\nu}p^2 - p_\mu p_\nu)\frac{4}{3}\cdot\frac{1}{\varepsilon} + \ldots \quad , \tag{6.42}$$

where we have used (6.38) and (2.32). One can see by comparing (6.42) with (6.39) that the residues of the poles have opposite signs (more on this later). The sum of (6.42) and (6.39) gives the total one-loop correction to the vacuum polarization.

Next, the correction to the fermion line (ignoring tadpoles) is given by

$\Sigma(p) =$ 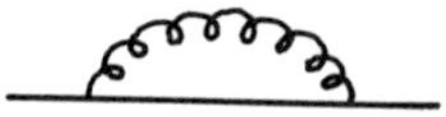 (6.43)

and it is a matrix both in spinor space and in the representation space of the fermions. It is easy to see that

$$\Sigma(p) = T^A_f T^A_f \Sigma^{QED}(p) \quad , \tag{6.44}$$

where $\Sigma^{QED}$ is given by (2.19). Furthermore the $d_f x d_f$ matrix in front of (6.44) is diagonal and given by

$$T^A T^A = \frac{N}{d_f} C_f \quad , \tag{6.45}$$

where we have not shown the representation indices. Hence we have

$$\Sigma(p) = -i\frac{N}{d_f} C_f \frac{g^2}{16\pi^2}[\not{p} + 4m]\frac{1}{\varepsilon} + \ldots \quad . \tag{6.46}$$

The one loop correction to the fermion vertex consists, ignoring tadpoles, of two diagrams. The first

$$\Gamma^A_{1\rho}(p,q) = \qquad (6.47)$$

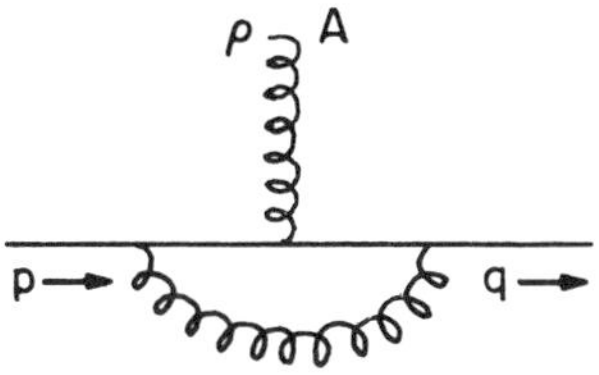


has the same character as in the Abelian case. Indeed we find

$$\Gamma^A_{1\rho}(p,q) = T^B_f T^A_f T^B_f \Gamma^{QED}_\rho(p,q) \quad . \qquad (6.48)$$

The $d_f x d_f$ matrix can be rewritten as

$$T^B_f T^A_f T^B_f = [T^B_f, T^A_f] T^B_f + T^A_f T^B_f T^B_f \qquad (6.49)$$

$$= i f^{BAC} T^C_f T^B_f + \frac{N}{d_f} C_f T^A_f \quad . \qquad (6.50)$$

Now, using the antisymmetry of C and B, we rewrite it as

$$T^B_f T^A_f T^B_f = \frac{1}{2} f^{BAC} f^{DCB} T^D_f + \frac{N}{d_f} T^A_f \qquad (6.51)$$

$$= \frac{1}{2} C_{ad} T^A_f + \frac{N}{d_f} C_f T^A_f \quad , \qquad (6.52)$$

where we have used (6.37), (6.38) and (6.45). Using the results of Section 2, we obtain

$$\Gamma^A_{1\rho}(p,q) = -ig\mu^\varepsilon T^A_f \gamma_\rho \left(\frac{1}{2} C_{ad.} + C_f \frac{N}{d_f}\right) \frac{g^2}{16\pi^2} \frac{1}{\varepsilon} + \ldots \quad . \qquad (6.53)$$

We recognize that the residue of the second term is exactly the same as that of (6.46) because these do not involve the non-Abelian character of the theory, and therefore satisfy the old QED Ward identity. If the Taylor-Slavnov identities are to hold, the ratios $Z_1/Z_2$ and $Z_8/Z_6$ must be equal. But the latter receives no contribution from the fermions at the one loop level, as we see from (6.17) and (6.20); hence the fermion wave function renormalization contribution must be exactly canceled by a term in $Z_1$. This is exactly what happens.

The second contribution to the vertex is definitely non-Abelian in character:

$$\Gamma^A_{2\rho}(p,q) = \tag{6.54}$$

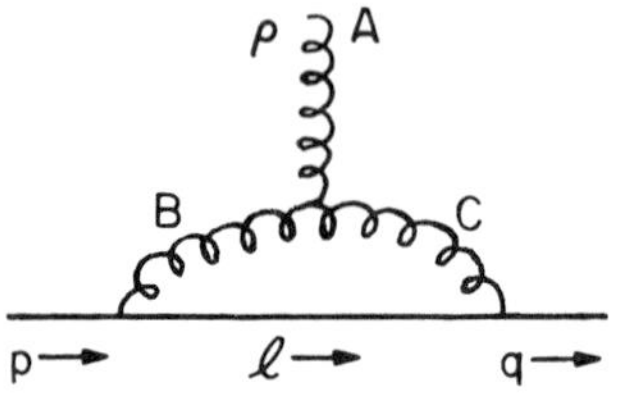


The rules give

$$\Gamma^A_{2\rho}(p,q) = -g^3\mu^{3\varepsilon} f^{ABC} T^B_f T^C_f \int \frac{d^{2\omega}\ell}{(2\pi)^{2\omega}} \gamma_\mu \frac{1}{\not\ell + m} \gamma_\nu \frac{[(\ell+p)_\mu \delta_{\nu\rho} + (q-2p-\ell)_\nu \delta_{\mu\rho} + (p+q-2\ell)_\rho \delta_{\mu\nu}]}{(p-\ell)^2(\ell-q)^2} \tag{6.55}$$

The group-theoretic factors are easily taken care of since

$$f^{ABC} T^B_f T^C_f = \frac{i}{2} f^{ABC} f^{BCD} T^D_f \tag{6.56}$$

$$= \frac{i}{2} C_{ad} T^A_f \quad , \tag{6.57}$$

using (6.37) and (6.38). Introduce Feynman parameters and appropriately shift the loop variable to obtain

$$\Gamma^A_{2\rho}(p,q) = -ig^3\mu^{3\varepsilon} C_{ad} T^A_f \int_0^1 dx \int_0^{1-x} dy \int \frac{d^{2\omega}\ell'}{(2\pi)^{2\omega}} \frac{N_\rho(\ell',p,q)}{[\ell'^2 + m^2(1-x-y) + q^2 x + p^2 y - (qx-py)^2]^3} , \tag{6.58}$$

with

$$N_\rho = 2\ell'_\rho \gamma_\nu \not{\ell}' \gamma_\nu + \hat{N}_\rho \quad , \qquad (6.59)$$

$$= 4\ell'_\rho \not{\ell}'(1-\varepsilon) + \hat{N}_\rho \qquad (6.60)$$

where $\hat{N}_\rho$ only contains terms linear in $\ell'$ which will be integrated out and terms that contain no $\ell'$. The latter will give a convergent loop integral. We find then

$$\Gamma^A_{2\rho}(p,q) = - ig\mu^\varepsilon T^A_f \frac{g^2 C_{ad}}{16\pi^2} \int_0^1 dx \int_0^{1-x} dy \{\gamma_\rho (1-\varepsilon)\Gamma(\varepsilon)\left(\frac{4\pi\mu^2}{m^2(1-x-y)+q^2x+p^2y-(qx-py)^2}\right)^\varepsilon +$$

$$+ \frac{\mu^2}{2} \frac{\hat{N}_\rho}{m^2(1-x-y)+q^2x+p^2y-(qx-py)^2}\} \quad . \qquad (6.61)$$

This formula is very complicated but the pole part is easily extracted

$$\Gamma^A_{2\rho}(p,q) = -ig\mu^\varepsilon \gamma_\rho T^A_f \frac{g^2 C_{ad}}{32\pi^2} \frac{1}{\varepsilon} + \ldots \quad , \qquad (6.62)$$

so that the total pole contribution to the fermion vertex function is

$$\Gamma^A_\rho(p,q) = - ig\mu^\varepsilon \gamma_\rho T^A_f \frac{g^2}{16\pi^2} (C_{ad} + C_f \frac{N}{d_f}) \frac{1}{\varepsilon} + \ldots \quad . \qquad (6.63)$$

Now that we have calculated the pole structure of the lowest order corrections to the fermion and gauge fields and to the fermion vertex, we can read off from (6.39), (6.42), (6.46) and (6.63) the corresponding renormalization constants. They are

$$Z_1 = 1 - \frac{g^2}{16\pi^2} ([C_{ad} + C_f \frac{N}{d_f}] \frac{1}{\varepsilon} + F_1) + \ldots \qquad (6.64)$$

$$Z_2 = 1 - \frac{g^2}{16\pi^2} (C_f \frac{N}{d_f} \frac{1}{\varepsilon} + F_2) + \ldots \quad , \quad (6.65)$$

$$Z_3 = 1 + \frac{g^2}{16\pi^2} ((\frac{5}{3} C_{ad} - \frac{4}{3} C_f) \frac{1}{\varepsilon} + F_3) + \ldots \quad , \quad (6.66)$$

where $F_1$, $F_2$, $F_3$ are the arbitrary finite parts of the counterterms. These expressions are all gauge dependent. [For instance, in the axial gauge, the effective Lagrangian has no ghosts and the same structure as in QED, resulting in the identity $Z_1 = Z_2$, which is patently untrue in the Feynman gauge.] Let us remark that the fermion contribution to the vector quartic self-interaction must diverge if the Slavnov-Taylor identities (6.14) are to hold because we explicitly see that the ratio $Z_3^{1/2} Z_1/Z_2$ contains a fermion contribution. Contrast this situation with the box diagram of QED which is finite. How can the same diagrams be finite in one case (QED) and diverge in the other (YM)? The answer, of course, is that the diagrams are not the same: in QED the box diagram's divergence vanishes only upon symmetrization of the external photon lines denoted only by their vector indices; in the Yang-Mills case, the symmetrization of the external lines can be performed in two ways, by symmetrizing on both vector and group indices, which as in QED gives no divergence <u>or</u> by antisymmetrizing on both vector and group indices. It is this new contribution which diverges (see problem). The very same reasoning can be applied to the fermion contribution to the triple gauge vertex (6.18).

Problems.

A. Compute the ghost contribution to the vacuum polarization using the altered Feynman rule (5.5). Show explicitly that it only affects the $p_\mu p_\nu$ (longitudinal) part of $\Pi_{\mu\nu}$. Find $Z_\alpha$ and compare it with the $Z_\alpha$ obtained by the conventional rules of Appendix C.

B. Perform the parametric integrals for $\Pi_{\mu\nu}$ in $2\omega$ dimensions starting from (6.28) and (6.32). Then expand about $\omega = 2$; discuss the significance of the extra poles in $\varepsilon$.

*C. Compute in the Feynman gauge the ghost field renormalization constant $Z_6$, as well as the ghost-gauge vertex renormalization constant $Z_7$ to one loop, and verify that $Z_7/Z_6$ is indeed equal to $Z_1/Z_2$.

**D. Evaluate the most divergent parts of the diagrams

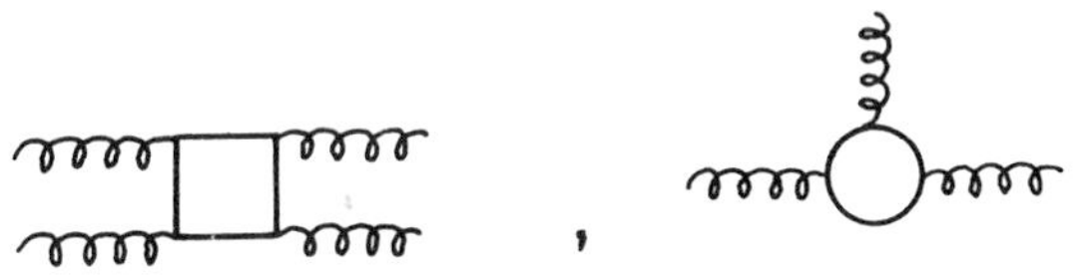

in the Feynman gauge and verify the Slavnov-Taylor identities for the fermion part.

7. Yang-Mills Theory: Slavnov-Taylor Identities

We have seen that the effective Action which constitutes the starting point of the quantum Yang-Mills theory is no longer gauge invariant because of the gauge fixing term and the ghosts (in a covariant gauge). The terms which break the gauge invariance are

$$\mathcal{L}_{extra} = \frac{1}{2\alpha} G^A G^A - i\eta^{*A} \frac{\delta g^A}{\delta A^B_\mu} (\mathcal{D}_\mu)^{BC} \eta^C \quad , \tag{7.1}$$

where $\eta^B$ and $\eta^{*B}$ are the ghost fields and $G^A$ the gauge conditions. It turns out to be more convenient to use real Grassmann fields $\omega^A$ and $\rho^A$ rather than complex ones; they are related by

$$\eta^A = \frac{1}{\sqrt{2}} (\omega^A + i\rho^A) \tag{7.2}$$

$$\eta^{*A} = \frac{1}{\sqrt{2}} (\omega^A - i\rho^A) \quad . \tag{7.3}$$

Then the gauge noninvariant terms are rewritten in terms of $\omega^A$ and $\rho^A$ in the form

$$\mathcal{L}_{extra} = \frac{1}{2\alpha} G^A G^A + \omega^A \frac{\delta G^A}{\delta A^B_\mu} (\mathcal{D}_\mu)^{BC} \rho^C \quad . \tag{7.4}$$

Consider now their behavior under a gauge transformation

$$\delta A^B_\mu = \mathcal{D}^{BC}_\mu \Lambda^C \quad , \tag{7.5}$$

where $\Lambda^C$ is the gauge parameter. We have

$$\delta\mathcal{L}_{extra} = \frac{1}{\alpha} G^A \frac{\delta G^A}{\delta A^B_\mu} (\mathcal{D}_\mu)^{BC} \Lambda^C + \delta\omega^A \frac{\delta G^A}{\delta A^B_\mu} (\mathcal{D}_\mu \rho)^B + \omega^A \frac{\delta G^A}{\delta A^B_\mu} \delta(\mathcal{D}_\mu \rho)^B \quad , \tag{7.6}$$

assuming a gauge condition linear in $A_\mu$. Then we notice that the first and second terms could cancel one another provided we chose the gauge parameter cleverly. Indeed by taking

$$\Lambda^C = \zeta\rho^C \qquad , \qquad (7.7)$$

and

$$\delta\omega^A = -\frac{1}{\alpha} G^A\zeta \qquad , \qquad (7.8)$$

the first two terms of the variation (7.6) cancel against one another. Furthermore, since $\rho^C$ is a real Grassmann field, the parameter $\zeta$ must itself be a real Grassmann number so as to make $\Lambda^C$ normal, i.e.,

$$\zeta^2 = 0 \qquad . \qquad (7.9)$$

Then if it can be arranged that

$$\delta(\mathcal{D}_\mu\rho)^B = 0 \qquad (7.10)$$

the invariance of the total effective Action will follow, since the transformation on the $A_\mu$ field is in the form of a (field dependent) gauge transformation. Explicitly

$$\delta(\mathcal{D}_\mu\rho)^B = \partial_\mu\delta\rho^B + f^{BCE}(\mathcal{D}_\mu\zeta\rho)^E\rho^C + f^{BCE}A^E_\mu\delta\rho^C \qquad . \qquad (7.11)$$

The terms involving one derivative give

$$\partial_\mu[\delta\rho^B + \frac{1}{2}\zeta f^{BCE}\rho^E\rho^C] \qquad ; \qquad (7.12)$$

thus if we set

$$\delta\rho^B = \frac{1}{2}\zeta f^{BCE}\rho^C\rho^E \qquad , \qquad (7.13)$$

we are left with

$$\delta(\mathcal{D}_\mu\rho)^B = \zeta f^{BCE}f^{EFG}A^G_\mu\rho^F\rho^C + \frac{1}{2}\zeta f^{BEG}A^G_\mu f^{EFC}\rho^F\rho^C \qquad . \qquad (7.14)$$

The identification

$$f^{ABC} = i(T^A)^{BC} \qquad (7.15)$$

enables us to rewrite (7.14) in the form

$$\delta(\mathcal{D}_\mu \rho)^B = -\zeta (T^F T^C)^{GB} A_\mu^G \rho^F \rho^C + \frac{i}{2} \zeta f^{FCE} (T^E)^{GB} A_\mu^G \rho^F \rho^C \quad . \qquad (7.16)$$

Now, using the antisymmetry under the interchange of F and C, and the Lie algebra obeyed by the T matrices, we verify that this vanishes with the choice (7.13). To summarize, the effective Lagrangian (5.1) is invariant under the combined transformations

$$\delta A_\mu^B = \frac{1}{g} \zeta (\mathcal{D}_\mu \rho)^B$$

$$\delta \omega^A = -\frac{1}{\alpha g} \partial \cdot A^A \zeta$$

$$\delta \rho^A = \frac{1}{2} \zeta f^{ABC} \rho^B \rho^C$$

$$\delta \Psi = -i \zeta \rho^A T_f^A \Psi \qquad , \qquad (7.17)$$

where we have explicitly shown the dependence on the coupling constant, set $G^A = \partial \cdot A^A$, and included the fermions. The effective Action is invariant under these BRS (for Becchi, Rouet and Stora, op cit) transformations only when the coupling constants are all the same. Since the BRS transformations are unaffected by changing space-time dimensions, they will go through dimensional regularization unscathed with the result that both $\mathcal{L}_{eff}$ and $\mathcal{L}_{ct}$ will be invariant. Hence as a first consequence all the bare couplings will be the same, that is in (6.3) we now can say that

$$g_0 = g_0' = g_0'' = g_0''' = g_0'''' \qquad (7.18)$$

with the consequence that (6.14) holds and that there are, besides $Z_\alpha$ and $Z_m$, only four independent renormalization constants, $Z_1$, $Z_2$, $Z_3$, $Z_6$ say, three of which we have already calculated.

As in the Abelian case we can use the BRS invariance to derive identities between the Yang-Mills Green's functions. In order to avoid complicated nonlinear terms, it is convenient to introduce sources that couple to the changes in the fields under a BRS transformation, that is we augment the usual set of sources $J^B_\mu$, $\sigma^B$, $\xi^B$, $\chi$, $\bar{\chi}$ that couple to the fundamental fields as

$$-\langle J^B_\mu A^B_\mu + \omega^B \sigma^B + \rho^B \xi^B + i\bar{\chi}\Psi + i\bar{\Psi}\chi \rangle \tag{7.19}$$

by other sources $\tau^A_\mu$, $u^A$, $\lambda$, $\bar{\lambda}$ which couple to the nonlinear BRS changes as

$$-\langle \frac{1}{g} \tau^A_\mu (\mathcal{D}_\mu \rho)^A + \frac{1}{2} f^{ABC} u^A \rho^B \rho^C + \bar{\lambda} \rho^A T^A_f \Psi - \bar{\Psi} T^A_f \rho^A \lambda \rangle \quad . \tag{7.20}$$

Our starting point is the generating functional

$$Z[J^B_\mu, \sigma^B, \xi^B, \chi, \bar{\chi}; \tau^B_\mu, u^B, \bar{\lambda}, \lambda] \quad , \tag{7.21}$$

given by

$$e^{-Z} = \int \mathcal{D} A_\mu \mathcal{D} \Psi \mathcal{D} \bar{\Psi} \, \mathcal{D} \omega \, \mathcal{D} \rho \; e^{-S'_{eff}} \quad , \tag{7.22}$$

where $S'_{eff}$ is $S_{eff}$ of Section 5, augmented by the sources (7.19) and (7.20). We perform a BRS transformation on the integration variables of (7.22); on the one hand such a change of integration variable should leave Z unaffected; on the other, its effect on the integrand can be explicitly traced - any deviation from (7.22) resulting from this change should vanish, yielding the Yang-Mills version of the Ward identities.

First of all, one can show that the measure of integration is unaltered by a BRS transformation (see problem). Secondly we have just shown that $S_{eff}$ is BRS invariant, hence only the source terms will change. The terms linear in the fields will give the extra terms in the exponent

$$-\langle J^B_\mu \delta A^B_\mu + \delta\omega^B \sigma^B + \delta\rho^B \xi^B + i\bar{\chi}\delta\Psi + i\delta\bar{\Psi}\chi\rangle \quad . \qquad (7.23)$$

The terms in (7.20) give no extra contribution because the BRS transformations are nilpotent: two BRS transformations give zero:

$$\delta(f^{BCD}\rho^C\rho^D) = 0 \qquad (7.24)$$

$$\delta(\rho^A T^A_f \Psi) = 0 \qquad (7.25)$$

and (7.10). This is not hard to prove (see problem). Thus the only effect of our transformation is to add to the exponent of the integrand the terms (7.23). They are all linear in the Grassmann variable $\zeta$ so that this extra exponential is easily expanded - the term linear in $\zeta$ must therefore vanish since it is the only change in the integrand, giving the identities

$$\int \mathcal{D}(\ldots)\langle \frac{1}{g} J^B_\mu (\mathcal{D}_\mu \rho)^B - \frac{1}{\alpha g} \partial\cdot A^B \sigma^B + \frac{1}{2} f^{BCD}\rho^C\rho^D\xi^B + \bar{\chi}\rho^A T^A_f \Psi - \bar{\Psi}\rho^A T^A_f \chi\rangle e^{-S'_{eff}} = 0 \; . \qquad (7.26)$$

This functional equation is highly nonlinear and therefore awkward to work with. The reason for the extra sources now becomes clear - we can replace these nonlinear terms by functional derivatives since by construction

$$A^B_\mu = -\frac{\delta}{\delta J^B_\mu} \qquad (7.27)$$

$$\frac{1}{g}(\mathcal{D}_\mu \rho)^B = \frac{\delta}{\delta \tau^B_\mu} \qquad (7.28)$$

$$\frac{1}{2} f^{ABC}\rho^B\rho^C = \frac{\delta}{\delta u^A} \qquad (7.29)$$

$$\rho^A T^A_f \Psi = \frac{\delta}{\delta \bar{\lambda}} \quad , \qquad (7.30)$$

all equations valid when acting on $e^{-S'_{eff}}$. Thus we now write the identities in the form of a linear functional differential equation

$$<J^B_\mu \frac{\delta}{\delta\tau^B_\mu} + \sigma^B \frac{1}{\alpha g} \partial_\mu \frac{\delta}{\delta J^B_\mu} + \xi^B \frac{\delta}{\delta u^B} + \bar{\chi} \frac{\delta}{\delta\bar{\lambda}} + \chi^T \frac{\delta}{\delta\lambda}> e^{-Z} = 0 \quad . \tag{7.31}$$

It is more useful to write these identities on the effective quantum Action $\Gamma$, which generates the proper vertices of the theory. To this effect we perform a functional Legendre transformation

$$J^B_\mu \rightarrow A^B_{\mu c\ell} \tag{7.32}$$

$$\sigma^B \rightarrow \omega^B_{c\ell} \tag{7.33}$$

$$\xi^A \rightarrow \rho^A_{c\ell} \tag{7.34}$$

$$\chi \rightarrow \Psi_{c\ell} \tag{7.35}$$

$$\bar{\chi} \rightarrow \bar{\Psi}_{c\ell} \quad , \tag{7.36}$$

$$Z \rightarrow \Gamma \quad , \tag{7.37}$$

where now $A^B_{\mu c\ell}$, $\omega^B_{c\ell}$, $\rho^B_{c\ell}$, $\Psi_{c\ell}$, $\bar{\Psi}_{c\ell}$ are the new sources which play the role of the fields in the classical approximation. Explicitly

$$\Gamma = Z - <J^B \cdot A^B_{c\ell} + i\bar{\chi}\Psi_{c\ell} + i\bar{\Psi}_{c\ell}\chi + \sigma^B \omega^B_{c\ell} + \xi^B \rho^B_{c\ell}> \quad . \tag{7.38}$$

We have not changed the nonlinear sources so that $\Gamma$ is also a functional of $\tau^A_\mu$, $u^A$, $\lambda$ and $\bar{\lambda}$. Then we have

$$J^B_\mu = -\frac{\delta\Gamma}{\delta A^B_{\mu c\ell}} \quad , \quad \bar{\chi} = i\,\frac{\delta\Gamma}{\delta\Psi_{c\ell}} \quad , \quad \text{etc...} \tag{7.39}$$

with the result that the Ward-Slavnov-Taylor identities now read

$$< \frac{\delta\Gamma}{\delta A^B_{\mu c\ell}} \frac{\delta\Gamma}{\delta\tau^B_\mu} - \frac{1}{\alpha g}\,\partial\cdot A^B_{\mu c\ell}\,\frac{\delta\Gamma}{\delta\omega^B_{c\ell}} - \frac{\delta\Gamma}{\delta\rho^B_{c\ell}}\frac{\delta\Gamma}{\delta u^B} - i\,\frac{\delta\Gamma}{\delta\Psi_{c\ell}}\frac{\delta\Gamma}{\delta\bar{\lambda}} + i\,\frac{\delta\Gamma}{\delta\lambda}\frac{\delta\Gamma}{\delta\bar{\Psi}_{c\ell}} > = 0. \quad (7.40)$$

This is still a formidable looking formula, but it is (almost) the best one can provide in this very complicated theory (see problem). The nonlinear source terms are simple only in the classical approximation where we can set

$$\frac{\delta\Gamma}{\delta\tau^B_\mu} = \frac{1}{g}\,(\mathcal{D}^{c\ell}_\mu \rho_{c\ell})^B + \mathcal{O}(\hbar),\ \text{etc.....} \quad . \quad (7.41)$$

In the Abelian case we encountered in the variation of $\Psi$ a nonlinear term $(\eta-\eta^*)\Psi$ which we replaced in (3.16) by the classical sources $(\eta_{c\ell}-\eta^*_{c\ell})\Psi_{c\ell}$; we could do this only because in that case the ghost field did not interact with the fermion field. This is no longer true in our case, leading to far more complicated expressions for the Ward identities. However, in an axial gauge where ghosts are absent, the Yang-Mills identities look the same as in the QED case with the result that $Z_1 = Z_2$.

These identities, cumbersome as they may be, are essential for proving the renormalizability of the theory. They reduce the number of independent ultraviolet divergences, thus allowing for their absorption by a redefinition of the input parameters. However, in the spirit of our approach we do not concern ourselves with such technicalities. Lastly let us mention that it does not seem possible to generalize the BRS transformations when a fictitious mass is present both for the gauge particle and the ghosts, as was possible in QED. Thus it would seem that IR divergences in Yang-Mills should also be handled by dimensional regularization along the lines sketched in Section 4.

Problems.

A. Derive the Yang-Mills Ward identities in the axial gauge, including fermions.

B. Show that the Yang-Mills functional measure is left invariant by a BRS transformation.

C. Show that the BRS transformations are nilpotent.

D. Show that

$$\partial_\mu \frac{\delta\Gamma}{\delta\tau^A_\mu} = - \frac{\delta\Gamma}{\delta\omega^A_{c\ell}} \ .$$

*E. Examine in detail how the QED Ward identity which relates the longitudinal part of the vertex to the inverse fermion propagator is altered in the Yang-Mills case in the covariant gauge $\partial\cdot A^B = 0$.

## 8. Yang-Mills Theory: Asymptotic Freedom

With the help of the BRS transformation we have shown that the Yang-Mills theory contains only one coupling constant, g. We can therefore use any of the formulae (6.9) - (6.13) to relate the bare coupling $g_0$ to the renormalized coupling g. Since it happens that we have calculated $Z_1$, $Z_2$ and $Z_3$ in the Feynman gauge, we start from (6.9)

$$g_0 = g\mu^{\varepsilon} \frac{Z_1}{Z_2 Z_3^{1/2}} \quad . \tag{8.1}$$

Using (6.64), (6.65) and (6.66), this equation now reads [by expanding in g at fixed ε]

$$g_0 = g\mu^{\varepsilon}[1 - \frac{g^2}{16\pi^2} [(\frac{11}{6} C_{ad.} - \frac{2}{3} C_f)\frac{1}{\varepsilon} + \frac{1}{2} F_3 - F_1 + F_2]] \quad . \tag{8.2}$$

If we take the mass independent renormalization prescription where we set all the finite parts of the counterterms to zero, we find immediately that

$$\mu \frac{\partial g}{\partial \mu} = (g \frac{\partial}{\partial g} - 1)(- \frac{g^3}{16\pi^2})(\frac{11}{6} C_{ad.} - \frac{2}{3} C_f) \tag{8.3}$$

or

$$\mu \frac{\partial g}{\partial \mu} = \beta(g) \tag{8.4}$$

$$\beta(g) = - \frac{g^3}{16\pi^2} (\frac{11}{3} C_{ad.} - \frac{4}{3} C_f) \quad . \tag{8.5}$$

Thus as long as

$$\frac{11}{3} C_{ad.} - \frac{4}{3} C_f > 0 \tag{8.6}$$

we see that the coupling constant decreases with μ! This is the statement of asymptotic freedom [discovered independently by H. D. Politzer, Phys.

Rev. Letters 26, 1346 (1973) and D. Gross and F. Wilczek, Phys. Rev. Letters 26, 1343 (1973)]. This discovery is of enormous significance. On the one hand it says that if at a given distance $\mu^{-1}$ the coupling constant is perturbative, it will decrease in value at shorter and shorter distances with the result that one can use perturbation theory with confidence at these scales. On the other hand, as the distance increases so does the coupling constant, leaving the domain where perturbative calculations can be trusted. The situation is shown below

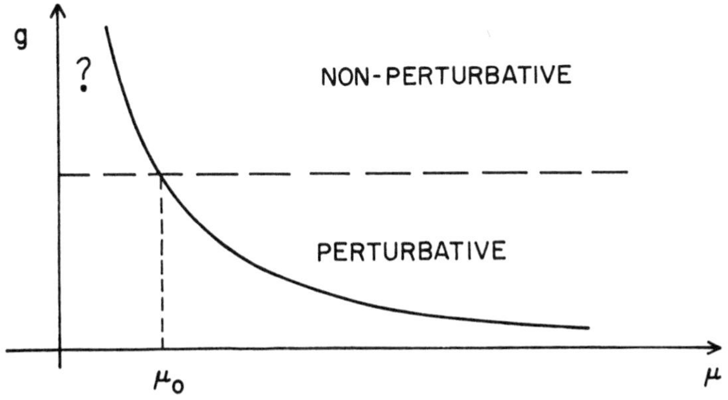


In the perturbative regime, we can integrate (8.4) with the result

$$g^2(\mu) = g^2(\mu_0)\left[1 + \frac{g^2(\mu_0)}{8\pi^2}\left(\frac{11}{3} C_{ad.} - \frac{4}{3} C_f\right)\ln\frac{\mu}{\mu_0}\right]^{-1} \quad . \tag{8.7}$$

This formula can in turn by expanded as long as

$$\frac{g^2(\mu_0)}{8\pi^2}\left(\frac{11}{3} C_{ad.} - \frac{4}{3} C_f\right)\ln\frac{\mu}{\mu_0} << 1 \quad . \tag{8.8}$$

Then $g^2(\mu)$ can be understood to represent an effective coupling constant, equal in lowest order to $g^2(\mu_0)$ and corrected by $\mathcal{O}(g^4(\mu_0))$

$$g^2(\mu) \approx g^2(\mu_0) - \frac{g^4(\mu_0)}{8\pi^2}\left(\frac{11}{3} C_{ad.} - \frac{4}{3} C_f\right)\ln\frac{\mu}{\mu_0} + \ldots \quad . \tag{8.9}$$

This behavior can, of course, be duplicated by diagrams which correct the fundamental vertex, but the full corrections to the vertex are more complicated. Typically, the corrections coming from integrating renormalization group equations give only the leading logarithms. Thus (8.7) will give to the next order only the $\ln^2 \frac{\mu}{\mu_0}$ term.

When $\mu < \mu_0$, i.e., at large distances, the perturbative calculations are not to be trusted. This means that the states on which perturbative calculations have been performed have a meaning only at short distances where the theory becomes asymptotically free, but in the large distance regime their coupling becomes strong with the result that one does not expect these perturbative states to be able to leave the interaction region to become asymptotic states [unless for some unknown reason the coupling becomes weak again at yet a larger distance: since no computation has so far been carried out in this regime we are free to speculate]. If so, the only way perturbative states could escape and form asymptotic states, would be as composite states which are neutral with respect to the long range gauge force. Such neutral or singlet combinations would still suffer multipole interactions but these are short range and do not alter the nature of the composite states. Thus there arises the <u>confinement</u> hypothesis (infrared slavery) which states that in an asymptotically free theory only singlets under the gauge force could serve as asymptotic state (otherwise they could not have escaped the effect of the large coupling).

Let us discuss this hypothesis in the context of Quantum Chromodynamics (QCD), <u>the</u> candidate theory for explaining Strong Interaction Physics. In this case the gauge group is SU(3) and the perturbative states are Dirac fermions, transforming according to the fundamental $\underset{\sim}{3}$ representation of $SU_3$, identified with quarks (q), and eight vector particles called gluons. From group theory it is easy to see that

$$C_{ad.} = 3 \text{ for } SU(3)$$

$$C_f = \frac{1}{2} \text{ for each Dirac fermion} \sim \underset{\sim}{3} .$$

Thus

$$\beta(g) = -\frac{g^3}{16\pi^2}\left(11 - \frac{2}{3} n_f\right) \qquad , \qquad (8.10)$$

where we have included $n_f$ flavors of quarks [experimentally $n_f = 5$, but is expected to be 6 on some theoretical grounds]: QCD is asymptotically free. Thus according to our previous discussion, we expect the asymptotic states of the theory to be not the quarks or the gluons, but rather <u>composites</u> made up of quarks, antiquarks and gluons, to be identified with strongly interacting particles found in the laboratory such as protons, neutrons, $\pi$-mesons, etc. We identify them by simply forming singlets under the color force! There are integer spin composites made up of quark-antiquark pairs $(q\bar{q})$, with either spin 0 or 1 in their lowest states when the spins are aligned or antialigned. Grouped theoretically, this corresponds to

$$\underset{\sim}{3} \otimes \underset{\sim}{\bar{3}} = \underset{\sim}{1} + \underset{\sim}{8} \qquad .$$

Fermionic asymptotic states can be formed from

$$(\underset{\sim}{3} \otimes \underset{\sim}{3} \otimes \underset{\sim}{3})_A = \underset{\sim}{1} \quad ,$$

corresponding to the baryons (qqq) and antibaryons ($\overline{q}\overline{q}\overline{q}$) with lowest spins 3/2 and 1/2 depending on quark spin alignments. Comparison of these composites, $q\bar{q}$, qqq and $\overline{q}\overline{q}\overline{q}$ with observed states is entirely satisfactory. Not so successful has been the detection of states made up of gluons. Since

$$(\underset{\sim}{8} \otimes \underset{\sim}{8})_S = \underset{\sim}{1} + \ldots$$

one should discover an asymptotic state made up of two gluons. No such state has been found presumably because of the value of its mass and decay parameters. Also missing are singlet states like $q\bar{q}$ gluon, where ($q\bar{q}$) form an $\underset{\sim}{8}$. The eventual acceptance of QCD as the theory of strong interactions rests on the ability to calculate strong binding processes in a gauge theory. However, nature has provided us with a short scale probe: the off-shell photons produced either in a time-like mode by $e^+e^-$ annihilation or in a space-like mode coming from a projectile electron. Such probes have verified that the proton looks to the space-like probe as if it were made up of three quarks (parton model), as evidenced by the famous scaling properties encountered in electroproduction experiments. Although severe infrared divergence problems usually complicate comparison between theory and experiment, space-like probes connect directly to the deep Euclidean region where we have seen that the structure of the field theory is at its simplest. On the negative side, the lack of understanding of the theory in its strong coupling regime makes it difficult to agree on a standard physical identification for the gauge coupling constant, unlike in QED where e is almost directly measured. Here the analog of an ideal identification would be to normalize with respect to, say, the $\pi NN$ coupling constant, but alas we simply do not know how to relate it to the quark-quark coupling constant. Thus devious (and very clever) means of comparison with experiment had to be devised, but we do not discuss them here as they would be ample subject for another book!

Finally we note that there exists a parameter with dimensions which should specify the theory since if one knew the value of $g^2$ at some scale $\mu_1$, presumably from clever comparison with experiment, one would know at which scale perturbation theory would cease to be valid. Traditionally one introduces a scale parameter $\Lambda$ defined such that

$$g^2(\mu) = \frac{8\pi^2}{(\frac{11}{3} C_{ad.} - \frac{4}{3} C_f)\ln(\frac{\mu}{\Lambda})} \quad . \tag{8.11}$$

Comparison with (8.7) yields

$$\ln \Lambda = \ln \mu_0 - \frac{8\pi^2}{g^2(\mu_0)[\frac{11}{3} C_{ad.} - \frac{4}{3} C_f]} \quad . \tag{8.12}$$

To this order, knowledge of $\mu_0$ and $g^2(\mu_0)$ gives the scale $\Lambda$, but this formula does not yield $\Lambda$ unambiguously since it is only approximated, coming from integrating (8.5).

Appendix A - Gaussian Integration

Functional Gaussian integrals will be understood to be the product of many regular Gaussian integrals. The simplest is

$$G(a) = \int_{-\infty}^{+\infty} dx\, e^{-ax^2} \qquad , \qquad (A-1)$$

which, using Poisson's trick of taking the square and of expressing the integrand in polar coordinates, is seen to be

$$G(a) = \sqrt{\frac{\pi}{a}} \qquad . \qquad (A-2)$$

We can generalize it to N degrees of freedom. Let

$$G(A) = \int_{-\infty}^{+\infty} dx_1 dx_2 .. dx_N \; e^{-x_i a_{ij} x_j} \qquad , \qquad (A-3)$$

where A is the real symmetric N x N matrix with matrix elements $a_{ij}$. We write

$$x_i a_{ij} x_j = X^T A X \qquad \text{with } A^T = A \qquad . \qquad (A-4)$$

A can be diagonalized by means of a rotation

$$A = R^T D R, \qquad R^T R = R R^T = 1 \qquad , \qquad (A-5)$$

with D a diagonal matrix with entries $d_1, d_2, .. d_N$. Then

$$G(A) = \int dx_1 .. dx_N \; e^{-X^T R^T D R X} \qquad (A-6)$$

$$= \int dy_1 .. dy_N \; e^{-Y^T D Y} \qquad , \qquad (A-7)$$

with y = RX. The Jacobian is one (prove it). In the y variables, G(A) is separable in the N-fold product of

$$G_N(A) = G(d_1) G(d_2) ... G(d_N) \qquad (A-8)$$

$$= \pi^{N/2} (d_1 d_2 .. d_N)^{-1/2} \qquad (A-9)$$

$$= \Pi^{N/2}(\det A)^{-1/2} \quad , \qquad \text{(A-10)}$$

provided that all the eigenvalues of A are positive. In a similar way we can prove that if $z_i$ are N complex variables

$$\int \prod_{i}^{N} dz_i dz_i^* \; e^{-z^+ C z} = (2\pi)^N (\det C)^{-1} \quad , \qquad \text{(A-11)}$$

where C is an hermitian N x N matrix with positive eigenvalues. Formally, one then defines Gaussian path integrals by taking the limit $N \to \infty$.

These formulae are valid when the determinant does not vanish. If it does, it means that some $d_i$ is equal to zero, leading to an infinity from integrating over an infinite interval. But can we get a sensible answer even if the determinant vanishes? Ideally we would like to divide out the culprit infinite integral. Can this procedure be formalized? Suppose the symmetric N x N matrix A has n zero eigenvalues. In the y variables <u>define</u> the restricted Gaussian integral

$$G_{rest}(A) = \int dy_1 \ldots dy_{N-n} \; e^{-x^T(y)Ax(y)} \quad , \qquad \text{(A-12)}$$

where we integrate only over the variables corresponding to a nonzero eigenvalue of A. This representation of $G_{rest}(A)$ is awkward since it depends on the right system of coordinates "y". To make up for this, invent new variables $y_{N-n+1}, .. y_N$ and rewrite (A-12) as

$$G_{rest}(A) = \int dy_1 .. dy_{N-n} dy_{N-n+1} .. dy_N \delta(y_{N-n+1}) .. \delta(y_N) e^{-x^T(y)Ax(y)} . \qquad \text{(A-13)}$$

Now change variables from y to x, using the Jacobi formula

$$dy_1 .. dy_N = dx_1 .. dx_N \det\left|\frac{\partial y}{\partial x}\right| \quad , \qquad \text{(A-14)}$$

to obtain the final expression

$$G_{rest}(A) = \int (\prod_1^N dx_i) \det\left|\frac{\partial y}{\partial x}\right| \prod_{a=N-n+1}^{N} \delta(y_a) e^{-x^T A x} \quad . \qquad \text{(A-15)}$$

This integral is perfectly well defined. The $y_a$ are some arbitrary functions of x, and the extra factors $\det\left|\frac{\partial y}{\partial x}\right| \Pi\delta(y)$ in the measure effectively restrict the integration from an N-dimensional space to an N-n dimensional one. As the construction has shown, $G_{rest}(A)$ does not depend on the specific form of $y_a(x)$ $a \geq N-n$. It goes without saying that the $y_a(x)$ should be cleverly chosen so as to do the job, i.e., restrict the integration region; if they do not, the Jacobian $\det\left|\frac{\partial y}{\partial x}\right|$ is seen to be singular. (This procedure of altering the measure will be used for path integrals of gauge theories. It will lead to the famous Faddeev-Popov ghosts in covariant gauges.)

Finally we prove one more expression. Consider now

$$F[A,\omega] \equiv \int \prod_1^N dx_i \; e^{-x^T A x + \omega^T x} \quad . \qquad \text{(A-16)}$$

We rewrite the exponent by completing the squares

$$x^T A x - \omega^T x = (x - \tfrac{1}{2} A^{-1}\omega)^T A (x - \tfrac{1}{2} A^{-1}\omega) - \tfrac{1}{4} \omega^T A^{-1} \omega \quad , \qquad \text{(A-17)}$$

provided that $A^{-1}$ exists. Letting

$$x' = x - \tfrac{1}{2} A^{-1}\omega \qquad \text{(A-18)}$$

so that $dx_i' = dx_i$, we find

$$F[A,\omega] = e^{-\frac{1}{4}\omega^T A^{-1}\omega} \int \prod_i^N dx_i' \; e^{-x'^T A x'} \qquad \text{(A-19)}$$

$$= \pi^{N/2} e^{-\frac{1}{4}\omega^T A^{-1}\omega} (\det A)^{-\frac{1}{2}} \quad . \qquad \text{(A-20)}$$

Again the path integral result is formally obtained in the limit $N \to \infty$.

## Appendix B - Integration over Arbitrary Dimensions

Consider the integral

$$I_N = \int d^N \ell F(\ell) \quad , \quad \text{(B-1)}$$

where $F(\ell)$ is an arbitrary integrand, depending only on the length of $\ell_\mu$ ($\mu = 1, ..N$); and N is an integer. Introduce polar coordinates in N dimensions

$$(\ell_1, \ldots \ell_N) \rightarrow (L, \phi, \theta_1, \ldots, \theta_{N-2}) \quad ,$$

with

$$L^2 = \ell_\mu \ell_\mu \quad . \quad \text{(B-2)}$$

Then

$$d^N \ell = L^{N-1} dL \; d\phi \; \sin\theta_1 d\theta_1 \; \sin^2\theta_2 d\theta_2 \ldots \sin^{N-2}\theta_{N-2} d\theta_{N-2}, \quad \text{(B-3)}$$

with the variables assuming values over the intervals

$$0 < L < \infty \quad ; \quad 0 < \phi < 2\pi \quad ; \quad 0 < \theta_i < \pi \quad i = 1,..N-2. \quad \text{(B-4)}$$

It is easy to obtain

$$I_N = 2\pi \prod_{k=1}^{N-2} \int_0^\pi \sin^k\theta_k d\theta_k \int_0^\infty L^{N-1} dL \; F(L) \quad . \quad \text{(B-5)}$$

The use of the well-known formula

$$\int_0^{\pi/2} (\sin t)^{2x-1} (\cos t)^{2y-1} dt = \frac{1}{2} \frac{\Gamma(x)\Gamma(y)}{\Gamma(x+y)} \quad , \; \text{Re } x, \text{Re } y > 0, \quad \text{(B-6)}$$

with $y = 1/2$, yields

$$\int_0^{\pi} dt\ \sin^k t = 2\int_0^{\pi/2} dt\ \sin^k t dt = \frac{\Gamma(\frac{k+1}{2})\Gamma(\frac{1}{2})}{\Gamma(\frac{k+2}{2})}$$

$$= \sqrt{\pi}\ \frac{\Gamma(\frac{k+1}{2})}{\Gamma(\frac{k+2}{2})} \quad . \tag{B-7}$$

Putting it back in (B-5), we obtain

$$I_N = \frac{\pi^{N/2}}{\Gamma(N/2)} \int_0^{\alpha} dx\ x^{\frac{N-2}{2}}\ F(x) \quad , \tag{B-8}$$

where $\chi = L^2$.

In general, F(x) will be of the form

$$F(x) = (x+a^2)^{-A} \quad , \qquad A = 2, 3, .. \tag{B-9}$$

leading to

$$\int_0^{\infty} dx \frac{x^{\frac{N-2}{2}}}{(x+a^2)^A} = (a^2)^{-A+N/2} \int_0^{\infty} dy\ y^{\frac{N-2}{2}}\ (1+y)^{-A} \quad . \tag{B-10}$$

Comparing with the expression for the beta function

$$B(N/2,A-N/2) = \frac{\Gamma(\frac{N}{2})\Gamma(A - \frac{N}{2})}{\Gamma(A)} = \int_0^{\infty} dy\ y^{\frac{N}{2} - 1}\ (1+y)^{-A} \quad , \tag{B-11}$$

which is valid for Re $\frac{N}{2} > 0$ and Re$(A - \frac{N}{2}) > 0$, we obtain our final result

$$\int \frac{d^N \ell}{(\ell^2+a^2)^A} = \pi^{N/2}\ \frac{\Gamma(A-N/2)}{\Gamma(A)}\ \frac{1}{(a^2)^{A-N/2}} \quad . \tag{B-12}$$

We have derived this expression for N integer and Re$(A - \frac{N}{2}) > 0$, Re $\frac{N}{2} > 0$. Now we take it to be true for noninteger N by means of (B-12), by analytic continuation.

Now, by letting $\ell = \ell' + p$, and relabeling $b^2 = a^2 + p^2$, we can write (B-12) in the form

$$\int \frac{d^N\ell}{(\ell^2+2p\cdot\ell+b^2)^A} = \pi^{N/2}\,\frac{\Gamma(A-N/2)}{\Gamma(A)}\,\frac{1}{(b^2-p^2)^{A-N/2}} \quad . \tag{B-13}$$

Next, by successive differentiation of (B-13) with respect to $p_\mu$, it is easy to obtain the formulae

$$\int d^N\ell \, \frac{\ell_\mu}{(\ell^2+2p\ell+a^2)^A} = \pi^{N/2}\,\frac{\Gamma(A-N/2)}{\Gamma(A)}\,\frac{(-p_\mu)}{(a^2-p^2)^{A-N/2}} \tag{B-14}$$

and

$$\int d^N\ell \, \frac{\ell_\mu \ell_\nu}{(\ell^2+2p\cdot\ell+a^2)^A} = \frac{\pi^{N/2}}{\Gamma(A)(a^2-p^2)^{A-N/2}} \times$$

$$\times \left[\Gamma(A-\tfrac{N}{2})p_\mu p_\nu + \tfrac{1}{2}\,\delta_{\mu\nu}\Gamma(A-1-\tfrac{N}{2})(a^2-p^2)\right] \quad . \tag{B-15}$$

These formulae are derived in Euclidean space and in each case the RHS is taken to be the correct continuation of the LHS to noninteger values of N.

Glossary of Dimensional Regularization Formulae

$$\int \frac{d^{2\omega}\ell}{(2\pi)^{2\omega}} \frac{1}{(\ell^2+M^2+2\ell\cdot p)^A} = \frac{\Gamma(A-\omega)}{(4\pi)^{\omega}\Gamma(A)} \frac{1}{(M^2-p^2)^{A-\omega}} \qquad \text{(B-16)}$$

$$\int \frac{d^{2\omega}\ell}{(2\pi)^{2\omega}} \frac{\ell_\mu}{(\ell^2+M^2+2\ell\cdot p)^A} = -\frac{\Gamma(A-\omega)}{(4\pi)^{\omega}\Gamma(A)} \frac{p_\mu}{(M^2-p^2)^{A-\omega}} \qquad \text{(B-17)}$$

$$\int \frac{d^{2\omega}\ell}{(2\pi)^{2\omega}} \frac{\ell_\mu \ell_\nu}{(\ell^2+M^2+2\ell\cdot p)^A} = \frac{1}{(4\pi)^{\omega}\Gamma(A)} \cdot$$

$$\cdot \left[p_\mu p_\nu \frac{\Gamma(A-\omega)}{(M^2-p^2)^{A-\omega}} + \frac{1}{2}\delta_{\mu\nu} \frac{\Gamma(A-1-\omega)}{(M^2-p^2)^{A-1-\omega}}\right] \qquad \text{(B-18)}$$

$$\int \frac{d^{2\omega}\ell}{(2\pi)^{2\omega}} \frac{\ell_\mu \ell_\nu \ell_\rho}{(\ell^2+M^2+2\ell\cdot p)^A} = \frac{-1}{(4\pi)^{\omega}\Gamma(A)} \left[ p_\mu p_\nu p_\rho \frac{\Gamma(A-\omega)}{(M^2-p^2)^{A-\omega}}\right.$$

$$\left. + \frac{1}{2}(\delta_{\mu\rho}p_\nu + \delta_{\nu\rho}p_\mu + \delta_{\mu\nu}p_\rho) \cdot \frac{\Gamma(A-1-\omega)}{(M^2-p^2)^{A-1-\omega}}\right] \qquad \text{(B-19)}$$

$$\int \frac{d^{2\omega}\ell}{(2\pi)^{2\omega}} \frac{\ell_\mu \ell_\nu \ell_\rho \ell_\sigma}{(\ell^2+M^2+2\ell\cdot p)^A} = \frac{1}{(4\pi)^{\omega}\Gamma(A)} \left[p_\mu p_\nu p_\rho p_\sigma \frac{\Gamma(A-\omega)}{(M^2-p^2)^{A-\omega}}\right.$$

$$+ \frac{1}{2}[\delta_{\mu\nu}p_\rho p_\sigma + \delta_{\nu\sigma}p_\mu p_\rho + \delta_{\rho\sigma}p_\mu p_\nu + \delta_{\mu\rho}p_\nu p_\sigma + \delta_{\nu\rho}p_\mu p_\sigma + \delta_{\mu\sigma}p_\rho p_\nu] \frac{\Gamma(A-1-\omega)}{(M^2-p^2)^{A-1-\omega}}$$

$$\left. + \frac{1}{4}[\delta_{\mu\nu}\delta_{\rho\sigma} + \delta_{\nu\rho}\delta_{\mu\sigma} + \delta_{\mu\rho}\delta_{\nu\sigma}] \frac{\Gamma(A-2-\omega)}{(M^2-p^2)^{A-2-\omega}}\right] \qquad \text{(B-20)}$$

## APPENDIX C: EUCLIDEAN SPACE FEYNMAN RULES IN COVARIANT GAUGE, (in $2\omega$ dimensions)

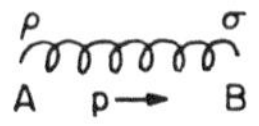


$$\frac{\delta^{AB}}{p^2}\left[\delta_{\rho\sigma} - (1-\alpha)\frac{p_\rho p_\sigma}{p^2}\right]$$

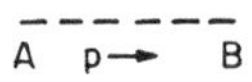


$$-i\,\frac{\delta^{AB}}{p^2}$$

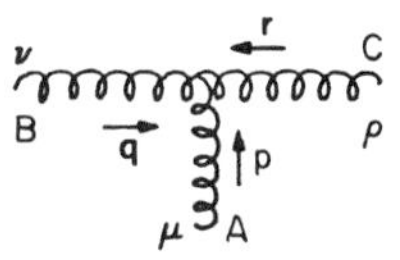


$$-ig\mu^{2-\omega}\, f^{ABC}\left[(r-q)_\mu\,\delta_{\nu\rho} + (q-p)_\rho\,\delta_{\mu\nu} + (p-r)_\nu\,\delta_{\rho\mu}\right]$$

$$\left[p + q + r = 0\right]$$

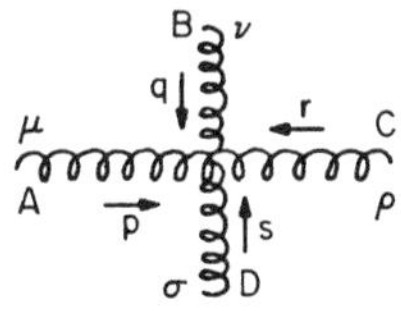


$$-g^2\mu^{4-2\omega}\left[f^{ABE} f^{CDE}\,(\delta_{\mu\rho}\delta_{\nu\sigma} - \delta_{\nu\rho}\delta_{\mu\sigma}) + \right.$$

$$\left. f^{CBE} f^{ADE}\,(\delta_{\mu\rho}\delta_{\nu\sigma} - \delta_{\nu\mu}\delta_{\rho\sigma}) + f^{DBE} f^{CAE}\,(\delta_{\sigma\rho}\delta_{\nu\mu} - \delta_{\nu\rho}\delta_{\mu\sigma})\right]$$

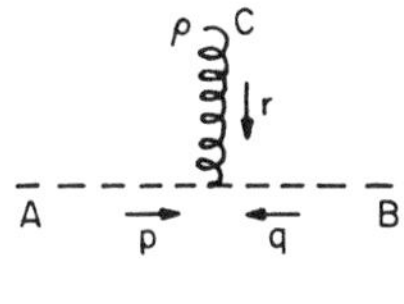


$$-g\mu^{2-\omega}\, f^{ABC}\, q_\rho$$

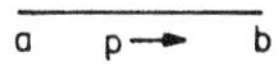


$$-i\,\frac{\delta^a_b}{\not p + m}$$ fermion line

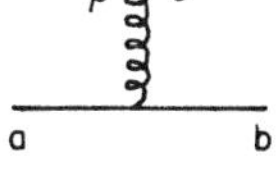


$$-ig\mu^{2-\omega}\,\gamma_\rho\,(T^C_f)^a_b$$

The purpose of this annotated bibliography is to offset the introductory character of the text by providing the reader with a list of references where the material is discussed extensively, both in length and in depth. The author apologizes for any omission, but this bibliography is intended to be pedagogical rather than complete.

There are many fine textbooks on Quantum Field Theory. Here we give a few, listed alphabetically by author:

V. B. Berestetskii, E. M. Lifshitz, L. P. Pitaevskii (1971) "Relativistic Quantum Theory", part I and II (Oxford: Pergamon Press).

J. D. Bjorken, S. D. Drell (1965) "Relativistic Quantum Mechanics" and "Relativistic Quantum Fields" (New York: McGraw-Hill).

N. N. Bogoliubov, D. V. Shirkov (1959) "An Introduction to the Theory of Quantized Fields" (New York: Wiley-Interscience).

L. D. Faddeev, A. A. Slavnov (1980) "Gauge Fields, Introduction to Quantum Theory" (Reading, Massachusetts: Benjamin/Cummings, Advanced Book Program).

E. M. Lifshitz, L. P. Pitaevskii (1974) "Relativistic Quantum Theory", part 2 (Oxford: Pergamon Press).

F. Mandl (1959) "Introduction to Quantum Field Theory" (New York: Wiley-Interscience).

C. Nash (1978) "Relativistic Quantum Fields" (New York: Academic Press).

S. Schweber (1961) "An Introduction to Relativistic Quantum Field Theory" (New York: Harper and Row).

J. C. Taylor (1976) "Gauge Theory of Weak Interactions" (Cambridge: Cambridge University Press).

See also the extraordinary lectures by S. Coleman, delivered at the International School of Subnuclear Physics "Ettore Majorana":

1971 part A: "Dilatations" (on the breaking of conformal invariance by quantum effects).

1971 part B: "Renormalization: a Review for Non-Specialists".

1973 "Secret Symmetry" (on the Higgs mechanism and gauge theories).

1977 "The Uses of Instantons" (self-explanatory).

More advanced readers will enjoy:

E. S. Abers and B. W. Lee (1973) "Gauge Theories", Phys. Reports 9C, 1-141.

R. Balian and J. Zinn-Justin (1975) "Methods in Field Theory" (Amsterdam: North Holland), which contains lectures delivered at the 1975 Les Houches summer school.

We present in the following additional references for each of the chapters in the text:

Chapter I.

Sec. 1: An extensive treatment of the relation between invariances of the Lagrangian and conservation laws can be found in E. L. Hill (1951) Rev. Mod. Phys. 23, 253-260.

Sec. 2, 3: The standard references for the Lorentz group are: I. M. Gel'fand, R. A. Minlos, Z. Ya. Shapiro (1963) "Representations of the Rotation and Lorentz Groups and their Applications" (Oxford: Pergamon Press), and I. M. Gel'fand, M. I. Graev, N. Ya. Vilenkin (1966) "Generalized Functions" vol. 5 (New York: Academic Press). Many of the original papers on the representation theory of the Lorentz and Poincaré groups are reprinted in F. J. Dyson (1966) "Symmetry Groups in Nuclear and Particle Physics" (New York: Benjamin). Another beautiful exposition of the representation theory can be found in the lecture notes of F. Gürsey (1970) on Field Theory, edited by M. Günaydin (unpublished).

Sec. 4, 5, 6, 7: The transformation properties of fields are discussed by T. W. Kibble (1961) J. Math. Phys. 2, 212-221.

Sec. 8: Supersymmetry first appeared in the context of string models: P. Ramond (1971) Phys. Rev. D3, 2415-2418, A. Neveu and J. Schwarz (1971) Nucl. Phys. B31, 86-112; and also as a generalization of the Poincaré group Yu. A. Gol'fand, E. P. Likhtman (1971) JETP Letters 13, 323-326, D. V. Volkov, V. P. Akulov (1973) Phys. Lett. 46B, 109-110, J. Wess and B. Zumino (1974) Nucl. Phys. B70, 39-50. For an extensive review of supersymmetry: P. Fayet, S. Ferrara (1977) Phys. Reports 32C, 249-334.

Chapter II.

Sec. 1: For more details on canonical transformations see H. Goldstein (1950) "Classical Mechanics" (Reading, Massachusetts: Addison-Wesley) and N. Mukunda, E. C. G. Sudarshan (1974) "Classical Dynamics: A Modern Perspective" (New-York: Wiley-Interscience).

Sec. 2, 3: Original papers on the Path Integrals are to be found in the reprint volume "Selected Papers on Quantum Electrodynamics" (1958), J. Schwinger ed. (New York: Dover). The uses of the Path Integral are detailed in R. P. Feynman, A. R. Hibbs (1965) "Quantum Mechanics and Path Integrals" (New York: McGraw-Hill).

Chapter III.

Sec. 1, 2, 3, 4: The treatment follows closely that of E. Abers and B. Lee, op. cit.; D. Lurie (1968) "Particles and Fields" (New York: Interscience)

Sec. 5, 6: For the ζ-function evaluation of determinants see D. B. Ray, I. M. Singer (1971) Adv. Math. 7, 145; J. S. Dowker, R. Critchley (1976) Phys. Rev. D13, 3224-3232; E. Corrigan, P. Goddard, H. Osborn, S. Templeton (1979) Nucl. Phys. B159, 469-496.

Chapter IV.

Sec. 3: Standard references on dimensional regularization (besides the one quoted in the text) are C. G. Bollini, J. J. Giambiagi (1972) Nuovo Cimento 12B, 20-26; J. F. Ashmore (1972) Nuovo Cimento Lett. 4, 289-290; for a review see G. Leibbrandt (1975) Rev. Mod. Phys. 47, 849-876.

Sec. 5: A lucid explanation of renormalization is to be found in Coleman's Erice lectures op. cit. See also J. C. Collins (1974) Phys. Rev. D10, 1213-1218; for proofs of renormalizability see W. Zimmermann (1970) in "Lectures on Elementary Particles and Quantum Field Theory" S. Deser, M. Grisaru, H. Pendleton eds. (Cambridge, Massachusetts: MIT Press), and E. Abers, B. Lee, op. cit. On the renormalization group: E. C. G. Stuckelberg, A. Petermann (1953) Helv. Physica Acta 26, 499-520; M. Gell-Mann, F. Low, op. cit.; C. C. Callan (1975) in "Methods in Field Theory" op. cit. For the difference between different renormalization group equations, see S. Weinberg (1973) Phys. Rev. D8, 3497-3509.

Sec. 8, 9: R. J. Eden, P. V. Landshoff, D. I. Olive, J. C. Polkinghorne (1966) "The Analytic S-Matrix" (Cambridge: Cambridge University Press), and E. M. Lifshitz, L. P. Pitaerskii, op. cit.

Chapter V.

The fundamental reference for this chapter is F. A. Berezin (1966) "The Method of Second Quantization" (New York: Academic Press), and inevitably S. Coleman (1977) "The Uses of Instantons" op. cit.

Chapter VI.

Sec. 1, 2: In addition to the Yang-Mills paper op. cit. see R. Utiyama (1956) 1597-1607; T. W. Kibble (1961) op. cit. and M. Gell-Mann, S. L. Glashow (1961) Ann. Phys. (N. Y.) 15, 437-460.

Sec. 3: For instantons: S. Coleman "The Uses of Instantons" _op. cit._; J. C. Taylor (1976) _op. cit._ Path-ordered exponentials were first introduced by S. Mandelstam (1962) Ann. Phys. (N.Y.) 19, 1-24.

Chapter VII

Sec. 1: A beautiful exposition of systems with constrained Hamiltonians is to be found in P. A. M. Dirac (1964) "Lectures on Quantum Mechanics" (New York: Yeshiva University) and (1965) "Lectures on Quantum Field Theory" (New York: Yeshiva University).

Sec. 2: See the text by L. D. Faddeev, A. A. Slavnov (1980) _op. cit._; L. D. Faddeev (1975) in "Methods in Field Theory" _op. cit._

Chapter VIII.

Sec. 1: The necessity for ghost fields in covariant gauges was first recognized by R. P. Feynman (1963) Acta Physica Polonica 24, 697-722; B. S. De Witt (1967) Phys. Rev. 162, 1195-1239.

Sec. 2, 4: Quantum Electrodynamics has been extensively discussed in many textbooks. Some additional ones are A. I. Akhiezer, V. B. Berestetskii (1965) "Quantum Electrodynamics" (New York: Wiley-Interscience); R. P. Feynman (1961) "Quantum Electrodynamic" (New York: Benjamin); G. Källen (1972) "Quantum Electrodynamics" (Berlin: Springer-Verlag); for dimensionally regularized QED, see for instance J. C. Collins, A. J. McFarlane (1973) Phys. Rev. D10, 1201-1212.

Sec. 3: J. C. Ward (1950) Phys. Rev. 78, 182-183; Y. Takahashi (1957) Nuovo Cimento 6, 370-375.

Sec. 4: For use of dimensional regularization to tag infrared divergences, see R. Gastmans, R. Mendermans (1973) Nucl. Phys. B63, 277-284; W. J. Marciano (1975) Phys. Rev. D12, 3861-3871.
Sec. 5: Yang-Mills theories were proved renormalizable by G. 't Hooft (1971) Nucl. Phys. B33, 173-199; see also E. Abers, B. W. Lee, _op. cit._

Sec. 7: A. A. Slavnov (1972) Theor. & Math. Phys. 10, 99-104; J. C. Taylor (1971) Nucl. Phys. B33, 436-444

Sec. 8: The phenomenon of asymptotic freedom was noted earlier by G. 't Hooft (1972) at the Conference on Lagrangian Field Theory, Marseille; for more details and reference see H. D. Politzer (1974) "Asymptotic Freedom: An Approach to Strong Interactions", Phys. Reports 14C, 129-180. For applications of QCD: R. D. Field (1978) "Applications of Quantum Chromodynamics", lectures given at the La Jolla Institute Summer Workshop.

# INDEX